Ordinary Differential Equations

Introduction and Qualitative Theory

Third Edition

MONOGRAPHS AND TEXTBOOKS IN PURE AND APPLIED MATHEMATICS

Recent Titles

Walter Ferrer and Alvaro Rittatore, Actions and Invariants of Algebraic Groups (2005)

Christof Eck, Jiri Jarusek, and Miroslav Krbec, Unilateral Contact Problems: Variational Methods and Existence Theorems (2005)

M. M. Rao, Conditional Measures and Applications, Second Edition (2005)

A. B. Kharazishvili, Strange Functions in Real Analysis, Second Edition (2006)

Vincenzo Ancona and Bernard Gaveau, Differential Forms on Singular Varieties: De Rham and Hodge Theory Simplified (2005)

Santiago Alves Tavares, Generation of Multivariate Hermite Interpolating Polynomials (2005)

Sergio Macías, Topics on Continua (2005)

Mircea Sofonea, Weimin Han, and Meir Shillor, Analysis and Approximation of Contact Problems with Adhesion or Damage (2006)

Marwan Moubachir and Jean-Paul Zolésio, Moving Shape Analysis and Control: Applications to Fluid Structure Interactions (2006)

Alfred Geroldinger and Franz Halter-Koch, Non-Unique Factorizations: Algebraic, Combinatorial and Analytic Theory (2006)

Kevin J. Hastings, Introduction to the Mathematics of Operations Research with *Mathematica*®, Second Edition (2006)

Robert Carlson, A Concrete Introduction to Real Analysis (2006)

John Dauns and Yiqiang Zhou, Classes of Modules (2006)

N. K. Govil, H. N. Mhaskar, Ram N. Mohapatra, Zuhair Nashed, and J. Szabados, Frontiers in Interpolation and Approximation (2006)

Luca Lorenzi and Marcello Bertoldi, Analytical Methods for Markov Semigroups (2006)

M. A. Al-Gwaiz and S. A. Elsanousi, Elements of Real Analysis (2006)

Theodore G. Faticoni, Direct Sum Decompositions of Torsion-Free Finite Rank Groups (2007)

R. Sivaramakrishnan, Certain Number-Theoretic Episodes in Algebra (2006)

Aderemi Kuku, Representation Theory and Higher Algebraic K-Theory (2006)

Robert Piziak and P. L. Odell, Matrix Theory: From Generalized Inverses to Jordan Form (2007)

Norman L. Johnson, Vikram Jha, and Mauro Biliotti, Handbook of Finite Translation Planes (2007)

Lieven Le Bruyn, Noncommutative Geometry and Cayley-smooth Orders (2008)

Fritz Schwarz, Algorithmic Lie Theory for Solving Ordinary Differential Equations (2008)

Jane Cronin, Ordinary Differential Equations: Introduction and Qualitative Theory, Third Edition (2008)

Ordinary Differential Equations

Introduction and Qualitative Theory

Third Edition

Jane Cronin

CRC Press
Taylor & Francis Group
Boca Raton London New York

CRC Press is an imprint of the
Taylor & Francis Group, an **informa** business
A CHAPMAN & HALL BOOK

CRC Press
Taylor & Francis Group
6000 Broken Sound Parkway NW, Suite 300
Boca Raton, FL 33487-2742

First issued in paperback 2019

© 2008 by Taylor & Francis Group, LLC
CRC Press is an imprint of Taylor & Francis Group, an Informa business

No claim to original U.S. Government works

ISBN-13: 978-0-8247-2337-8 (hbk)
ISBN-13: 978-0-367-38796-9 (pbk)

Library of Congress Cataloging-in-Publication Data

Cronin, Jane, 1922-
    Ordinary differential equations : introduction and qualitative theory / Jane Cronin-Scanlon. -- 3rd ed.
        p. cm. -- (Pure and applied mathematics ; 292)
    Rev. of ed.: Differential equations. 2nd ed., rev. and expanded. c1994.
    Includes bibliographical references and index.
    ISBN 978-0-8247-2337-8 (alk. paper)
    I. Cronin, Jane, 1922- Differential equations. II. Title. III. Series.

QA372.C92 2007
515'.352--dc22   2007038844

**Visit the Taylor & Francis Web site at
http://www.taylorandfrancis.com**

**and the CRC Press Web site at
http://www.crcpress.com**

Contents

Preface to Third Edition

There are two principal changes in this edition: the treatment of two-dimensional systems has been considerably extended; and the discussion of periodic solutions in small parameter problems has been extended and unified.

The reason for a more extended study of the venerable subject of two-dimensional systems is the growing use of nonlinear systems in applications in biology. Many models in biology are two-dimensional systems or involve two-dimensional systems. Also the treatment of biological models which are singularly perturbed systems sometimes requires analysis of two-dimensional subsystems.

We give a detailed analysis of equilibrium points (rest points or singular points or critical points) which is based on the treatment given by Lefschetz [1962]. The main goal is to obtain the portrait of the orbits in a neighborhood of the singular point: following Lefschetz, we describe the general appearance of these portraits and then obtain constructive (calculational) methods for determining the portrait in a given case, especially by using critical directions. We also introduce the Bendixson index and obtain the Bendixson formula for relating the index to the number of hyperbolic and elliptic sectors of the equilibrium point.

If any equilibrium point, say $(0, 0)$, of a system

$$dx/dt = f(x, y)$$
$$dy/dt = g(x, y)$$

is such that the lowest order terms of f and g have different orders, then it becomes more complicated to apply the calculational methods described above. A simple but important case of this problem is that in which the matrix

$$\begin{bmatrix} f_x(0, 0) & f_y(0, 0) \\ g_x(0, 0) & g_y(0, 0) \end{bmatrix}$$

has exactly one nonzero eigenvalue. This problem was studied by Bendixson [1901] and, using a different approach, Lefschetz [1962]. We follow Lefschetz and describe a straightforward calculational method for determining the portrait of the equilibrium point, that is, whether it is a node, saddle point, or saddle-node. As will be indicated, this method can also be applied to more complicated cases.

Studying the periodic solutions of an equation

$$dx/dt = F(t, x, \varepsilon)$$

where ε is a small parameter and the equation is well-understood if $\varepsilon = 0$, is a very old problem which plays a significant role in subjects ranging from celestial mechanics to biology. As might be expected, there is a huge literature on the problem, and our purpose here is simply to give an extended and unified treatment of the qualitative aspects. Our treatment is entirely based on a method of Poincaré and makes possible the study of both nonautonomous and autonomous problems. It can be regarded as an extension of the discussion given in Coddington and Levinson [1955]. But we also borrow heavily from Roseau [1966]. The extension includes the use of topological degree and treatment of Hopf bifurcations. As part of the unification, we show that use of the averaging method to study periodic solutions is a special case of use of this method of Poincaré.

This book owes much to a number of sources: most especially Lefschetz [1962], also Coddington and Levinson [1955], Malkin [1959], Roseau [1966], and Farkas [1994].

I wish to thank Dr. Joseph McDonough who read in detail a large part of this book, contributed very useful comments, and pointed out needed corrections.

I am also greatly indebted to Dottie Phares who prepared the LaTeX version of the manuscript with great skill and understanding and also made very helpful suggestions about the preparation of the manuscript.

Jane Cronin

Preface to Second Edition

The main purpose of the changes in this edition is to make the book more effective as a textbook. To this end, the following steps have been taken.

First, the material in Chapter 1 has been rearranged and augmented to make it more readable. Motivations for studying existence theorems and extension theorems are given. The exercises at the end of the chapter have been extended to illustrate the theory more extensively, and the examples from chemistry and biology introduced at the end of the chapter are treated in detail in the solutions manual. Also the exposition and organization of the chapter have been reworked with a view to clarification.

In Chapter 2, a treatment of the Sturm-Liouville theory has been added. The theory has been treated in outline form with specific references for proofs. The purpose of the discussion is to make clear the relation between initial value problems and boundary value problems, to describe how the Sturm-Liouville theory is rooted in elementary linear algebra, and to show that the Sturm-Liouville problem and its solution form a concrete example in functional analysis. (Unlike the other material in this book, the Sturm-Liouville problem is an infinite-dimensional problem and really belongs in a course in functional analysis.)

The text in all the chapters has been reviewed and in some parts clarified and improved. (The proof of the instability theorem has been substantially simplified by allowing use of Lyapunov theory.)

Finally, a solutions manual is available for instructors. The solutions manual (of more than 100 pages) includes detailed solutions for almost all the exercises at the ends of the chapters and also a detailed analysis of the examples given at the end of Chapter 1: the Volterra population equations, the Hodgkin-Huxley nerve conduction equations, the Field-Noyes model of the Belousov-Zhabotinsky reaction, and the Goodwin model of cell activity. The analysis given for these examples that uses the theory from Chapter 1 shows the student the importance of the existence and extension theorems and also provides an early introduction to the use of geometric methods. As the theory is developed in later chapters, it is applied to these examples in the exercises at the ends of the chapters. Thus fairly extensive analyses of these examples are given in the solutions manual.

The solutions in the manual are written in detail so that the instructor can easily judge the length of an outside assignment and so that the students can read them. Thus if time limitations prevent discussion in class of a problem (an unfortunately frequent event in my experience) copies of the solutions can be distributed to students.

I want to thank the faculty and students whose comments have helped greatly in the preparation of this book.

I am deeply indebted to Dottie Phares who prepared the LaTeX version of this book with great understanding and skill.

Jane Cronin

Preface to First Edition

This book has two objectives: first, to introduce the reader to some basic theory of ordinary differential equations which is needed regardless of the direction pursued in later studies and, second, to give an account of some qualitative theory of ordinary differential equations which has been developed in the last couple of decades and which may be useful in problems which have arisen in qualitative studies in chemical kinetics, biochemical systems, physiological problems, and other biological problems. The only prerequisites for reading the book are the first semester of a course in advanced calculus and a semester of linear algebra.

Chapters 1, 2, 3, 4 and the first part of Chapter 6 are a suitable basis for a first course in ordinary differential equations at the advanced undergraduate or beginning graduate level. The discussion differs from the conventional presentation only in that there is heavier emphasis on stability theory and there is no treatment of eigenvalue problems. The reason for emphasizing stability lies in the growing importance of stability in studies of problems in chemistry and biology. While neutral stability may be significant in physics, the unaccounted-for disturbances in chemical and biological systems are sufficiently important to suggest that in these studies, a stronger stability condition, that is, some kind of asymptotic stability, is needed. Eigenvalue problems have been treated in admirable detail in many books, and we have chosen a more limited course (the search for periodic solutions) and have extended only this to the nonlinear case.

Not all of the material in Chapters 1 through 4 needs to be considered. It is certainly not necessary to study in detail all the existence theorems given in Chapter 1, although the functional analysis approach used in some of the proofs helps to orient a student who has some experience in that subject. On the other hand, the extension theorems should be emphasized because they are of crucial importance in understanding the difficulties of working with solutions of nonlinear equations. Because the linear theory is so important in itself and for later work, the material in Chapter 2 should be given strong emphasis. But the treatment of two-dimensional systems in the latter part of Chapter 3 can be merely indicated or omitted. In Chapter 4, the basic results are important, but whether all the details of proofs, for example the proof of the instability theorem, should be included is questionable. The proof of the Poincaré-Bendixson theorem in Chapter 6 is given in full detail, and the use of the Jordan curve theorem in the proof is described with some care.

Chapter 5 is an introduction to the Lyapunov second method. The reason for using the method and the basic theorem are discussed in detail, but only indications are given of the important applications that have been made of the method. A more complete discussion may be found in the lucid account given by LaSalle and Lefschetz [1963].

The second part of Chapter 6 treats a theorem of George Sell which is a kind of extension of the Poincaré-Bendixson theorem to the n-dimensional case. The Poincaré-Bendixson theorem itself cannot be generalized to the n-dimensional case, but Sell's theorem says roughly that if an n-dimensional autonomous system has a bounded solution with suitable asymptotic stability, then that solution approaches an asymptotically stable periodic solution. From the viewpoint of pure mathematics, this is much less impressive than the Poincaré-Bendixson theorem because the strong hypothesis is imposed that an asymptotically stable solution exists. From the point of view of applications in chemistry and biology, where it seems highly likely that only solutions with some kind of asymptotic stability are significant, Sell's result is very important, for it suggests that the search for periodic solutions should be supplanted by a search for asymptotically stable solutions.

Chapter 7 is a fairly detailed treatment of the classical problem of branching or bifurcation of periodic solutions if a small parameter is varied. There is an enormous literature on various aspects of this problem, and our aim here is to provide a general discussion of the qualitative aspects of the problem, that is, the existence and stability of periodic solutions is treated. For the nonautonomous case, a general discussion along the lines given by Malkin [1959] is given. The treatment is somewhat more streamlined than Malkin's and is completed by applying topological degree theory to obtain explicit (i.e., computable) sufficient conditions for the existence and stability of periodic solutions. For the autonomous case, the treatment given by Coddington and Levinson [1952] and the Hopf bifurcation theorem are discussed and extended by using degree theory.

I would like to express my gratitude to the U.S. Army Research Office for partial support during the writing of this book.

In preliminary form, the book was used as a text in several courses in differential equations. I would like to express my thanks to the many students whose comments improved the text considerably, to Mary Anne Jablonski and Lynn Braun for their efficiency and partient good humor when they typed the various versions of the text, and to Edmund Scanlon who made the drawings for this book.

Jane Cronin

Introduction

In Freshman and Sophomore calculus, the study of differential equations is begun by developing a number of more or less computational techniques such as separation of variables. For example, suppose we consider the equation

$$\frac{dx}{dt} = tx + 5x$$

We "separate variables" and integrate the equation by carrying out the following steps:

$$\frac{dx}{dt} = (t + 5)x$$

$$\frac{dx}{x} = (t + 5)dt$$

$$\ell n|x| = \frac{t^2}{2} + 5t + C$$

$$|x| = K \exp\left(\frac{t^2}{2} + 5t\right)$$

where C is an arbitrary constant of integration and $K = e^C$ so that K is an arbitrary positive constant. These formal steps are more or less familiar to the student who has taken Sophomore calculus or an elementary course in differential equations. A rigorous justification for these steps is less familiar (Exercise 1 in Chapter 1).

It is natural to think that a more advanced study of differential equations consists of the extension and refinement of such techniques as separation of variables. Actually, however, we must take a different direction. The reason is this: although methods like separation of variables are effective when they can be applied, they are applicable only to a very limited class of differential equations, and despite the strenuous efforts of many mathematicians over a long period of time (the eighteenth and nineteenth centuries), there is no indication that these methods can be extended beyond a very limited class of differential equations. The direction that we follow instead is to establish existence theorems for differential equations, that is, to prove that solutions of a differential equation exist even though we may not be able to compute them explicitly. (The proofs of the existence theorems will suggest methods for finding approximations to solutions but will not, in general, show us how to write the solutions explicitly.) The existence theorems apply to wide classes of differential equations. Thus we give up hope of obtaining explicitly written solutions but gain the advantage of being able to study large classes of equations.

Having obtained the existence of solutions, we proceed then to study some of the properties of the solutions. First, we consider such basic characteristics as differentiability of the solution and its continuity with respect to a parameter in the differential equation. Next, we study the theory of linear systems which is essential for later work. After these basic studies have been made, there are several courses which can be followed. We choose to study some so-called qualitative properties of solutions: stability, periodicity, and almost periodicity. There are several reasons for pursuing this direction. The most important reason is that such study yields a coherent, aesthetically pleasing theory which has important applications in the physical and life sciences.

Qualitative theory of solutions of differential equations originates in the giant developments due to Poincaré and Lyapunov. One of the marks of true mathematical genius is the ability to ask the right questions. The greatly gifted mathematician formulates "good questions," that is, the questions whose study leads to the development of mathematical theory which is beautiful, profound and useful to people in other fields. In their work on differential equations, Poincaré and Lyapunov displayed this to an extraordinary degree. Both perceived that much of the future development of differential equations would be in the direction of qualitative studies. Indeed, both initiated (independently and from different viewpoints) qualitative studies. Today, when topological notions are familiar to beginning students of mathematics, the idea of a qualitative study may not seem striking. At the time when Poincaré and Lyapunov were working, the introduction of such ideas required tremendous intellectual force and originality. It is scarcely necessary to point out the extent to which Poincaré's work has influenced twentieth century mathematics. Lyapunov's work has had a lesser influence on the development of pure mathematics, but his work on stability foreshadowed twentieth century developments in applied mathematics, the physical sciences, and the life sciences to an uncanny extent. To see why this is true, it is necessary to keep in mind that Lyapunov's and Poincaré's work on differential equations was motivated by problems in celestial mechanics. By the 1920s the efforts to develop a theory for the burgeoning subject of radio circuitry led to intensive study of new classes of nonlinear differential equations, and it was realized that the qualitative theory developed by Poincaré [1892–1899] was applicable to these equations. Lyapunov's stability theory (Lyapunov [1892], LaSalle and Lefschetz [1961]), however, remained largely disregarded. It was useful but did not play a crucial role in the study of radio circuit problems (see Andronov and Chaikin [1949]), and its use in celestial mechanics was limited. This limitation stemmed partly from the fact that the stability problems in celestial mechanics were extremely difficult. For example, to prove that certain solutions of the three-body problem are stable in the sense of Lyapunov was only proved in 1961 by Leontovich [1962] who used profound results due to Kolmogorov and Arnold. Indeed, Wintner [1947, p. 98] dismissed Lyapunov's definition of stability as unrealistic because it was too strong a condition. Another reason for the limited use of Lyapunov stability theory in celestial mechanics is the fact that the concept of asymptotic stability introduced by Lyapunov is not applicable in Hamiltonian systems. (The systems of differential equations which occur in celestial mechanics are special cases of Hamiltonian systems.) Thus the Lyapunov theory lay dormant, attended only by some Russian mathematicians, until the advent of control theory in the years following World War II. Efforts to develop a mathematical control theory led to the realization that the Lyapunov theory was well-designed for such studies and there was a widespread growth of interest in stability theory.

More recently, it has become clear that if a biological problem can be formulated in terms of a system of ordinary differential equations, then stability theory must play an important role in the study of the system. The reason for this is that since biological systems tend to be quite complicated, the differential equation which is used to describe the system is only a rather crude approximate description. It must be assumed that disturbances of the system (as described by the differential equation) are constantly occurring. This suggests that only those solutions of the differential equation which have strong stability properties are biologically significant, that is, describe phenomena which actually occur in the biological system.

In choosing to study the qualitative properties described above, we are disregarding very important theory which is useful in the study of physical and biological problems. The crucially important subject of numerical or computational methods for solving differential equations will be omitted. (As will be pointed out later, our choice of a proof for the basic existence theorem is influenced by the wish to indicate the underlying idea of one numerical procedure.)

Chapter 1

Existence Theorems

What This Chapter Is About

We shall be concerned with the existence and uniqueness of solutions and the size of domains of solutions. (We use the word "solution" here to refer to a solution of the initial value problem. See page 13 for a description of the initial value problem.) Since we shall obtain no practical procedures for calculating solutions, it is natural for the student to feel dubious about the value of this material. Would it not be better to omit this theoretical stuff and get on with the real problem of finding explicit solutions? Besides, those who have solved differential equations by using a computer know full well that this is accomplished without any talk about existence, uniqueness, and so forth.

These are reasonable, serious questions deserving careful answers. First of all, using a computer to solve a differential equation means that a particular program, code, or software is used. The design of the software is based on numerical analysis of differential equations, and this numerical analysis is, in turn, based on the theory to be described here. So if one uses a computer program, this means that somebody else has thrashed out the theory beforehand.

Second, the topics in this chapter are essential to both the structure of the theory of differential equations and the applications of differential equations in physics, engineering, and so forth. As we proceed in later chapters, we shall see the importance of this material to theory. Here we shall merely point out why these topics are important for applications. Before doing so, however, a few more words to the student whose thoughts may run like this: Do I have to slog through more than 40 pages laden with mathematical symbols, except for 3 small pictures (a description of Chapter 1), before I get to solving anything? Mercifully, the answer is no. There are just two important results in this chapter: the existence (under very reasonable hypotheses) of unique solutions (Existence Theorem 1.1) and the fact that one must be careful about the domain of the solution. It may not be as large an interval on the t-axis as one might hope. (See the example of equation (1.35).) Sooner or later, one may need more of the analysis in this chapter. But it may be sufficient for a while simply to be aware of what topics are studied there.

Certainly it is logical to establish the existence of a solution before attempting to find an explicit formula or approximation for the solution, especially since the question of existence turns out to be somewhat more complicated than might at first be expected. (See Exercises 6, 10, and 11 for untoward outcomes which can occur.) From the point of view of applications, the condition of uniqueness is essential. A solution of a differential equation which models a physical system is used to predict the behavior of the system. If there is more than one solution which satisfies the given initial condition, then prediction is not possible because one does not know which solution to use. Thus, if a differential equation is proposed as a model for a physical system and if it can be shown that the solutions of the differential solution are not unique, then the differential equation has questionable value as a model of the physical system. Hence a prudent step before initiating any numerical analysis is to verify that the solutions of the differential equation exist and are unique.

Equally important is the size of the domain of the solution. Very often, the solutions of differential equations that are obtained in a first course in differential equations are defined for all real values of the independent variable. (This frequently occurs because the equation studied is linear.) But it would be wrong to conclude that this is generally the case. As soon as one ventures outside the realm of linear equations, the domain of a solution, that is, the set of values of the independent variable for which the solution is defined, becomes a serious question. Even for a very simple nonlinear equation, the domain of the solution may be unexpectedly small. As will be shown in Exercise 11 at the end of this chapter, it may be very important to investigate the domain of the solution before undertaking a numerical analysis.

Finally, we want to point out the relative importance of various parts of this chapter. The basic existence and uniqueness theorem is Existence Theorem 1.1 (Picard theorem) of this chapter. Anyone who expects to work with differential equations needs to be well-acquainted with this theorem, its results, and its limitations. The differentiability theorem and the existence theorem for an equation with a parameter describe properties of solutions which are also very important for later work.

Existence Theorems 1.2 and 1.4 are somewhat more sophisticated. They are interesting for a reader with some background in functional analysis, but unnecessary for later parts of this book. Existence Theorem 1.3 has more general interest. A weaker hypothesis (continuity) is used to prove existence of solutions (not necessarily unique). The method of proof is the Euler-Cauchy method, which is the basis for numerical studies of differential equations. Also the theorem yields a quick approach to the problem of estimating how large the domain of the solution is.

The second major topic in this chapter concerns the size of the domain of the solution (Extension Theorems 1.1, 1.2, and 1.3). This is a crucially important topic and requires careful attention both for later theory and for applications. Study of the domain of the solution is especially important because the student's previous experience with differential equations may be misleading. Judging on the basis of the first course in differential equations, the student may be inclined, quite reasonably, to conclude that if the functions which appear in the differential equation are defined and have, say, continuous second derivatives for all real values of the variables, then the same will be true of the solutions. This is far from true, and getting information about the domains of solutions is often not easy as we shall see.

Existence Theorem by Successive Approximations

First we introduce some notation and terminology. Throughout, all quantities will be real unless otherwise specified. Whenever we speak of Euclidean n-space (including the line or the plane), we will mean real Euclidean n-space. The domains and ranges of all functions will be subsets of real Euclidean spaces.

It is not purposeful to give a formal definition of a differential equation, we shall merely say that a *differential equation* or *system of differential equations* is a statement of equality relating a set of functions and their derivatives. We will consider differential equations in which there is just one independent variable relative to which the derivatives are taken. The differential equation is then said to be an *ordinary* differential equation. We study ordinary differential equations of the form

$$x_1' = f_1(t, x_1, \ldots, x_n)$$
$$x_2' = f_2(t, x_1, \ldots, x_n)$$
$$\ldots$$
$$x_n' = f_n(t, x_1, \ldots, x_n) \tag{1.1}$$

where $x_1, \ldots, x_n$ are real-valued functions of t; $x_1', \ldots, x_n'$ are the derivatives of $x_1, \ldots, x_n$; and $f_1, \ldots, f_n$ are real-valued functions of $t, x_1, \ldots, x_n$. Notice that by considering only equations of the form (1.1), we are implicitly using the assumption that our equations can be solved for the derivative. Thus we are excluding from our study equations of the form, for example,

$$a(t)x' + F(t, x) = 0$$

where $F(t, x)$ is a real-valued function and the function $a(t)$ is zero for some of the values of t being considered (i.e., the equation cannot be solved for x').

It is useful, indeed almost essential, to write equation (1.1) in vector form. For this, let x denote the n-vector

$$\begin{bmatrix} x_1 \\ \vdots \\ x_n \end{bmatrix}$$

let x' denote the n-vector

$$\begin{bmatrix} x_1' \\ \vdots \\ x_n' \end{bmatrix}$$

and let $f(t, x)$ denote the n-vector

$$\begin{bmatrix} f_1(t, x_1 \ldots, x_n) \\ \vdots \\ f_n(t, x_1, \ldots, x_n) \end{bmatrix}$$

For our convenience, we recall the following definitions. A vector function $f(t, x)$ is *continuous* on a set A if each of its components is continuous on A. A vector function

$$x(t) = \begin{bmatrix} x_1(t) \\ \vdots \\ x_n(t) \end{bmatrix}$$

is *differentiable* on (a, b) if each of these components $x_1(t), \ldots, x_n(t)$ is differentiable on (a, b). The *derivative of the vector function* $x(t)$ is

$$\frac{dx}{dt} = \begin{bmatrix} \frac{dx_1}{dt} \\ \vdots \\ \frac{dx_n}{dt} \end{bmatrix}$$

A vector function $x(t)$ is *integrable* on $[a, b]$ if each of the components is integrable on $[a, b]$. The *integral* of $x(t)$ on $[a, b]$ is

$$\int_a^b x(t)dt = \begin{bmatrix} \int_a^b x_1(t)dt \\ \vdots \\ \int_a^b x_n(t)dt \end{bmatrix}$$

The *norm* of the vector

$$x = \begin{bmatrix} x_1 \\ \vdots \\ x_n \end{bmatrix}$$

denoted by $|x|$, is $\sum_{i=1}^n |x|$. The norm $|x|$ is the most convenient for our use, but there are other norms in Euclidean n-space. The most familiar is $\|x\| = \{\sum_{i=1}^n |x_i|^2\}^{1/2}$. But it is easy to see that for all vectors x,

$$\|x\| \leq |x|$$

and

$$|x| \leq \sqrt{n}\, \|x\|$$

(See Exercise 2.) If inequalities of this kind hold, the norms give rise to the same topology and they are said to be *equivalent*.

Now (1.1) can be written in vector notation as

$$x' = f(t, x) \tag{1.2}$$

and will be referred to as a differential equation or as a system of differential equations. The choice of terminology will depend on whether we are primarily concerned with the components of x or x itself. If x is an n-vector, then (1.2) is sometimes called an *n-dimensional system*. (Unless otherwise stated, n will always denote the dimension of the system.) A differential equation such as (1.2) in which only first derivatives appear is called a *first-order equation*. If a derivative of nth-order of a component of x occurs in the differential equation but no derivative of order higher than n occurs, the differential equation is called an nth-*order differential equation or* nth-*order system of differential equations*. However, we need not institute a separate study for nth-order equations because nth-order systems are easily represented as first-order systems. To describe this representation, we show how it works for an example. Consider the nth-order equation

$$x^{(n)} = g(t, x, x^{(1)}, \ldots, x^{(n-1)}) \tag{1.3}$$

where $x^{(j)}$ denotes the jth derivative $(j = 1, \ldots, n)$ of a real-valued function of t. Equation (1.3) can be represented as a system of first-order equations, that is, as a system of the form (1.2) if we let

$$x_1 = x$$
$$x_2 = x'$$
$$x_3 = x''$$
$$\cdots$$
$$x_{n-1} = x^{(n-2)}$$
$$x_n = x^{(n-1)}$$

In this notation, equation (1.3) becomes the first-order system

$$x_1' = x_2$$
$$x_2' = x_3$$
$$\cdots$$
$$x_{n-1}' = x_n$$
$$x_n' = g(t, x_1, x_2, \ldots, x_n)$$

We shall be concerned with the existence of solutions of differential equations, and our first step is to make precise the concept of solution.

Definition   The *projection of a point* $(t, x_1, \ldots, x_n)$ on the t-axis is t. The *projection of a set* is the collection of the projections of the points in the set.

Definition   Let f be an n-vector function defined and continuous on a connected open set D in (t, x)-space (i.e., the Euclidean space of points $(t, x_1, \ldots, x_n)$, where $t, x_1, \ldots, x_n$ are real numbers), and let $I = (a, b)$ be an open interval on the t-axis (I may be finite or infinite, that is, a may be a number or $-\infty$ and b may be a number or $+\infty$) such that the projection of D on the t-axis contains I. Let $x(t)$ be an n-vector function defined and differentiable on (a, b) such that $(t, x(t)) \in D$ for all $t \in I$. Then $x(t)$ is a *solution on* I *of the* differential equation

$$x' = f(t, x)$$

if for each $t \in I$,

$$\frac{d}{dt}x(t) = f[t, x(t)]$$

(Sometimes it will be convenient to refer to solutions defined on an interval J which includes one or both of its endpoints. This will mean that either the solution is defined on an open interval which contains J or that one-sided derivatives are considered at the included endpoints.)

Notice that we do not require that the domain of the solution $x(t)$ be in any sense maximal. That is, there may exist an interval (c, d) which contains (a, b) properly and a solution $y(t)$ on (c, d) of the differential equation such that $y|(a, b) = x$, where $y|(a, b)$ denotes the function $y(t)$ on the domain (a, b). The solution $y(t)$ is then called an *extension* of the solution $x(t)$, and we say that $x(t)$ is *extended* (to the solution $y(t)$). Later we will study the question of whether such extensions exist.

Before introducing our first existence theorem, we make more precise the question of whether there exists a solution. The reason for doing this is that if there is a solution of the differential equation, there is frequently an infinite set of solutions. For example, the differential equation

$$x' = t$$

has the set of solutions

$$x(t) = \frac{t^2}{2} + c$$

where c is an arbitrary constant. Consequently, instead of seeking just any solution of the differential equation, we search for a solution which satisfies a special condition: That is, we look for a solution which has a given value x^0 when t has a given value t_0. Such a solution is denoted by $x(t, t_0, x^0)$. We say that the solution $x(t, t_0, x^0)$ *satisfies the initial condition* that it has the value x^0 when $t = t_0$. Writing the initial condition in the form of an equation, we have

$$x(t_0, t_0, x^0) = x^0$$

Our existence theorems are designed to answer the question: Does there exist a solution of the differential equation which satisfies a given initial condition? (This question is sometimes called the *initial value problem*.)

For the basic existence theorem, we use the concept of a Lipschitz condition on a function.

Definition   Suppose function $f(t, x)$ has domain D in (t, x)-space, and suppose there exists a constant $k > 0$ such that if $(t, x^1), (t, x^2) \in D$, then

$$|f(t, x^1) - f(t, x^2)| \le k|x^1 - x^2|$$

Then f satisfies a *Lipschitz condition* with respect to x in D, and k is a *Lipschitz constant* for f.

It is clear that if f satisfies a Lipschitz condition with respect to x in D, then for each fixed t, $f(t, x)$ is a continuous function of x. However, for a fixed x, $f(t, x)$ need not be a continuous function of t. For example, let

$$f(t, x) = 1 \qquad \text{for } t \geq 0, \ x \text{ real}$$
$$f(t, x) = 0 \qquad \text{for } t < 0, \ x \text{ real}$$

This function satisfies a Lipschitz condition with respect to x (let k be any positive constant) but is not continuous as a function of t.

A simple sufficient condition that f satisfy a Lipschitz condition is obtained as follows. Suppose that D is an open set in the (t, x)-plane (here x denotes a 1-vector) such that if (t, x^1) and (t, x^2) are in D, then $(t, x^1 + \theta(x^2 - x^1))$, where $0 \leq \theta \leq 1$, is in D, and suppose that $\partial f / \partial x$ exists and is bounded on D. Then by the Mean Value Theorem

$$f(t, x^2) - f(t, x^1) = \left\{ \frac{\partial f}{\partial x}[t, x^1 + \theta_1(x^2 - x^1)] \right\} (x^2 - x^1)$$

where $\theta_1 \in (0, 1)$. Thus, if $|\partial f / \partial x| \leq M$ on D, then f satisfies a Lipschitz condition with respect to x on D with a Lipschitz constant equal to M. Similar considerations can be made, if x is an n-vector $(n > 1)$, by using the mean value theorem for functions of several variables.

But f may satisfy a Lipschitz condition even if $\partial f / \partial x$ is not defined, as the following example shows: Let

$$f(t, x) = |x|$$

for all (t, x) in the (t, x)-plane (here x denotes a 1-vector).

Finally a function may be uniformly continuous on D, an open set, and yet not satisfy a Lipschitz condition as shown in the following example. Let

$$f(t, x) = |x|^\alpha$$

where α is a constant such that $\alpha \in (0, 1)$, for all (t, x) in the (t, x)-plane. First $f(t, x)$ is continuous and hence uniformly continuous on

$$\{(t, x) | x \in [-1, 1], \ t \text{ real}\}$$

but if $x > 0$,

$$|f(t, x) - f(t, 0)| = x^\alpha = x(x^{\alpha-1}) = \frac{1}{x^{1-\alpha}} |x - 0|$$

As x approaches zero, $\frac{1}{x^{1-\alpha}}$ increases without bound. Thus $f(t, x)$ does not satisfy a Lipschitz condition.

Now we are ready to prove the basic existence theorem.

Existence Theorem 1.1 (Picard Theorem). *Let D be an open set in (t, x)-space. Let $(t_0, x^0) \in D$ and let a, b be positive constants such that the set*

$$R = \{(t, x) \mid |t - t_0| \leq a, \ |x - x^0| \leq b\}$$

is contained in D. Suppose function f is defined and continuous on D and satisfies a Lipschitz condition with respect to x in R. Let

$$M = \max_{(t,x)\in R} |f(t,x)|$$

$$A = \min\left[a, \frac{b}{M}\right]$$

Then the differential equation

$$x' = f(t,x) \tag{1.2}$$

has a unique solution $x(t, t_0, x^0)$ on $(t_0 - A, t_0 + A)$ such that $x(t_0, t_0, x^0) = x^0$. This solution $x(t, t_0, x^0)$ is such that

$$|x(t, t_0, x^0) - x^0| \le MA$$

for all $t \in (t_0 - A, t_0 + A)$.

Before proceeding to the proof of Existence Theorem 1.1, we make a few remarks concerning the hypotheses of the theorem and indicating which questions it answers and which questions it leaves unresolved. First, as we will see later, continuity of f is sufficient to ensure the existence of a solution. The additional condition that f satisfy a Lipschitz condition is really needed only to prove the uniqueness of solution. In the proof of Existence Theorem 1.1, we will use the Lipschitz condition on f to prove existence because the Lipschitz condition makes it possible to use the method of successive approximations. We obtain in this way an elementary, fairly short, and rather elegant proof. But in Existence Theorem 1.3, we will obtain existence of a solution by using only the continuity of f.

From the point of view of pure mathematics, it is vitally important to sort out exactly which hypotheses are needed for existence and which for uniqueness. This question is, however, somewhat less important than might first be thought. The reason is that much of the theory of ordinary differential equations is developed for systems of equations which satisfy hypotheses strong enough so that solutions are unique. In Chapter 2, we will see how useful the uniqueness is for obtaining information about solutions of linear systems. The theory developed for autonomous systems in Chapter 3 rests largely on the uniqueness condition. From the point of view of applications, uniqueness is all important because without uniqueness, the system of ordinary differential equations and its solutions cannot be used to make quantitative predictions about the behavior of a physical system. (Of course, the presence of chaos can be investigated.) Thus for many purposes, it is reasonable to impose at the beginning hypotheses strong enough to assure uniqueness of solution.

Existence Theorem 1.1 leaves unanswered two important kinds of questions. First, we often need to know whether a solution satisfies additional conditions: For example, is the solution continuous or differentiable with respect to the initial value x^0? Second, it is often important to be able to estimate how large a domain the solution has.

Existence Theorem 1.1 merely says that the interval $(t^0 - A, t_0 + A)$ is contained in the domain of the solution $x(t, t_0, x^0)$. The solution $x(t, t_0, x^0)$ might have a much larger domain, and we need to obtain criteria for determining if, for example, the domain includes all $t > \bar{t}$, some fixed value. These questions will be dealt with later in this chapter.

Now we proceed to the proof of Existence Theorem 1.1.

Lemma 1.1 *A necessary and sufficient condition that the function* $x(t, t_0, x^0)$, *continuous in t, which satisfies the condition*

$$x(t_0, t_0, x^0) = x^0$$

is a solution of (1.2) *on the interval* $(t_0 - r, t_0 + r)$ *where* $r > 0$, *is that* $x(t, t_0, x^0)$ *satisfy the equation:*

$$x(t, t_0, x^0) = x^0 + \int_{t_0}^{t} f[s, x(s, t_0, x^0)] \, ds \qquad (1.4)$$

for $t \in (t_0 - r, t_0 + r)$.

Proof If $x(t, t_0, x^0)$ satisfies (1.4), then since f and x are continuous, we can differentiate (1.4). The result is equation (1.2). If $x(t, t_0, x^0)$ satisfies (1.2), that is, if

$$\frac{dx}{dt} = f[t, x(t, t_0, x^0)]$$

then taking the definite integral from t_0 to t, where $t \in (t_0 - r, t_0 + r)$ on both sides of equation, we obtain:

$$\int_{t_0}^{t} \frac{dx}{ds} \, ds = x(t, t_0, x^0) - x(t_0, t_0, x^0)$$

$$= x(t, t_0, x^0) - x^0 = \int_{t_0}^{t} f[s, x(s, t_0, x^0)] \, ds \qquad \square$$

The remainder of the proof of the theorem consists of showing that the sequence

$$x_0(t) = x^0$$

$$x_1(t) = x^0 + \int_{t_0}^{t} f[s, x_0(s)] \, ds$$

$$x_{m+1}(t) = x^0 + \int_{t_0}^{t} f[s, x_m(s)] \, ds \quad (m = 1, 2, \ldots)$$

converges on $[t_0 - A, t_0 + A]$ to a function which is a solution on $(t_0 - A, t_0 + A)$ of (1.4) and then showing that this solution is unique. We will actually show that the sequence converges on $[t_0, t_0 + A]$ and that the limit function is a solution on $[t_0, t_0 + A]$ of (1.4). Similar treatment can be made on $[t_0 - A, t_0]$.

Lemma 1.2   *For each m, the function $x_m(t)$ is defined and continuous on $[t_0, t_0 + A]$ and if $t \in [t_0, t_0 + A]$, then*

$$|x_m(t) - x^0| \le M|t - t_0|$$

Proof   The proof is by induction. If $m = 0$, the statement is obviously true. If the statement is true for $m = q$, then for $t \in [t_0, t_0 + A]$,

$$|x_q(t) - x^0| \le MA \le b$$

Therefore $f[t, x_q(t)]$ is defined for $t \in [t_0, t_0 + A]$. Since $f[t, x_q(t)]$ is a continuous function of t, then

$$x_{q+1}(t) = x^0 + \int_{t_0}^{t} f[s, x_q(s)] \, ds$$

is defined and continuous. Also

$$|x_{q+1}(t) - x^0| = \left| \int_{t_0}^{t} f[s, x_q(s)] \, ds \right| \le M(t - t_0) \qquad\qquad \Box$$

Lemma 1.3   *The sequence $\{x_m(t)\}$ converges uniformly on $[t_0, t_0 + A]$ to a continuous function $x(t)$.*

Proof   We will prove that the series

$$x_0(t) + \sum_{n=0}^{\infty} [x_{n+1}(t) - x_n(t)] \tag{1.5}$$

converges uniformly on $[t_0, t_0 + A]$. For $t \in [t_0, t_0 + A]$, let

$$d_n(t) = |x_{n+1}(t) - x_n(t)|$$

Then for each n,

$$d_n(t) = \left| \int_{t_0}^{t} \{ f[s, x_n(s)] - f[s, x_{n-1}(s)] \} \, ds \right|$$

$$\le \int_{t_0}^{t} |f[s, x_n(s)] - f[s, x_{n-1}(s)]| \, ds$$

$$\le k \int_{t_0}^{t} |x_n(s) - x_{n-1}(s)| \, ds$$

$$= k \int_{t_0}^{t} d_{n-1}(s) \, ds$$

where k is a Lipschitz constant for f on R.

Next we obtain an estimate for $d_n(t)$ by induction. By Lemma 1.2, if $t \in [t_0, t_0 + A]$,

$$d_0(t) = |x_1(t) - x_0(t)| \leq M|t - t_0|$$

Assume that if $t \in [t_0, t_0 + A]$,

$$d_n(t) \leq \frac{M}{k} \frac{k^{n+1}(t - t_0)^{n+1}}{(n+1)!}$$

Then

$$d_{n+1}(t) \leq k \int_{t_0}^{t} d_n(s) \, ds \leq k \frac{M}{k} \frac{k^{n+1}}{(n+1)!} \int_{t_0}^{t} (s - t_0)^{n+1} \, ds$$

$$= \frac{M}{k} \frac{k^{n+2}}{(n+1)!} \frac{1}{n+2} (t - t_0)^{n+2}$$

Thus if $t \in [t_0, t_0 + A]$,

$$\sum_{n=0}^{\infty} d_n(t) \leq \frac{M}{k} \sum_{n=0}^{\infty} \frac{k^{n+1}(t - t_0)^{n+1}}{(n+1)!}$$

$$\leq \frac{M}{k} \sum_{n=0}^{\infty} \frac{k^{n+1} A^{n+1}}{(n+1)!}$$

$$= \frac{M}{k} \{\exp[kA] - 1\}$$

Thus the uniform convergence of (1.5) follows from the Weierstrass M-test or by a simple direct argument. □

The proof of Lemma 1.3 uses the convergence of the exponential series. An even simpler proof which uses the convergence of the geometric series runs as follows. First impose, if necessary, the additional condition that A is small enough so that

$$kA = r < 1$$

Then we have

$$\max_{t \in [t_0, t_0 + A]} |x_{n+1}(t) - x_n(t)|$$

$$\leq \max \int_{t_0}^{t} |f(s, x_n(s)) - f(s, x_{n-1}(s))| \, ds$$

$$\leq \max \int_{t_0}^{t} k|x_n(s) - x_{n-1}(s)| \, ds$$

$$\leq kA \max_{s \in [t_0, t_0 + A]} |x_n(s) - x_{n-1}(s)|$$

$$= r \max |x_n(s) - x_{n-1}(s)|$$

and hence

$$\max |x_{n+1}(t) - x_n(t)| \le r^n \max |x_1(s) - x^0|$$

The uniform convergence follows at once.

As we will see later, there is no significant advantage in using the exponential series in the proof of Lemma 1.3. The advantage of not requiring that kA be less than 1 is illusory. It turns out that it is not particularly important to try to maximize the interval of existence of the solution when proving the basic existence theorems. The question which requires detailed study is how far the solution can be extended. We consider this question in the extension theorems later in this chapter.

Lemma 1.4   *The function $x(t)$ is a solution of (1.4) such that $x(t_0) = x^0$.*

Proof   First we show that for $t \in [t_0, t_0 + A]$,

$$|x(t) - x^0| \le b$$

and hence that for all $t \in [t_0, t_0 + A]$, $f[t, x(t)]$ is defined. If $t \in [t_0, t_0 + A]$ and if $\varepsilon > 0$, then if m is sufficiently large,

$$|x(t) - x^0| \le |x(t) - x_m(t)| + |x_m(t) - x^0| < \varepsilon + M(t - t_0)$$

Therefore

$$|x(t) - x^0| \le M(t - t_0) \le MA \le b$$

By the Lipschitz condition on f, we have for $\varepsilon > 0$,

$$\left| \int_{t_0}^{t} \{f[s, x(s)] - f[s, x_m(s)]\} \, ds \right|$$

$$\le \int_{t_0}^{t} |f[s, x(s)] - f[s, x_m(s)]| \, ds$$

$$\le k \int_{t_0}^{t} |x(s) - x_m(s)| \, ds$$

$$\le k\varepsilon(t - t_0) \text{ if } m \text{ is sufficiently large}$$

$$\le k\varepsilon A$$

Therefore

$$\lim_{m \to \infty} \int_{t_0}^{t} f[s, x_m(s)] \, ds = \int_{t_0}^{t} f[s, x(s)] \, ds$$

Taking the limit in m on both sides of the equation

$$x_{m+1}(t) = x^0 + \int_{t_0}^{t} f[s, x_m(s)] \, ds$$

we obtain

$$x(t) = x^0 + \int_{t_0}^t f[s, x(s)] \, ds$$

$\square$

Lemma 1.5   *The solution $x(t)$ of (1.4), which satisfies the initial condition*

$$x(t_0) = x^0$$

is the only solution of (1.2) which satisfies this initial condition.

Proof   Suppose there exist solutions $x(t)$ and $\bar{x}(t)$ of (1.4) on an interval $(t_0 - r, t_0 + r)$, where r is a positive number, such that $x(t_0) = \bar{x}(t_0) = x_0$. We obtain an estimate on $|x(t) - \bar{x}(t)|$ for $t \in [t_0, t_0 + r - \delta]$, where $0 < \delta < r$ and δ is fixed, as follows. Since $x(t), \bar{x}(t)$ are continuous on $[t_0, t_0 + r - \delta]$ for fixed δ, there exists $B > 0$ such that if $t \in [t_0, t_0 + r - \delta]$, then

$$|x(t) - \bar{x}(t)| \le B$$

But

$$
\begin{aligned}
|x(t) - \bar{x}(t)| \quad &\le \int_{t_0}^t |f[s, x(s)] - f[s, \bar{x}(s)]| \, ds \\
&\le k \int_{t_0}^t |x(s) - \bar{x}(s)| \, ds
\end{aligned}
$$

Therefore

$$|x(t) - \bar{x}(t)| \le k B(t - t_0)$$

Assume that $|x(t) - \bar{x}(t)| \le [k^m/m!] \, B(t - t_0)^m$, for m a positive integer. Then by (4), $|x(t) - \bar{x}(t)| \le [k^{m+1}/(m+1)!] \, [B(t - t_0)^{m+1}]$, which is the $(m+2)$th term in the (convergent) series for $Be^{k(t-t_0)}$. Therefore $|x(t) - \bar{x}(t)| < \varepsilon$; hence $x(t) = \bar{x}(t)$ for $t \in [t_0, t_0 + r - \delta]$. Since δ is arbitrarily small, $x(t) = \bar{x}(t)$ for $t \in (t_0, t_0 + r)$. A similar argument holds for $t \in (t_0 - r, t_0)$.

We emphasize that the uniqueness result given by Lemma 1.5, although easy to prove, is crucially important both in later development of the theory and in applications of ordinary differential equations.

This completes the proof of Existence Theorem 1.1. (For sketches of typical solutions, see Figure 1.1.) $\square$

In Figure 1.1, if $\frac{b}{M} < a$, the solution is defined on

$$\left[t_0 - \frac{b}{M}, t_0 + \frac{b}{M} \right] \subseteq [t_0 - a, t_0 + a]$$

and $A = \frac{b}{M}$, and solution $x(t)$ satisfies the condition

$$|x(t) - x^0| \le M|t - t_0|$$

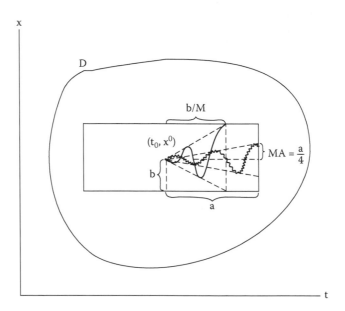

Figure 1.1

A possible solution curve, if $\frac{b}{M} < a$, is indicated with a solid line in Figure 1.1. If $\frac{b}{M} \geq a$, the solution is defined on $[t_0 - a, t_0 + a]$. If, for example, $M = \frac{1}{4}$, then $MA = \frac{a}{4}$. A possible solution curve, if $\frac{b}{M} \geq a$ is sketched with a "furry" line in Figure 1.1.

Corollary 1.1   *The solution $x(t, t_0, x^0)$ is a continuous function of (t, x^0). That is, given $\varepsilon > 0$, then there exists $\delta > 0$ such that if (t_1, x_1^0), (t_2, x_2^0) are points in the interior of the set R and if*

$$\left| \left(t_1, x_1^0 \right) - \left(t_2, x_2^0 \right) \right| < \delta$$

then

$$\left| x \left(t_1, t_0, x_1^0 \right) - x \left(t_2, t_0, x_2^0 \right) \right| < \varepsilon$$

Proof   See Exercise 8.   ☐

Corollary 1.1 is an important result in applying differential equations to describe physical and biological systems because the specification of x^0 is only an approximate description of a physical or biological condition.

Differentiability Theorem

It is important sometimes to use the fact that if f has continuous second derivatives, then the solution is a differentiable function of x^0. For example, we use it in Chapters 7 and 8 in applying the Poincaré method to study periodic solution.

Differentiability Theorem. *Let D be an open set in (t, x)-space. Let $(t_0, x^0) \in D$ and suppose function $f(t, x)$ is defined and has continuous second partial derivatives with respect to all variables at each point of D. Then there exist positive numbers a, b such that if $|\bar{x} - x^0| \leq b$, the differential equation*

$$x' = f(t, x) \tag{1.6}$$

has a unique solution $x(t, t_0, \bar{x})$ on $(t_0 - a, t_0 + a)$ with

$$x(t_0, t_0, \bar{x}) = \bar{x}$$

and such that at each point of the set,

$$S = \{(t, x)/|t - t_0| < a, |x - x^0| < b\}$$

the function $x(t, t_0, \bar{x})$ has a continuous third partial derivative with respect to t and has continuous first partial derivatives with respect to each component of $\bar{x}$.

Proof By Existence Theorem 1.1, there is a solution $x(t, t_0, \bar{x})$ of (1.2), that is,

$$\frac{\partial x}{\partial t}(t, t_0, \bar{x}) = f[t, x(t, t_0, \bar{x})] \tag{1.7}$$

Since f is continuous, then $\partial x/\partial t$ is a continuous function of $(t, \bar{x})$. Since f has continuous first and second partial derivatives, then using the chain rule to differentiate (1.6), we find that $\partial^2 x/\partial t^2$ and $\partial^3 x/\partial t^3$ are continuous functions of $(t, \bar{x})$.

By Lemma 1.1, we know that if $x(t, t_0, \bar{x})$ is a solution of (1.2) such that $x(t_0, t_0, \bar{x}) = \bar{x}$, then

$$x(t, t_0, \bar{x}) = \bar{x} + \int_{t_0}^{t} f[s, x(s, t_0, \bar{x})] \, ds \tag{1.8}$$

Now if such a solution $x(t, t_0, \bar{x})$ exists and is differentiable with respect to the components $\bar{x}_1, \ldots, \bar{x}_n$ of $\bar{x}$, then we can differentiate both sides of (1.7) with respect to $\bar{x}_i$ and obtain

$$\frac{\partial x}{\partial \bar{x}_i} = I_i + \int_{t_0}^{t} \left[\frac{\partial f}{\partial x}[s, x(s, t_0, \bar{x})] \right] \left[\frac{\partial x}{\partial \bar{x}_i} \right] ds \tag{1.9}$$

where

$$I_i = \begin{bmatrix} 0 \\ \vdots \\ 1 \\ \vdots \\ 0 \end{bmatrix}$$

that is, I_i has a 1 in the ith position and 0 elsewhere, and

$$\left[\frac{\partial f}{\partial x} [s, x(s, t_0, \bar{x})] \right]$$

is the $n \times n$ matrix whose entry in the (i, j) position is

$$\frac{\partial f_i}{\partial x_j} [s, x(s, t_0, \bar{x})]$$

This suggests that in order to obtain a solution of (1.2) which is differentiable with respect to the components of $\bar{x}$, the system of equations described by (1.8) and (1.9) should be solved. This is the procedure used. We set up the system of integral equations

$$x(t) = \bar{x} + \int_{t_0}^{t} f[s, x(s)] \, ds$$

$$y^{(i)}(t) = I_i + \int_{t_0}^{t} \left[\frac{\partial f}{\partial x} [s, x(s)] \right] [y^{(i)}(s)] \, ds \qquad (1.10)$$

$$(i = 1, \dots, n)$$

Since f has continuous second partial derivatives with respect to x, it follows that $\frac{\partial f_i}{\partial x_j}(t, x)$ satisfies a Lipschitz condition in x. Solving (1.10) for $x(t)$ and $y^{(1)}(t), \dots, y^{(n)}(t)$ is equivalent to solving (1.8) and (1.9) for $x(t, t_0, \bar{x})$ and $\frac{\partial x}{\partial \bar{x}_i}$, $i = 1, \dots, n$.

Let $(x(t, t_0, \bar{x}), y(t, t_0, \bar{x}))$ denote a solution of (1.10) obtained by applying the successive approximation method used to prove Existence Theorem 1.1. Then $(x(t, t_0, \bar{x}), y^{(1)}(t, t_0, \bar{x}), \dots, y^{(n)}(t, t_0, \bar{x})$ is the limit (in uniform convergence) on a set

$$\mathcal{R} = \{(t, x)/|t - t_0| \le a, |x - \bar{x}| \le b\}$$

of the sequence $\{(x_m(t, t_0, \bar{x}), y_m^{(1)}(t, t_0, \bar{x}), \dots, y_m^{(n)}(t, t_0, \bar{x})\}$ where

$$(x_0(t, t_0, \bar{x}), y_0^{(1)}(t, t_0, \bar{x}), \dots, y_0^{(n)}(t, t_0, \bar{x})) = (\bar{x}, 1, \dots, 1) \qquad (1.11)$$

and

$$x_{m+1}(t, t_0, \bar{x}) = \bar{x} + \int_{t_0}^{t} f[s, x_m(s, t_0, \bar{x})] \, ds \qquad (1.12)$$

$$y_{m+1}^{(i)}(t, t_0, \bar{x}) = I_i + \int_{t_0}^{t} \left[\frac{\partial f}{\partial x}[s, x_m(s, t_0, \bar{x})] \right] \left[y_m^{(i)}(s, t_0, \bar{x}) \right] ds \qquad (1.13)$$

By (1.11), we have $y_0^{(i)} = \frac{\partial x_0}{\partial \bar{x}_i}$. Now suppose that for some fixed m and for all $(t, x) \in \text{Int } \mathcal{R}$, it is true that $\frac{\partial x_m}{\partial \bar{x}_i} = y_m^{(i)}$, for $i = 1, \ldots, n$. Differentiating (1.12) with respect to $\bar{x}_i$, we obtain

$$\begin{aligned}
\frac{\partial x_{m+1}}{\partial \bar{x}_i} &= I_i + \int_{t_0}^{t} \left[\frac{\partial f}{\partial x}[s, x_m(s, t_0, \bar{x})] \right] \left[\frac{\partial x_m}{\partial \bar{x}_i}(s, t_0, \bar{x}) \right] ds \\
&= I_i + \int_{t_0}^{t} \left[\frac{\partial f}{\partial x}[s, x_m(s, t_0, \bar{x})] \right] \left[y_m^{(i)}(s, t_0, \bar{x}) \right] ds \\
&= y_m^{(i)}(t, t_0, \bar{x})
\end{aligned}$$

Thus by induction, it follows that for all m and for all $(t, x) \in \text{Int } \mathcal{R}$, it is true that $\frac{\partial x_m}{\partial \bar{x}_i} = y_m^{(i)}$. But the sequence

$$\left\{ y_m^{(i)}(t, t_0, \bar{x}) \right\} = \left\{ \frac{\partial x_m}{\partial \bar{x}_i}(t, t_0, \bar{x}) \right\}$$

converges uniformly on $\mathcal{R}$ to $y^{(i)}(t, t_0, \bar{x})$. Hence by a standard convergence theorem from calculus, it follows that

$$y^{(i)}(t, t_0, \bar{x}) = \frac{\partial x}{\partial \bar{x}_i}(t, t_0, \bar{x}), \quad i = 1, \ldots, n$$

This completes the proof of the differentiability theorem. ☐

Existence Theorem for Equation with a Parameter

It is natural to suppose that if $f(t, x, \mu)$ is continuous on (t, x)-space and is also continuous in a parameter μ, then solutions of the equation

$$x' = f(t, x, \mu)$$

also depend continuously on the parameter μ. For later work, we will need a result of this kind.

Existence Theorem for Equation with a Parameter. *Consider the equation*

$$\frac{dx}{dt} = f(t, x, \mu) \qquad (1.14)$$

and suppose that f is defined and continuous on

$$(a, b) \times \bar{G} \times [-\tilde{\mu}, \tilde{\mu}]$$

where $a < b$, and $\bar{G}$ is the closure of a bounded open set G in R^n, and $\tilde{\mu} > 0$. Suppose that the function $f(t, x, \mu)$ satisfies a Lipschitz condition with respect to x in $(a, b) \times \bar{G} \times [-\tilde{\mu}, \tilde{\mu}]$ where the Lipschitz constant is independent of μ. Let $t_0 \in (a, b)$ and let $x(t, x^0, \mu)$ be the solution of (1.14) such that

$$x(t_0, x^0, \mu) = x^0$$

where $x_0 \in G$. Then $x(t, x^0, \mu)$ is a continuous function of (x^0, μ) and there exists $r > 0$ such that $x(t, x^0, \mu)$ is continuous in (x^0, μ) uniformly for $t \in [t_0 - r, t_0 + r]$.

Proof   Let $\mu_1, \mu_2 \in [-\tilde{\mu}, \tilde{\mu}]$ and suppose $x_1^0, x_2^0 \in G$. By Lemma 1.1, we have if $|t - t_0| \leq r$, where r is a sufficiently small positive number, then

$$
\begin{aligned}
\left| x\left(t, x_2^0, \mu_2\right) - x\left(t, x_1^0, \mu_1\right)\right| &= \left| x_2^0 - x_1^0 + \int_{t_0}^t \left\{ f\left[s, x\left(s, x_2^0, \mu_2\right), \mu_2\right] \right.\right. \\
&\quad \left.\left. - f\left[s, x\left(s, x_1^0, \mu_1\right), \mu_1\right]\right\} ds\right| \\
&= \left| x_2^0 - x_1^0 + \int_{t_0}^t \left\{ f\left[s, x\left(s, x_2^0, \mu_2\right), \mu_2\right] - f\left[s, x\left(s, x_1^0, \mu_1\right), \mu_2\right]\right.\right. \\
&\quad \left.\left. + f\left[s, x\left(s, x_1^0, \mu_1\right), \mu_2\right] - f\left[s, x\left(s, x_1^0, \mu_1\right), \mu_1\right]\right\} ds\right| \\
&\leq \left| x_2^0 - x_1^0\right| + k|t - t_0| \max_{|t-t_0|\leq r} \left| x\left(t, x_2^0, \mu_2\right) - x\left(t, x_1^0, \mu_1\right)\right| \\
&\quad + |t - t_0| \max_{|t-t_0|\leq r} \left| f\left[t, x\left(t, x_1^0, \mu_1\right), \mu_2\right] - f\left[t, x\left(t, x_1^0, \mu_1\right), \mu_1\right]\right|
\end{aligned}
$$

$$(1.15)$$

where k is the Lipschitz constant given by hypothesis. Now we choose r small enough so that

$$k|t - t_0| \leq \frac{1}{2}. \tag{1.16}$$

Applying (1.16) to (1.15), we obtain

$$
\begin{aligned}
\max_{|t-t_0|\leq r} \left| x\left(t, x_2^0, \mu_2\right) - x\left(t, x_1^0, \mu_1\right)\right| &\leq \left| x_2^0 - x_1^0\right| \\
&\quad + \frac{1}{2} \max_{|t-t_0|\leq r} \left| x\left(t, x_2^0, \mu_2\right) - x\left(t, x_1^0, \mu_1\right)\right| \\
&\quad + r \max_{|t-t_0|\leq r} \left| f\left[t, x\left(t, x_1^0, \mu_1\right), \mu_2\right] - f\left[t, x\left(t, x_1^0, \mu_1\right), \mu_1\right]\right|
\end{aligned}
$$

and hence

$$
\begin{aligned}
\max_{|t-t_0|\leq r} &\left| x\left(t, x_2^0, \mu_2\right) - x\left(t, x_1^0, \mu_1\right)\right| \\
&\leq 2\left| x_2^0 - x_1^0\right| + 2r \max_{|t-t_0|\leq r} \left| f\left[t, x\left(t, x_1^0, \mu_1\right), \mu_2\right]\right.\\
&\quad \left. - f\left[t, x\left(t, x_1^0, \mu_1\right), \mu_1\right]\right|
\end{aligned}
$$

$$(1.17)$$

But $f[t, x(t, x_1^0, \mu_1), \mu]$ is uniformly continuous on

$$\{t/|t - t_0| \le r\} \times \{\mu/ - \tilde{\mu} \le \mu \le \tilde{\mu}\}.$$

Hence if $|x_2^0 - x_1^0|$ and $|\mu_2 - \mu_1|$ are sufficiently small it follows from (1.17) that

$$\max_{|t - t_0| \le r} \left| x\left(t, x_2^0, \mu_2\right) - x\left(t, x_1^0, \mu_1\right) \right|$$

can be made less than ε. Thus we have shown that $x(t, x^0, \mu)$ is continuous at (x_1^0, μ_1) uniformly for $|t - t_0| \le r$. □

Note 1: The proof shows that the hypothesis concerning the Lipschitz condition of f can be weakened to a local Lipschitz condition. Stating the weaker condition is messy and this messiness outweighs the theoretical advantage.

Note 2: For an extended version of the existence theorem for an equation with a parameter, see Exercise 13.

Existence Theorem Proved by Using a Contraction Mapping

The technique used in the proof of Existence Theorem 1.1, the method of successive approximations, is used in many parts of analysis, and it is used in a generalized form to prove a standard existence theorem in functional analysis, the banach fixed point theorem or principle of contraction mappings (see appendix).

Our next step is to prove a slightly different version of the basic existence theorem by using the banach fixed point theorem. Using a result from functional analysis makes the proof simpler and shorter. After we obtain our second version of the existence theorem we will make a more extensive comparison of it with Existence Theorem 1.1.

As before, let D be an open set in (t, x)-space and suppose that $(t_0, x^0) \in D$. Let A, B be positive numbers such that the set

$$\mathcal{D} = \{(t, x)/|t - t_0| \le A, |x - x^0| \le B\}$$

is contained in D. Let

$$M = \max_{(t,x) \in \mathcal{D}} |f(t, x)|$$

Let a, b be positive numbers which are small enough so that if $|x^1 - x^0| \le b$, then the "cone"

$$\mathcal{K}_{x^1} = \{(t, x)/|t - t_0| \le a, |x - x^1| \le M|t - t_0|\}$$

is contained in $\mathcal{D}$. See Figure 1.2 where the constants a, b are such that $Ma \le B - b$ and the set $\mathcal{K}_{x^1}$, where $x^1 - x^0 = b$, is shaded vertically.

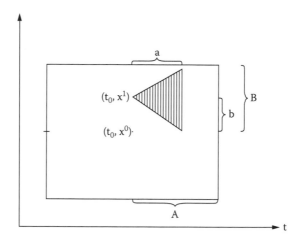

Figure 1.2

Let $\mathcal{M}$ be the space of continuous mappings m from the domain

$$\mathcal{R} = \{(t, x)/|t - t_0| \le a; |x - x^0| \le b\}$$

into R^n and let $\mathcal{E}$ be the set

$$\mathcal{E} = \{m \in \mathcal{M}/|m(t, x)| \le M|t - t_0|\}$$

It is easy to show that the function ρ, with domain $\mathcal{E} \times \mathcal{E}$, defined by

$$\rho(m_1, m_2) = \max_{(t,x)\in\mathcal{R}} |m_1(t, x) - m_2(t, x)|$$

where $m_1, m_2 \in \mathcal{E}$, is a metric on $\mathcal{E}$. By standard theorems from calculus, it follows that $\mathcal{E}$ is a complete metric space.

Now let S be the mapping from domain $\mathcal{E}$ into $\mathcal{M}$ such that if $m \in \mathcal{E}$, then Sm is defined by:

$$(Sm)(t, x) = \int_{t_0}^{t} f[s, x + m(s, x)]\, ds \tag{1.18}$$

We suppose that f satisfies a Lipschitz condition (with Lipschitz constant k) on $\mathcal{R}$.

Lemma 1.6   *If the positive number a is sufficiently small, mapping S is a contraction mapping from $\mathcal{E}$ into $\mathcal{E}$. That is, there exists a real number $q \in (0, 1)$ such that if $m_1, m_2 \in \mathcal{E}$, then $Sm_1, Sm_2 \in \mathcal{E}$ and*

$$\rho(Sm_1, Sm_2) \le q[\rho(m_1, m_2)]. \tag{1.19}$$

Proof   We must show:

1. S is well-defined, that is, the point $(s, x + m(s, x))$ is in the domain of mapping f so that the right-hand of (1.18) makes sense;

2. S takes $\mathcal{E}$ into $\mathcal{E}$, that is, if $m \in \mathcal{E}$, then $Sm \in \mathcal{E}$;

3. If a is sufficiently small, then S satisfies inequality (1.19).

Proof of 1. Let $(s, x) \in \mathcal{R}$ and let

$$K_x = \{(t, y)/|t - t_0| \leq a, |y - x| \leq M|t - t_0|\}$$

Then $(s, x + m(s, x))$ is contained in K_x because since $(s, x) \in \mathcal{R}$, then

$$|s - t_0| \leq a$$

and

$$|[x + m(s, x)] - x| = |m(s, x)| \leq M|s - t_0|$$

But $K_x \subset \mathcal{D}$ and $\mathcal{D}$ is a subset of the domain of f.

Proof of 2. By standard theorems from calculus, Sm is continuous on $\mathcal{R}$ and

$$|(Sm)(t, x)| \leq M|t - t_0|$$

Proof of 3. Since f satisfies a Lipschitz condition with Lipschitz constant k on $\mathcal{R}$, then

$$
\begin{aligned}
|(Sm_1)(t, x) - (Sm_2)(t, x)| \\
\leq \int_{t_0}^{t} |f[s, x + m_1(s, x)] - f[s, x + m_2(s, x)]|\, ds \\
\leq k \int_{t_0}^{t} |m_1(s, x) - m_2(s, x)|\, ds \\
\leq k[\rho(m_1, m_2)]|t - t_0| \\
\leq ka\rho(m_1, m_2)
\end{aligned}
$$

Thus

$$\rho[S(m_1) - S(m_2)] \leq ka\rho(m_1, m_2)$$

and if $ka < 1$, inequality (1.19) is satisfied. This completes the proof of Lemma 1.6.

$\square$

Remark.   Since k is a Lipschitz constant for f on the set $\mathcal{R}$, then if we choose a small enough so that $ka < 1$, the set $\mathcal{R}$ is either unchanged or made smaller. Hence k is still a Lipschitz constant for f on the (possibly smaller) set $\mathcal{R}$.

Lemma 1.6 permits us to apply the Banach Fixed Point Theorem or Principle of Contraction Mappings because Lemma 1.6 shows that $\mathcal{S}$ is a contraction mapping from the complete metric space $\mathcal{E}$ into itself. The Banach Fixed Point Theorem shows that mapping $\mathcal{S}$ has a unique fixed point $\bar{m}(t, x) \in \mathcal{E}$, that is,

$$\bar{m}(t, x) = \int_{t_0}^{t} f[s, x + \bar{m}(s, x)] \, ds \tag{1.20}$$

Let $x = \bar{x}$, where $|\bar{x} - x^0| \leq b$, be fixed in (1.20) and add $\bar{x}$ to each side of (1.20). We obtain

$$\bar{x} + \bar{m}(t, \bar{x}) = \bar{x} + \int_{t_0}^{t} f[s, \bar{x} + \bar{m}(s, \bar{x})] \, ds \tag{1.21}$$

Differentiation with respect to t on both sides of (1.21) shows that $\bar{x} + \bar{m}(t, x)$ is a solution $x(t)$ of (1.2) such that

$$x(t_0) = \bar{x} \tag{1.22}$$

Moreover $\bar{x} + \bar{m}(t, \bar{x})$ is a continuous function of $\bar{x}$.

By Lemma 1.1,

$$x(t) = \bar{x} + \int_{t_0}^{t} f[s, x(s)] \, ds$$

and hence

$$|x(t) - \bar{x}| \leq M|t - t_0|$$

Since $\bar{m}(t, x)$ is a unique fixed point in $\mathcal{E}$ of $\mathcal{S}$, it follows that $x(t)$ is a unique solution of (1.2) satisfying the initial condition (1.22). Hence we obtain the following somewhat extended version of our existence theorem.

Existence Theorem 1.2   *Let D be an open set in (t, x)-space. Let $(t_0, x^0) \in D$, and suppose function f is defined and continuous on D and satisfies a Lipschitz condition with respect to x on D. Then there exist positive numbers a, b such that if $|\bar{x} - x^0| \leq b$, the differential equation*

$$x' = f(t, x)$$

has a unique solution $x(t, t_0, \bar{x})$ on $(t_0 - a, t_0 + a)$ such that

$$x(t_0, t_0, \bar{x}) = \bar{x}$$

and this solution is a continuous function of $(t, \bar{x})$.

In comparing Existence Theorems 1.1 and 1.2, we notice first a fundamental similarity: both are proved by using successive approximations. Successive approximations are used directly in the proof of Existence Theorem 1.1. In Existence Theorem 1.2, the proof is obtained by applying the Banach Fixed Point Theorem, and the Banach Theorem, in turn, is proved by using successive approximations.

Existence Theorem 1.1 has, of course, the advantage that its proof is straightforward and used only results from calculus. Its second advantage is that the domain of the solution, that is, the interval $(t_0 - A, t_0 + A)$, is easily computed $(A = \min(a, b/M))$ whereas the domain of the solution obtained in Existence Theorem 1.2 is not described as explicitly and, moreover, it may be smaller than the domain obtained in Existence Theorem 1.1. For Existence Theorem 1.1, the numbers a and b simply indicate the dimensions of a rectangle "centered about" (t_0, x^0) and contained in D whereas for Existence Theorem 1.2, the numbers a and b have to be small enough so that the "cone" K_x is contained in $\mathcal{D}$ (as shown in Figure 1.2). The numbers a and b have to be calculated in each case. Also, as Figure 1.2 shows, the numbers a and b may be considerably smaller than the numbers a and b used in Existence Theorem 1.1. Finally, as shown in the proof of Lemma 1.6, the number a used in Existence Theorem 1.2 must be smaller than the reciprocal of the Lipschitz constant of f. The numbers a and b used in Existence Theorem 1.1 are completely independent of the Lipschitz constant of f. These advantages of Existence Theorem 1.1 are more apparent than real because, as pointed out earlier, it is not important at this stage to try to maximize the interval of existence of the solution. (We take up this question in the Extension Theorems later in this chapter.) Existence Theorem 1.2 has, on the other hand, two significant advantages. Its proof uses a result from functional analysis (the Banach Fixed Point Theorem) but the very use of this abstract theorem relates the proof of Existence Theorem 1.2 to other parts of analysis where the Banach Fixed Point Theorem is used. Another advantage of Existence Theorem 1.2 is that part of the conclusion is the fact that the solution depends continuously on the initial condition x^0 whereas with Existence Theorem 1.1 we gave a separate proof of continuity (Corollary 1.1).

Existence Theorem without Uniqueness

We used the Lipschitz condition of f heavily in the proofs of both Existence Theorem 1.1 and Existence Theorem 1.2. We needed the Lipschitz condition to prove both existence and uniqueness. The next question we consider is the following: Is it possible to reduce the hypothesis on f? As we will see, if the Lipschitz condition on f is replaced by a mere continuity condition on f, then we can still prove the existence of a solution but we lose the uniqueness. First we describe a simple example which shows that if f does not satisfy a Lipschitz condition, then the solution may not be unique. Let x be a scalar (or 1-vector) and consider the equation

$$\frac{dx}{dt} = x^{1/3} \tag{1.23}$$

Let $k \in (0, 1)$ and define the function $x_k(t)$ as follows:

$$x_k(t) = 0 \qquad \text{for } t \in (0, k)$$

$$x_k(t) = \left[\frac{2}{3}(t - k) \right]^{3/2} \qquad \text{for } t \in (k, 1)$$

If $t_0 \in (k, 1)$ then the derivative of $x_k(t)$ at t_0 is

$$\frac{dx_k}{dt} = \frac{3}{2} \left[\frac{2}{3}(t_0 - k) \right]^{1/2} \left(\frac{2}{3} \right) = [x_k(t_0)]^{1/3}$$

If $t_0 \in (0, k)$, the derivative at t_0 is

$$\frac{dx_k}{dt} = 0 = [x_k(t_0)]^{1/3}$$

If $t_0 = k$, the left-hand derivative is clearly 0, and it is a short calculation to show that the right-hand derivative is zero. Hence $x_k(t)$ is a solution of (1.23). But for fixed $t_0 \in (0, 1)$, the initial condition $x(t_0) = 0$ is satisfied by each $x_k(t)$ with $k \geq t_0$.

For an example in which there is more than one solution for every initial condition, see Hartman [1964, p. 18].

We are left with the problem of proving the existence of a solution if f is merely continuous, that is, if f does not satisfy a Lipschitz condition. We will prove this existence essentially by using techniques from calculus. There are several reasons for proving such an existence result. First, from the pure mathematics viewpoint, we want to obtain as clear a picture as possible of what conditions are needed to insure existence. Second, the technique to be used in the proof is the basis for a method often used in numerical analysis. Finally, the theorem we will prove makes possible a quick approach to the problem of estimating how large the domain of the solution is.

Existence Theorem 1.3   *Let $\bar{G}$ be the closure of a bounded open set G in R^n and let $f_1, \ldots, f_n$ be real-valued and continuous on $\bar{G}$. Let t_0 be a fixed real number and let*

$$x^0 = \left(x_1^0, \ldots, x_n^0 \right)$$

be a fixed point in G. Then there exist functions $x_1(t), \ldots, x_n(t)$ satisfying the following conditions:

1. Each $x_i(t)$, $i = 1, \ldots, n$, is defined on a domain which contains the interval

$$I = \left[t_0 - \frac{d}{M\sqrt{n}}, t_0 + \frac{d}{M\sqrt{n}} \right]$$

where $d = \inf_{q \in \bar{G} - G} \|x^0 - q\|$ and $\|x^0 - q\|$ denotes the Euclidean norm and M is an upper bound for the set

$$\{ |f_i(x_1, \ldots, x_n)| / (x_1, \ldots, x_n) \in \bar{G}, i = 1, \ldots, n \}$$

(Since $x^0 \in G$, then $d > 0$. Since $\bar{G}$ is bounded, M is finite.)

2. $(x_1(t_0), \ldots, x_n(t_0)) = x^0$.

3. $(x_1(t), \ldots, x_n(t))$ *is a solution on* $(t_0 - \frac{d}{M\sqrt{n}}, t_0 + \frac{d}{M\sqrt{n}})$ *of the system*

$$x_i' = f_i(x_1, \ldots, x_n), \quad i = 1, \ldots, n \tag{1.24}$$

Remark. Notice first that Existence Theorem 1.3 applies only to autonomous equations, that is, systems in which the functions $f_i(1 = i, \ldots, n)$ are independent of t. However, this limitation is only apparent because the problem of solving a nonautonomous system can be reduced to the problem of solving an autonomous system by using the following procedure. Suppose we seek a solution $(x_1(t), \ldots, x_n(t))$ of the system

$$x_i' = g_i(t, x_1, \ldots, x_n), \quad i = 1, \ldots, n \tag{1.25}$$

such that

$$x_i(t_0) = x_i^0, \quad i = 1, \ldots, n \tag{1.26}$$

where the point $(t_0, x_1^0, \ldots, x_n^0)$ is in the interior of the domain of the functions g_i $(i = 1, \ldots, n)$. Consider the autonomous system

$$\frac{dx_1}{d\tau} = g_1(t, x_1, \ldots, x_n)$$

$$\cdots$$

$$\frac{dx_n}{d\tau} = g_n(t, x_1, \ldots, x_n)$$

$$\frac{dt}{d\tau} = 1 \tag{1.27}$$

Suppose that system (1.27) has a solution $(\bar{x}_1(\tau), \ldots, \bar{x}_n(\tau), t(\tau))$ such that

$$x_i(t_0), = x_i^0, \quad i = 1, \ldots, n$$
$$t(t_0) = t_0$$

Since $dt/d\tau = 1$, then $t(\tau) = \tau + c$ where c is a constant. Since $t(t_0) = t_0$, then $c = 0$ and $t(\tau) = \tau$. Thus $(\bar{x}_1(t), \ldots, \bar{x}_n(t))$ is a solution of (1.25) which satisfies the initial condition (1.26).

Proof of Existence Theorem 1.3. The method of proof is called the Euler or Cauchy-Euler method, and it is the basis for several methods in numerical analysis. First, we explain the underlying idea of the method for the two-dimensional case.

The set $\bar{G}$ is "chopped up" into small rectangles and subsets of rectangles by using lines parallel to the x_1-axis and the x_2-axis. See Figure 1.3. Draw the line segment which starts at the point (x_1^0, x_2^0), which is in one of the rectangles, and has slope

$$\frac{f_2(x_1^0, x_2^0)}{f_1(x_1^0, x_2^0)}$$

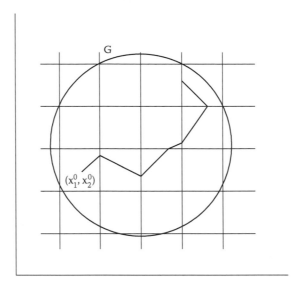

Figure 1.3

and extend the line segment until it reaches the boundary of the rectangle, say at the point (x_1^1, x_2^1). Take the line segment which starts at (x_1^1, x_2^1) and has slope

$$\frac{f_2(x_1^1, x_2^1)}{f_1(x_1^1, x_2^1)}$$

and extend the line segment until it reaches the boundary of the rectangle. Continue in this way and obtain a broken line as shown in Figure 1.3. Corresponding to each "chopping up" of $\bar{G}$ by lines parallel to the x_1-axis and the x_2-axis such a broken line is obtained. As smaller and smaller rectangles are used, the broken lines obtained become better and better approximations to a solution of the system

$$\begin{aligned} x_1' &= f_1(x_1, x_2) \\ x_2' &= f_2(x_1, x_2) \end{aligned}$$

The proof of Existence Theorem 1.3 consists in giving a rigorous, n-dimensional account of the procedure just described. It is interesting to notice that whereas the underlying idea of the method can be described almost instantaneously by using Figure 1.3, a detailed rigorous account, which follows, makes burdensome reading. This is a good example of work that often occurs in mathematics: the decision to use a simple but powerful and far-reaching idea is followed by the onerous labor of showing that the idea can actually be used.

First, since G is bounded, then $\bar{G}$ is compact. Hence the functions $f_1, \ldots, f_n$ are uniformly continuous on $\bar{G}$. That is, if $\varepsilon > 0$, then there exists $\delta > 0$ such that if

$$\sum_{i=1}^{n} \left| x_i^{(1)} - x_i^{(2)} \right| < \delta$$

then for $i = 1, \ldots, n$,

$$\left| f_i\left(x_1^{(1)}, \ldots, x_n^{(1)}\right) - f_i\left(x_1^{(2)}, \ldots, x_n^{(2)}\right) \right| < \varepsilon$$

Take a fixed $\varepsilon > 0$ and by using planes parallel to the coordinate planes in R^n, divide $\bar{G}$ into closed cubes of side with length δ/n and subsets of these cubes. Denote these cubes and subsets of cubes by $K_1, \ldots, K_s$. Choose the planes parallel to the coordinate planes in R^n so that x^0 is in the interior of some K_i, say K_1. Define the functions

$$x_i^{(1)}(t) = x_i^0 + \left[f_i\left(x_1^0, \ldots, x_n^0\right) \right](t - t_0)$$

where $i = 1, \ldots, n$ and $t \in [t_0, t_1]$ and

$$t_1 = \mathrm{lub}\{t/t \geq t_0, \left(x_1^{(1)}(\tau), \ldots, x_n^{(1)}(\tau)\right) \in K_1 \text{ for } \tau \in [t_0, t]\}$$

(Notice that $t_1 > t_0$ because x^0 is in the interior of K_1.) If $t_1 = \infty$, then since K_1 is bounded, it follows that for $i = 1, \ldots, n$

$$f_i\left(x_1^0, \ldots, x_n^0\right) = 0$$

Thus if $t_1 = \infty$, the solution promised in the statement of the theorem is for $i = 1, \ldots, n$

$$x_i(t) = x_i^0$$

for all real t. If $t_1 < \infty$, let

$$x_i^{(1)} = \lim_{t \to t_1} x_i^{(1)}(t) = x_i^0 + \left[f_i\left(x_1^0, \ldots, x_n^0\right) \right](t_1 - t_0)$$

If $(x_1^{(1)}, \ldots, x_n^{(1)}) \in \bar{G} - G$, proceed no further. If $(x_1^{(1)}, \ldots, x_n^{(1)}) \in G$, define

$$x_i^{(2)}(t) = x_i^{(1)} + \left[f_i\left(x_1^{(1)}, \ldots, x_n^{(1)}\right) \right](t - t_1) \qquad (i = 1, \ldots, n)$$

Then there exists $t^{(1)} > t_1$ and a K_2 (K_2 may be K_1) such that if $\tau \in [t_1, t^{(1)}]$, then

$$\left(x_1^{(2)}(\tau), \ldots, x_n^{(2)}(\tau)\right) \in K_2$$

Let

$$t_2 = \mathrm{lub}\{t/t \geq t_1, \left(x_1^{(2)}(\tau), \ldots, x_n^{(2)}(\tau)\right) \in K_2 \text{ for } \tau \in [t_1, t]\}$$

and let

$$x_i^{(2)} = \lim_{t \uparrow t_2} x_i^{(2)}(t) = x_i^{(1)} + \left[f_i\left(x_1^{(1)}, \ldots, x_n^{(1)}\right) \right](t_2 - t_1)$$

If

$$f_1\left(x_1^{(1)}, \ldots, x_n^{(1)}\right) = \cdots = f_n\left(x_1^{(1)}, \ldots, x_n^{(1)}\right) = 0$$

then $t_2 = \infty$ and for all $t \geq t_1$,

$$x_i^{(2)}(t) = x_i^{(1)} \quad (i = 1, \ldots, n)$$

We proceed by induction to define

$$x_i^{(k+1)}(t) = x_i^{(k)} + \left[f_i\left(x_1^{(k)}, \ldots, x_n^{(k)}\right) \right](t - t_k)$$

with domain $[t_k, t_{k+1}]$. (The induction process comes to an end after k steps if

$$\left(x_1^{(k)}, \ldots, x_n^{(k)}\right) \in \bar{G} - G$$

or if $t_k = \infty$.)

Next we "piece together" the domains $[t_0, t_1], [t_1, t_2], \ldots$ and the vector functions $(x_1^{(1)}(t), \ldots, x_n^{(1)}(t)), (x_1^{(2)}(t), \ldots, x_n^{(2)}(t)), \ldots$ and obtain a new vector function $(\bar{x}_1(t), \ldots, \bar{x}_n(t))$ with domain

$$\bigcup_{i \geq 0} [t_i, t_{i+1}]$$

and such that if $t \in [t_i, t_{i+1}]$, then

$$\bar{x}_j(t) = x_j^{(i+1)}(t)$$

for $j = 1, \ldots, n$. From the definition, it follows that each $\bar{x}_j(t)$ is continuous on

$$\bigcup_{i \geq 0} [t_i, t_{i+1}].$$

A sufficient condition that $(x_1^{(s)}, \ldots, x_n^{(s)}) \in \bar{G}$ is that

$$\sum_{j=0}^{s-1} \left\{ \sum_{i=1}^{n} \left[f_i\left(x_1^{(j)}, \ldots, x_n^{(j)}\right) \right]^2 (t_{j+1} - t_j)^2 \right\}^{1/2} \leq d \qquad (1.28)$$

Since

$$\left\{ \sum_{i=1}^{n} \left[f_i\left(x_1^{(j)}, \ldots, x_n^{(j)}\right) \right]^2 (t_{j+1} - t_j)^2 \right\}^{1/2} \leq \{ n M^2 (t_{j+1} - t_j)^2 \}^{1/2}$$

$$\leq M\sqrt{n}(t_{j+1} - t_j)$$

then a sufficient condition that (1.28) holds is

$$\sum_{j=0}^{s-1} M\sqrt{n}(t_{j+1} - t_j) \leq d$$

$$M\sqrt{n}(t_s - t_0) \leq d$$

$$(t_s - t_0) \leq \frac{d}{M\sqrt{n}}$$

Thus the domain of $(\bar{x}_1(t), \ldots, \bar{x}_n(t))$ contains the interval

$$\left[t_0, t_0 + \frac{d}{M\sqrt{n}} \right]$$

We proceed in an exactly similar way to consider $t < t_0$ and obtain finally the n-vector function $(\bar{x}_1(t), \ldots, \bar{x}_n(t))$ whose domain contains the interval

$$I = \left[t_0 - \frac{d}{M\sqrt{n}}, t_0 + \frac{d}{M\sqrt{n}} \right]$$

Lemma 1.7   *For $i = 1, \ldots, n$ and $t \in I$,*

$$\bar{x}_i(t) = x_i^0 + \int_{t_0}^t f_i\,[\bar{x}_1(s), \ldots, \bar{x}_n(s)]\,ds + \int_{t_0}^t g_i(s)\,ds \qquad (1.29)$$

where the function g_i is such that for all $s \in I$,

$$|g_i(s)| < \varepsilon$$

Proof

$$x_i^{(1)} - x_i^0 = \int_{t_0}^{t_1} f_i\left(x_1^0, \ldots, x_n^0 \right) ds$$

$$x_i^{(2)} - x_i^{(1)} = \int_{t_1}^{t_2} f_i\left(x_1^{(1)}, \ldots, x_n^{(1)} \right) ds$$

$$\cdots$$

$$x_i^{(k)} - x_i^{(k-1)} = \int_{t_{k-1}}^{t_k} f_i\left(x_1^{(k-1)}, \ldots, x_n^{(k-1)} \right) ds$$

Adding these equations together we obtain

$$x_i^{(k)} - x_i^0 = \sum_{j=0}^{k-1} \int_{t_j}^{t_{j+1}} f_i\left(x_1^{(j)}, \ldots, x_n^{(j)} \right) ds \qquad (1.30)$$

(where $x_i^{(0)} = x_i^0$ for $i = 1, \ldots, n$).

If $t_k \leq t \leq t_{k+1}$,

$$\int_{t_k}^t f_i\left(x_1^{(k)}, \ldots, x_n^{(k)} \right) ds = \left[f_i\left(x_1^{(k)}, \ldots, x_n^{(k)} \right) \right](t - t_k)$$

$$= x_i^{(k+1)}(t) - x_i^{(k)} = \bar{x}_i(t) - x_i^{(k)} \qquad (1.31)$$

From equations (1.30) and (1.31), we obtain if $t_k \le t \le t_{k+1}$,

$$\bar{x}_i(t) = x_i^{(k)} + \int_{t_k}^{t} f_i\left(x_1^{(k)}, \ldots, x_n^{(k)}\right) ds$$

$$= x_i^0 + \sum_{j=0}^{k-1} \int_{t_j}^{t_{j+1}} f_i\left(x_1^{(j)}, \ldots, x_n^{(j)}\right) ds$$

$$+ \int_{t_k}^{t} f_i\left(x_1^{(k)}, \ldots, x_n^{(k)}\right) ds$$

Hence if $t_k \le t \le t_{k+1}$,

$$\bar{x}_i(t) - x_i^0 = \sum_{j=0}^{k-1} \int_{t_j}^{t_{j+1}} f_i\left[x_1^{(j)}(s), \ldots, x_n^{(j)}(s)\right] ds$$

$$+ \int_{t_k}^{t} f_i\left[x_1^{(k)}(s), \ldots, x_n^{(k)}(s)\right] ds$$

$$+ \sum_{j=0}^{k-1} \int_{t_j}^{t_{j+1}} \left\{ f_i\left(x_1^{(j)}, \ldots, x_n^{(j)}\right) - f_i\left(x_1^{(j)}(s), \ldots, x_n^{(j)}(s)\right) \right\} ds$$

$$+ \int_{t_k}^{t} \left\{ f_i\left(x_1^{(k)}, \ldots, x_n^{(k)}\right) - f_i\left[x_1^{(k)}(s), \ldots, x_n^{(k)}(s)\right] \right\} ds$$

$$= \int_{t_0}^{t} f_i[\bar{x}_1(s), \ldots, \bar{x}_n(s)] ds + \int_{t_0}^{t} g_i(s) ds$$

where if $s \in [t_j, t_{j+1}], j = 1, \ldots, k-1$

$$|g_i(s)| = |f_i\left(x_1^{(j)}, \ldots, x_n^{(j)}\right) - f_i\left(x_1^{(j)}(s), \ldots, x_n^{(j)}(s)\right)| < \varepsilon$$

because $(x_1^{(j)}, \ldots x_n^{(j)})$ and $(x_1^{(j)}(s), \ldots, x_n^{(j)}(s))$ are in the same set K_j. A similar estimate holds for $|g_i(s)|$ if $s \in [t_k, t]$. This completes the proof of Lemma 1.7. $\square$

The n-vector function $(\bar{x}_1(t), \ldots, \bar{x}_n(t))$ was obtained corresponding to a number $\varepsilon > 0$. Next we take $\varepsilon = 1/m$, where m is a positive integer, and let $(\bar{x}_1^m(t), \ldots, \bar{x}_n^m(t))$ be the corresponding n-vector function. It follows from (1.28) that if $t \in I$,

$$|\bar{x}_i^m(t)| \le B + M|t - t_0| + \frac{1}{m}|t - t_0|$$

where $B = \max_{i=1,\ldots,n} |x_i^0|$. It also follows from (1.28) that if $t', t'' \in I$, then

$$\left|\bar{x}_i^m(t') - \bar{x}_i^m(t'')\right| \le M|t' - t''| + \frac{1}{m}|t' - t''|$$

Thus $\{\bar{x}_i^m(t)\}$ is a sequence of uniformly bounded, equicontinuous functions on I. Hence by Ascoli's theorem, there is a subsequence $\{(\bar{x}_1^{m_\ell}(t), \ldots, \bar{x}_n^{m_\ell}(t))\}$ such that for $i = 1, \ldots, n$ $\{\bar{x}_i^{m_\ell}(t)\}$ converges uniformly on I to a continuous function $u_i(t)$. Now we have

$$\bar{x}_i^{m_\ell}(t) = x_i^0 + \int_{t_0}^t f_i \left[\bar{x}_1^{m_\ell}(s), \ldots, \bar{x}_n^{m_\ell}(s)\right] ds + \int_{t_0}^t g_i^{m_\ell}(s)\, ds \qquad (1.32)$$

If $t \in I$,

$$\left| \int_{t_0}^t g_i^{m_\ell}(s)\, ds \right| \le \frac{1}{m_\ell}|t - t_0| \le \frac{1}{m_\ell} \frac{d}{\sqrt{n}\, M}$$

Hence $\int_{t_0}^t g_i^{m_\ell}(s)\, ds$ converges uniformly to zero on I. Also we have

$$\left| \int_{t_0}^t f_i \left[\bar{x}_1^{m_\ell}(s), \ldots, \bar{x}_n^{m_\ell}(s)\right] ds - \int_{t_0}^t f_i[u_1(s), \ldots, u_n(s)]\, ds \right|$$

$$\le |t - t_0| \max_{s \in [t_0, t]} |f_i \left[\bar{x}_1^{m_\ell}(s), \ldots, \bar{x}_n^{m_\ell}(s)\right] - f_i[u_1(s), \ldots, u_n(s)]|$$

$$< |t - t_0|\varepsilon$$

The last inequality holds for m_ℓ sufficiently large because for $i = 1, \ldots, n$, $\bar{x}_i^{m_\ell}(t)$ converges uniformly to $u_i(t)$ and f_i is uniformly continuous on $\bar{G}$. Thus the function

$$\int_{t_0}^t f_i \left[\bar{x}_1^{m_\ell}(s), \ldots, \bar{x}_n^{m_\ell}(s)\right] ds$$

converges uniformly on I to

$$\int_{t_0}^t f_i[u_1(s), \ldots, u_n(s)]\, ds$$

Thus from (1.32) it follows that for $t \in I = [t_0 - d/M\sqrt{n}, t_0 + d/M\sqrt{n}]$

$$u_i(t) = x_i^0 + \int_{t_0}^t f_i[u_1(s), \ldots, u_n(s)]\, ds \qquad (1.33)$$

Since $u_1(s), \ldots, u_n(s)$ are continuous and f is continuous, then we may differentiate the right-hand side of (1.33) and conclude that for each $t \in (t_0 - d/M\sqrt{n}, t_0 + d/M\sqrt{n})$,

$$\frac{du_i}{dt} = f_i[u_1(t), \ldots, u_n(t)] \quad i = 1, \ldots, n$$

This completes the proof of Existence Theorem 1.3.

Notice that the proof of Existence Theorem 1.3 is not constructive because when Ascoli's Theorem is applied, we can only conclude that there exists a subsequence $\{(\bar{x}_1^{m_\ell}(t), \ldots, \bar{x}_n^{m_\ell}(t))\}$ which converges to a solution. We do not exhibit the subsequence. However, if enough hypotheses are imposed so that the solution of (1.24) which satisfies the initial value is unique, then as the following corollary shows, the sequence $\{(\bar{x}_1^m(t), \ldots, \bar{x}_n^m(t))\}$ itself converges to the solution.

Corollary to Existence Theorem 1.3   *Suppose that equation* (1.24) *is such that if a solution satisfying the initial condition exists, it is unique (a sufficient condition for this is that f satisfies a Lipschitz condition). Then the sequence* $\{(\bar{x}_1^m(t), \ldots, \bar{x}_n^m(t))\}$ *obtained in the proof of Existence Theorem 1.3 converges to a solution of* (1.24).

Proof   The proof of Existence Theorem 1.3 shows that there is a subsequence of $\{(\bar{x}_1^m(t), \ldots, \bar{x}_n^m(t))\}$ which converges uniformly on interval I to a solution $(u_1(t), \ldots, u_n(t))$ of (1.24). Suppose that $\{(\bar{x}_1^m(t), \ldots, \bar{x}_n^m(t))\}$ itself does not converge uniformly to $(u_1(t), \ldots, u_n(t))$ on the interval I. Then there is an $\varepsilon_1 > 0$ and a subsequence

$$\left\{ \left(\bar{x}_1^{m_v}(t), \ldots, \bar{x}_n^{m_v}(t) \right) \right\}$$

of $\{(\bar{x}_1^m(t), \ldots, \bar{x}_n^m(t))\}$ such that for each m_v there is a number $t_v \in I$ such that

$$\sum_{i=1}^{n} \left| u_i(t_v) - \bar{x}_i^{m_v}(t_v) \right| > \varepsilon_1 \tag{1.34}$$

But the proof of Existence Theorem 1.3 shows that $\{(\bar{x}_1^{m_v}(t), \ldots, \bar{x}_n^{m_v}(t))\}$ is a uniformly bounded equicontinuous sequence of functions on I. Hence it contains a subsequence which converges uniformly to I to a function $(w_1(t), \ldots, w_n(t))$. By the same arguments used in the proof of Existence Theorem 1.3 it follows that $(w_1(t), \ldots, w_n(t))$ is a solution of (1.24) on the interval

$$\left(t_0 - \frac{d}{M\sqrt{n}}, \, t_0 + \frac{d}{M\sqrt{n}} \right)$$

such that

$$w_i(t_0) = x_i^0 \quad (i = 1, \ldots, n)$$

But inequality (1.34) shows that the solutions $(u_1(t), \ldots, u_n(t))$ and $(w_1(t), \ldots, w_n(t))$ are different functions. This is a contradiction to the hypothesis of uniqueness of solution.   □

For explicit computations of solutions or approximations to solutions, it would seem important to measure how fast the sequences converge which occur in the proofs of Existence Theorem 1.1, Existence Theorem 1.2, and the Corollary to Existence Theorem 1.3. Actually questions about rapidity of convergence and choice of a sequence which converges with maximal rapidity are very serious and would lead us deeply into the subject of numerical analysis. As stated in the introduction, we shall (with all due respect for this important subject) not enter a study of these questions.

Next we show how a quick proof of Existence Theorem 1.3 can be given if a technique from functional analysis, the Schauder fixed point theorem (see the appendix), is used.

Existence Theorem 1.4   *Suppose that f is continuous on D, an open set in* (t, x)-*space. Let* t_0, x^0, a, b, M, A *have the same meaning as in the statement of Existence*

Theorem 1.1. Then there exists a solution $x(t, t_0, x^0)$ of

$$x' = f(t, x)$$

on $(t_0 - A, t_0 + A)$ such that $x(t_0, t_0, x^0) = x^0$.

Proof   Let $J = [t_0 - A, t_0 + A]$ and let $C[J, R^n]$ denote the Banach space of continuous n-vector functions $\phi = (\phi_1, \ldots, \phi_n)$ with domain J such that

$$\|\phi\| = \sup_{t \in J} \sum |\phi_i(t)|$$

Let

$$\mathcal{F} = \{\phi \in C[J, R^n] / \phi(t_0) = x^0, |\phi(t) - x^0| \le b \text{ for } t \in J\}$$

Then it follows easily that $\mathcal{F}$ is a bounded, convex, closed subset of $C[J, R^n]$. Now we define the mapping

$$\mathcal{M} : \mathcal{F} \to C[J, R^n]$$

as follows: if $\phi(t) \in \mathcal{F}$

$$(\mathcal{M}\phi)(t) = x^0 + \int_{t_0}^t f[s, \phi(s)] \, ds$$

From standard theorems of calculus, it follows that if $\phi \in C[J, R^n]$, then $\mathcal{M}\phi \in C[J, R^n]$. Also

$$\mathcal{M}(\mathcal{F}) \subset \mathcal{F} \quad \text{because}$$

$$(\mathcal{M}\phi)(t_0) = x^0$$

and

$$|(\mathcal{M}\phi)(t) - x^0| \le \int_{t_0}^t |f(s, \phi(s))| \, ds \le M|t - t_0| \le MA \le M\frac{b}{M} = b$$

$\mathcal{M}(\mathcal{F})$ is an equicontinuous set because if $t, \bar{t} \in J$, then

$$|(\mathcal{M}\phi)(t) - (\mathcal{M}\phi)(\bar{t})| \le |\int_{\bar{t}}^t f(s, \phi(s)) \, ds| \le M|t - \bar{t}|$$

and M is independent of ϕ. Hence by Ascoli's theorem, $\mathcal{M}$ is a compact map of $\mathcal{F}$ into itself and by Schauder's theorem, $\mathcal{M}$ has a fixed point. That is, there is a function $\phi \in \mathcal{F}$ such that

$$\phi(t) = (\mathcal{M}\phi)(t) = x^0 + \int_{t_0}^t f[s, \phi(s)] \, ds$$

By Lemma 1.1, ϕ is the desired solution $x(t, t_0, x^0)$.   ∎

The brevity of the proof of Existence Theorem 1.4 is misleading because the proof uses the Schauder Fixed Point Theorem, the proof of which is by no means short. Note also that the proof of Existence Theorem 1.4 is nonconstructive. That is, we arrive at the conclusion that a solution exists but we obtain no hint about how to compute the solution. If f satisfies a Lipschitz condition, then the sequences constructed in the proofs of Existence Theorems 1.1 and 1.3 converge to solutions. As pointed out earlier, this convergence may be too slow for practical computation of a uniform approximation to a solution. But the solution is approximated in some theoretical sense whereas the proof of Existence Theorem 1.4 yields no approximation at all.

Extension Theorems

In Existence Theorems 1.1, 1.2, 1.3, and 1.4, we have obtained estimates on the size of the domain of the solution. For example, in Existence Theorem 1.1, the domain of the solution contained the interval $(t_0 - A, t_o + A)$. However, none of these estimates put a limitation or bound on the domain of the solution. That is, it might be possible that for a particular equation, the solution were defined for all real t even though the existence theorem only guaranteed the existence of a solution on a finite interval. (We will see later that such a result holds for "well-behaved" linear systems.) Consequently we want now to study further the question of how large the domain of the solution may be.

In the differential equations studied in an elementary course in differential equations, the solutions are often defined for all real t. But it would be quite wrong to conclude that this kind of result generally holds for solutions of differential equations. The following simple example shows a typical complication that may arise. Let x be a 1-vector and consider the equation

$$x' = x^2 \tag{1.35}$$

By separation of variables, we have

$$\frac{dx}{x^2} = dt$$

Integrating, we obtain

$$-\frac{1}{x} = t + c$$

or

$$x = \frac{-1}{t + c}$$

where c is an arbitrary real constant. Hence the solution $x(t)$ of (1.35) which satisfies the initial value

$$x(1) = -1$$

is

$$x(t) = -\frac{1}{t}$$

As $t \to 0$, the solution $x(t)$ decreases without bound and solution $x(t)$ is certainly not defined at $t = 0$. Actually no solution of (1.35) is defined for all real t (except for the solution $x(t) \equiv 0$). This is true in spite of the fact that the function $f(t, x)$ in (1.35) is just the simple expression x^2 so that $f(t, x)$ is defined on the entire (t, x)-plane and is, indeed, independent of t.

Our next step is to obtain some conditions under which extensions of domains of solutions can be made.

Definition If $x(t)$ is a solution on (a, b) of $x' = f(t, x)$ and $y(t)$ is a solution on (α, β) of $x' = f(t, x)$ and $(\alpha, \beta) \supset (a, b)$ and $y/(a, b) = x$, then the solution $y(t)$ is an *extension* of solution $x(t)$.

Definition If $x(t)$ is a solution on (a, b) of $x' = f(t, x)$ and if $x(t)$ is such that any extension $y(t)$ of $x(t)$, where $y(t)$ is a solution on (α, β) of $x' = f(t, x)$, has the property that $(\alpha, \beta) = (a, b)$, then $x(t)$ is a *maximal solution* of $x' = f(t, x)$.

Extension Theorem 1.1 *Suppose G is a bounded open set in (t, x)-space and f is continuous on $\bar{G}$ and f satisfies a Lipschitz condition in x in G. Suppose $x(t)$, with domain (α, β), where $\beta < \infty$, is a maximal solution of*

$$x' = f(t, x)$$

Let

$$p(t) = \inf_{(t_1, x^1) \in \partial G}\{|t - t_1| + |x(t) - x^1|\}$$

where ∂G denotes the boundary of G. Then $\lim_{t \to \beta} p(t) = 0$.

Remarks. (1) Extension Theorem 1.1 says roughly that if the solution cannot be extended any further, then it has gone out to the boundary of G.

(2) The hypothesis that f satisfies a Lipschitz condition in G is to ensure uniqueness. (We will point out in the proof where this uniqueness condition is needed.)

Proof of Extension Theorem 1.1. Suppose the conclusion of the theorem is not true. Then there exists $\bar{\varepsilon} > 0$ and a sequence $\{t_m\}$ such that $t_m \to \beta$ and for all m, $p(t_m) > \bar{\varepsilon}$, that is,

$$\inf_{(t_1, x^1) \in \partial G} \{|t_m - t_1| + |x(t_m) - x^1|\} > \bar{\varepsilon} \tag{1.36}$$

Suppose $M > 0$ is such that

$$\max_{\substack{(t,x) \in \bar{G} \\ i=1,\ldots,n}} |f_i(t, x)| \leq M$$

Given integer N, then there exists t_m such that

$$|\beta - t_m| < \frac{\bar{\varepsilon}}{N} \tag{1.37}$$

Let $y(t)$ be a solution of $x' = f(t, x)$ such that

$$y(t_m) = x(t_m)$$

The domain of $y(t)$ contains

$$[t_m - r, t_m + r]$$

where

$$r = \min\left\{a, \frac{b}{M}\right\} \tag{1.38}$$

and a, b are as described in Existence Theorem 1.1. Let $a = \frac{\bar{\varepsilon}}{N}$. Then we may set b equal to $(\frac{N-1}{N})\bar{\varepsilon} = (1 - \frac{1}{N})\bar{\varepsilon}$ because by (1.36) the set

$$\left\{(t, x) / |t_m - t| < \frac{\bar{\varepsilon}}{N}, \ |x(t_m) - x| < \left(1 - \frac{1}{N}\right)\bar{\varepsilon}\right\}$$

is contained in G. But if N is sufficiently large

$$a = \frac{\bar{\varepsilon}}{N} < \left(1 - \frac{1}{N}\right)\frac{\bar{\varepsilon}}{M} = \frac{b}{M} \tag{1.39}$$

By (1.37), (1.38), and (1.39), the domain of $y(t)$ contains β. By the Lipschitz condition (See Remark (2) above.) $y(t)$ yields an extension of $x(t)$. Since the domain of the extension contains β, then $x(t)$ is not maximal. This contradicts the hypothesis.

Extension Theorem 1.2   *Let G be an open set in (t, x)-space and let f be contin-uous on $\bar{G}$. Suppose f satisfies a Lipschitz condition in x in G. Let $x(t)$, with domain (α, β) and $\beta < \infty$, be a maximal solution of*

$$x' = f(t, x)$$

Then either $|x(t)|$ becomes unbounded as $t \to \beta$ (i.e., given $\varepsilon > 0$ and $M_1 > 0$, then there exists t such that $0 < \beta - t < \varepsilon$ and $|x(t)| > M_1$) or $\partial G \neq \phi$ and

$$\lim_{t \to \beta} p(t) = 0$$

where $p(t)$ is defined as in the statement of Extension Theorem 1.1.

Proof   Suppose $|x(t)|$ remains bounded as $t \to \beta$. That is, suppose there exists $\varepsilon_0 > 0$ and $c > 0$ such that for all $t \in (\beta - \varepsilon_0, \beta)$,

$$(t, x(t)) \in \{(t, x) / |t| + |x| < c\}$$

Let

$$B_1 = \{(t, x)/|t| + |x| < 2c\}$$

and

$$G_1 = G \cap B_1$$

Then by Extension Theorem 1.1,

$$\lim_{t \to \beta} \inf_{(t_1, x^1) \in \partial G_1} \{|t - t_1| + |x(t) - x^1|\} = 0 \tag{1.40}$$

But

$$\partial G_1 \subset \partial G \cup \partial B_1$$

Hence if $(t, x) \in \partial G_1$ and $(t, x) \notin \partial G$, then $(t, x) \in \partial B_1$. But if $(\bar{t}, \bar{x}) \in \partial B_1$, and if $t \in (\alpha, \beta)$, then from the definition of B_1, it follows that if $t \in (\beta - \varepsilon_0, \beta)$, then

$$|\bar{t} - t| + |\bar{x} - x(t)| \geq c$$

Hence from (1.40), we have

$$\lim_{t \to \beta} \inf_{(t_1, x^1) \in \partial G} \{|t - t_1| + |x(t) - x^1|\} = 0 \qquad \Box$$

Extension Theorem 1.3 *Suppose f is continuous on (t, x)-space and that for each point in (t, x)-space, there is a neighborhood N of (t, x) such that in N, f satisfies a Lipschitz condition in x. Suppose also that f is bounded, that is, suppose there exists $M > 0$ such that for all (t, x),*

$$|f(t, x)| < M$$

Then the domain of each maximal solution of

$$x' = f(t, x)$$

is the t-axis.

Proof Suppose a maximal solution $x(t)$ has domain (α, β) where $\beta < \infty$. Let $t_0 \in (\alpha, \beta)$. Then by Lemma 1.1,

$$x(t) = x(t_0) + \int_{t_0}^{t} f[s, x(s)] \, ds$$

and if $\alpha < t_2 < t_1 < \beta$,

$$|x(t_1) - x(t_2)| \leq M|t_1 - t_2|$$

Hence the functional values $x(t)$ for t near β and less than β satisfy a Cauchy condition and there exists $\lim_{t \to \beta} x(t)$. Call this limit $\bar{x}$ and let $y(t)$ be a solution of

$$x' = f(t, x)$$

such that $y(\beta) = \bar{x}$. By the Lipschitz condition, $y(t)$ yields an extension of $x(t)$. Since the domain of this extension contains β, then $x(t)$ is not maximal. A similar argument shows that $\alpha > -\infty$ leads to a contradiction.   ▯

(For another version of Extension Theorem 1.3, see Exercise 14.)

Exercises

1. Justify rigorously the method of separation of variables as applied to the equation: $(*)$ $g(x)\frac{dx}{dt} = f(t)$.

2. If

$$x = \begin{bmatrix} x_1 \\ \vdots \\ x_n \end{bmatrix}$$

prove that

$$\|x\| \le |x| \tag{1.41}$$

and

$$|x| \le \sqrt{n}\|x\| \tag{1.42}$$

3. Let x denote a scalar (i.e., a 1-vector). Prove that the equation

$$\frac{dx}{dt} = \left(\frac{1}{t^2 + 1}\right) e^{-x^2 \sin^2 t}$$

has a unique solution $x(t)$ such that $x(0) = 1$. Show that the domain of $x(t)$ is the real line.

4. Let a, b, c, d be real constants. Show that if $(x(t), x_2(t))$ is a solution of the 2-dimensional system

$$\frac{dx_1}{dt} = ax_1 + bx_2$$

$$\frac{dx_2}{dt} = cx_1 + dx_2$$

such that $x_1(0) = 5$, $x_2(0) = 0$, then for each t in the domain of the solution $(x_1(t), x_2(t))$ it is true that

$$[x_1(t)]^2 + [x_2(t)]^2 > 0$$

(That is, there is no value $\bar{t}$ such that $x(\bar{t}) = y(\bar{t}) = 0$.)

5. Given the system

$$\frac{d^3x}{dt^3} + \left(\frac{1}{1+t^2}\right)\frac{d^2x}{dt^2} + (\sin t)\frac{dx}{dt} + \frac{t}{x^2 + y^2 + 1} = 0 \qquad (1.43)$$

$$\frac{d^2y}{dt^2} + e^{-t}\frac{dy}{dt} + \cos(x + y) = 0$$

show that there exists a unique solution $(x(t), y(t))$ such that $x(0) = 1$, $\frac{dx}{dt}(0) = 0$, $\frac{d^2x}{dt^2}(0) = 5$, $y(0) = 0$, $\frac{dy}{dt}(0) = 3$. Find an interval I on the t-axis with midpoint $t = 0$ such that the domain of the solution contains I.

6. Find the solution of

$$\frac{dy}{dt} = 3t^2y^2$$

such that $y(1) = 0$.

7. Prove Lemma 1.5 without using the exponential series.

8. Prove Corollary 1.1. (Hint: use Lemma 1.1.)

9. In the Remark after the statement of Existence Theorem 1.3, it is shown that any nonautonomous system of n equations can be transformed into an autonomous system of $(n + 1)$ equations. This suggests that the theory of differential equations need only be developed for autonomous systems. (To apply such theory to a nonautonomous system, one would simply transform the nonautonomous system into an autonomous system.) Can you suggest why this possibility might not be feasible?

10. Find $x\left(\frac{3}{4}\right)$ where $x(t)$ is the solution of

$$\frac{dx}{dt} = x^3$$

such that $x(0) = 1$.

11. Find $x\left(\frac{1}{4}\right)$ where $x(t)$ is the solution of

$$\frac{dx}{dt} = x^{1/3}$$

such that $x\left(\frac{1}{2}\right) = 0$.

12. An extremely useful result which will be used in Exercise 13 and used often in Chapter 4, is the following:

Gronwall's Lemma. If u, v are real-valued nonnegative continuous functions with domain $\{t/t \geq t_0\}$ and if there exists a constant $M \geq 0$ such that for all $t \geq t_0$

$$u(t) \leq M + \int_{t_0}^{t} u(s)v(s)\,ds \tag{1.44}$$

then

$$u(t) \leq M \exp \int_{t_0}^{t} v(s)\,ds \tag{1.45}$$

Prove Gronwall's Lemma. (Hint: first assume $M > 0$. Then for all $t \geq t_0$

$$\frac{u(t)v(t)}{M + \int_{t_0}^{t} u(s)v(s)\,ds} \leq v(t).$$

Integrate both sides of this inequality from t_0 to t.)

13. The existence theorem for equation with a parameter is a local result. That is, we proved that in a sufficiently small domain

$$\{t/|t - t_0| \leq r\}$$

the solution is a continuous function of parameter μ and initial value x_0. Now suppose that if $\mu = 0$, the equation

$$\frac{dx}{dt} = f(t, x, \mu) \tag{1.14}$$

has a unique solution $x(t, \bar{x}^0, 0)$ whose domain contains the interval $[a, b]$. (In all solutions, the initial value will be assumed at $t = t_0$. Hence we will omit the t_0 in the notation for the solution.) Assume that f is continuous and satisfies a Lipschitz condition with Lipschitz constant k on an open set G in (t, x^0, μ)-space where

$$G \supset [a, b] \times \bigcup_{t \in [a,b]} \{x/|x - x(t, \bar{x}_0, 0)| \leq r\} \times [-\mu_0, \mu_0]$$

where r, μ_0 are given positive numbers. Assume also that at each point of G, the function f has a continuous partial derivative $\frac{\partial f}{\partial \mu}$.

Prove that if $|\mu|$ is sufficiently small, equation (1.14) has a unique solution $x(t, x^0, \mu)$ whose domain contains $[a, b]$ and $x(t, x^0, \mu)$ is continuous in (x^0, μ) uniformly for $t \in [a, b]$.

14. State and prove a version of Extension Theorem 1.3 which does not have a uniqueness hypothesis.

15. Prove: Suppose G is an open set in (t, x)-space and let f be continuous on $\bar{G}$ and be such that for each $(t, x) \in G$ there is a neighborhood N of (t, x) such that in N, f satisfies a Lipschitz condition in x. Let $x(t)$ be a solution with domain (α, β) of the equation

$$x' = f(t, x) \tag{1.2}$$

such that $x(t)$ cannot be extended beyond β and suppose there exists a closed bounded set $A \subset R^n$ such that $[\alpha, \beta] \times A \subset G$ and such that for all $t \in (\alpha, \beta)$

$$x(t) \in A$$

Then $\beta = \infty$.

Examples

Now we describe some differential equations which are used to model biological, physiological, and chemical systems, and indicate how the theory in this chapter is used in the study of these equations. As theory is developed in subsequent chapters, it will be applied to further study of these equations.

1. The Volterra Equations for Predator-Prey Systems

For discussions of the derivation of the Volterra equations and other more general cases, see Maynard Smith [1974] and May [1973].

If x denotes the population density of the prey and y the population density of the predator, the Volterra equations describe the rates of change of x and y as follows:

$$x' = ax - Ax^2 - cxy \tag{1.46}$$
$$y' = -dy + exy$$

where a, A, c, d, and e are positive constants. These equations are sufficiently explicit so that a detailed analysis of their solutions can be made as indicated in Maynard Smith [1974]. Here we will make just a few general observations of a kind that are applicable for more general classes of equations.

Since x, y denote populations and hence are nonnegative and remain finite for all $t > t_0$, some fixed value, we need to show that any biologically significant solution of (1.46) has such properties. That is, we must show that if the solution $(x(t), y(t))$ is such that $(x(t_1), y(t_1))$ is in the first quadrant, then for all $t > t_1$, $(x(t), y(t))$ is defined, is in the first quadrant, and the set

$$\{(x(t), y(t))/t > t_1\}$$

is bounded.

(a) Show that no solution "escapes" the first quadrant, that is, that if $(x(\bar{t}), y(\bar{t}))$ is in the first quadrant, then there is no value $\tilde{t} > \bar{t}$ such that $(x(\tilde{t}), y(\tilde{t}))$ is not in the first quadrant.

(b) Show that if $\frac{a}{A} < \frac{d}{e}$ and if

$$K = \frac{a}{A} + \delta < \frac{d}{e}$$

where δ is a sufficiently small positive number, then no solution escapes the rectangle R with vertices

$$(0, 0), \ (K, 0), \ (K, B), \ (0, B)$$

where B is any fixed positive number. Show also that any solution which passes through a point in the first quadrant ultimately enters the rectangle R.

These results show that each solution which passes through a point in the first quadrant ultimately enters and thereafter remains in rectangle R. Hence by Exercise 15, any solution which passes through a point in the first quadrant is defined for all t greater than some t_0.

It should be emphasized that we describe here only some mathematical aspects of the study of the above equations. The more important question of the biological significance of the mathematical results will not be dealt with here at all.

2. The Hodgkin-Huxley Equations

One of the most successful mathematical models used in biological sciences is the system of differential equations obtained by Hodgkin and Huxley [1952] in their study of nerve conduction (for which they received a Nobel prize in 1959). For a lucid account of some of the physiological background of their work, see FitzHugh [1969].

Besides their value as a model in nerve conduction, the Hodgkin-Huxley equations have proved to be very valuable in modeling other physiological systems. They have provided a paradigm, sometimes called Hodgkin-Huxley like equations, for models, among others, of cardiac components (see Cronin [1987]) and brain components (see Traub and Miles [1991]).

The Hodgkin-Huxley equations describe the relationships among the potential difference across the membrane surface of the nerve axon and the current arising from the flow of ions (mostly sodium and potassium) across the membrane and the current caused by the fact that the membrane has a capacitance. For the standard temperature 6.3°C, the H-H equations are the following four equations:

$$\frac{dV}{dt} = \frac{I}{C} - \frac{I_i}{C}$$

or

$$\frac{dV}{dt} = \frac{I}{C} - \frac{1}{C}[\bar{g}_{Na}m^3h(V - V_{Na}) + \bar{g}_K n^4(V - V_K) + \bar{g}_L(V - V_L)]$$

and

$$\frac{dm}{dt} = \alpha_m(V)[1 - m] - \beta_m(V)m = \frac{m_\infty(V) - m}{\tau_m(V)}$$

$$\frac{dh}{dt} = \alpha_h(V)[1 - h] - \beta_h(V)h = \frac{h_\infty(V) - h}{\tau_h(V)}$$

$$\frac{dn}{dt} = \alpha_n(V)[1 - n] - \beta_n(V)n = \frac{n_\infty(V) - n}{\tau_n(V)} \qquad \text{(H-H)}$$

where

t = time

V = potential across membrane

I = total current through membrane, per unit area

I_i = ionic current through membrane, per unit area

$\quad = \bar{g}_{Na}m^3h(V - V_{Na}) + \bar{g}_K n^4(V - V_K) + \bar{g}_L(V - V_L)$

C = capacitance per unit area of membrane (which is a constant)

$\bar{g}_{Na}$ = sodium conductance constant (positive)

$\bar{g}_K$ = potassium conductance constant (positive)

$\bar{g}_L$ = leakage conductance constant (positive)

V_{Na} = 115 mV

V_K = −12 mV

V_L = 10.5989 mV

m = sodium activation in H-H model

h = sodium inactivation in H-H model

n = potassium activation in H-H model

The second expression for $\frac{dm}{dt}$, that is, $\frac{m_\infty(V) - m}{\tau_m(V)}$, is obtained as follows: Since

$$\frac{dm}{dt} = \alpha_m(V) - [\alpha_m(V) + \beta_m(V)]m = \frac{\frac{\alpha_m(V)}{\alpha_m(V) + \beta_m(V)} - m}{\frac{1}{\alpha_m(V) + \beta_m(V)}}$$

we take

$$m_\infty(V) = \frac{\alpha_m(V)}{\alpha_m(V) + \beta_m(V)}$$

and

$$\tau_m(V) = \frac{1}{\alpha_m(V) + \beta_m(V)}$$

The second expressions for $\frac{dh}{dt}$ and $\frac{dn}{dt}$ are obtained similarly. The variables m, h, n are "phenomenological variables" which describe changes in conductance of sodium and potassium. Their ranges are the interval $[0, 1]$. For a discussion of the meaning of m, h, n, see Hodgkin and Huxley [1952] or Cronin [1987].

$$\alpha_m(V) = \frac{0.1(25 - V)}{e^{0.1(25-V)} - 1}$$

$$\beta_m(V) = 4e^{-\frac{V}{18}}$$

$$\alpha_h(V) = 0.07e^{-\frac{V}{20}}$$

$$\beta_h(V) = \frac{1}{e^{0.1(30-V)} + 1}$$

$$\alpha_n(V) = \frac{0.01(10 - V)}{e^{0.1(10-V)} - 1}$$

$$\beta_n(V) = 0.125e^{-\frac{V}{80}}$$

These equations were obtained from "space-clamped" data, that is, from experimental data in which V, m, h, n depend on time but not on position along the axon.

Besides giving a mathematical description which summarizes experimental data, the H-H equations make a number of valid predictions. For example, the equations are derived empirically from data in which V is controlled and I and I_i are measured (sometimes called voltage-clamped data). But the equations reproduce or predict data from current-clamped experiments, that is, experiments in which I is controlled and V is measured. However the H-H equations are obtained by empirical considerations and partly as a consequence of this, their status is quite different from the status of differential equations in mechanics or electrical circuit theory which are, derived from first principles such as Newton's laws or the Kirchhoff's laws.

The H-H equations present very serious mathematical problems as we will see later. However it is easy to show that the solutions are appropriately bounded.

(a) Show that if I is a bounded continuous function of t, there exists a positive number K such that if $(V(t), m(t), h(t), n(t))$ is a solution of (H-H) such that for $t = t_1$,

$$(V(t), m(t), h(t), n(t)) \in$$
$$\{(V, m, h, n)/ - K \le V \le K, m \in [0, 1], h \in [0, 1], n \in [0, 1]\} \qquad (1.47)$$

then (1.47) holds for all $t \ge t_1$. (That is, the solution is defined for all $t \ge t_1$ and (1.47) is true for all $t \ge t_1$.)

It is not difficult to show by combining (H-H) and a few statements from electrical theory that a system of partial differential equations can be derived which describes V, m, h, n as functions of space, that is, position on the nerve axon, as well as time.

(See Cronin [1987], pp. 61–62.) These equations have the following form:

$$\frac{\partial^2 V}{\partial x^2} - a\frac{\partial V}{\partial t} = F(V, m, h, n)$$

$$\frac{\partial m}{\partial t} = G_1(V, m)$$

$$\frac{\partial h}{\partial t} = G_2(V, h)$$

$$\frac{\partial n}{\partial t} = G_3(V, n) \tag{1.48}$$

where F, G_1, G_2, G_3 are "well-behaved" functions and a is a positive constant. Study of these equations is crucially important because certain of their solutions can be interpreted as descriptions of the impulse which travels along the axon when stimulus is applied. Such solutions, called travelling wave solutions, are of the form

$$V(x - \omega t), \ m(x - \omega t), \ h(x - \omega t), \ n(x - \omega t)$$

where ω is a constant. The search for travelling wave solutions of (1.48) reduces to the problem of solving a system of ordinary differential equations as the following problem shows.

(b) Let $\xi = x - \omega t$. Show that

$$\frac{\partial^2 V}{\partial x^2}(x - \omega t) = \frac{d^2 V(\xi)}{d\xi^2}$$

and

$$\frac{\partial V}{\partial t} = -\omega \frac{dV}{d\xi}$$

and find the system of ordinary differential equations obtained by making these substitutions in (1.48).

One of the major successes of the Hodgkin-Huxley theory is that by studying travelling wave solutions, Hodgkin and Huxley were able to make a theoretical estimate of the velocity of nerve conduction which agreed quite well with the experimentally observed velocity.

Since the study of (H-H) and (1.48) presents serious mathematical difficulties, simpler systems which seem to retain many of the important properties of (H-H) or (1.48) have been studied instead. One such system which has received considerable study is the FitzHugh-Nagumo equation

$$V_{xx} - V_t = F(V) + R$$
$$R_t = \varepsilon(V - bR)$$

where ε, b are positive constants and $F(V)$ is a function such that $F(0) = 0$ and $F'(0)$ is positive, for example, $F(V)$ can be cubic. For a description of some of the work on the FitzHugh-Nagumo equation as well as other topics in nerve conduction theory, see Scott [1975] and Cronin [1987].

3. The Field-Noyes Model for the Belousov-Zhabotinsky Reaction

The Belousov-Zhabotinsky reaction is a chemical reaction which exhibits temporal oscillations. A detailed chemical mechanism for the reaction was developed by Field, Köros and Noyes [1972] and this mechanism was described in terms of a differential equation by Field and Noyes [1974]. The mathematical properties of the differential equation have been studied by Hastings and Murray [1975] and Murray [2003]. The Field-Noyes model can be written as

$$\frac{dx}{dt} = k_1 Ay - k_2 xy + k_3 Ax - 2k_4 x^2$$

$$\frac{dy}{dt} = -k_1 Ay - k_2 xy + k_5 fz \qquad \text{(F-N)}$$

$$\frac{dz}{dt} = k_3 Ax - k_5 z$$

where A, f, k_1, k_2, k_3, k_4, k_5 are positive constants, and x, y, z denote concentrations of certain molecules and ions.

(a) Show that no solution of (F-N) escapes the first octant.

(b) Find a closed bounded set E in the first octant such that no solution of (F-N) escapes it and such that every solution of (F-N) which passes through a point in the first octant ultimately enters and remains in the set E; more precisely, if t_1 is such that $(x(t_1), y(t_1), z(t_1))$ is in the first octant, then there is a number $t_2 > t_1$ such that for all $t \geq t_2$, solution $(x(t), y(t), z(t))$ is defined and

$$(x(t), y(t), z(t)) \in E$$

The Belousov-Zhabotinsky reaction also exhibits spatial structure and this can be studied mathematically by using a partial differential equation, that is, a reaction-diffusion equation. See Tyson [1976], and Kopell and Howard [1973].

4. The Goodwin Equations for a Chemical Reaction System

Goodwin [1963, 1965] has introduced and studied a differential equation which describes a chemical reaction system of a type which occurs in the study of cells in biology. The chemical system is assumed to have n constituents and the concentrations $x_1, \ldots, x_n$ of the constituents are related by the equations

$$\frac{dx_1}{dt} = \frac{K}{1 + \alpha x_n^\rho} - b_1 x_1$$

$$\frac{dx_2}{dt} = g_1 x_1 - b_2 x_2 \qquad \text{(1.49)}$$

$$\cdots$$

$$\frac{dx_n}{dt} = g_{n-1} x_{n-1} - b_n x_n$$

where ρ is a positive integer, and α, K, $g_1, \ldots, g_{n-1}$, $b_1, \ldots, b_n$ are positive constants. The term $K/(1 + \alpha x_n^\rho)$ refers to the fact that the reaction dx_1/dt is inhibited by feedback metabolite x_n in a reaction in which the stoichiometric coefficient of x_n is ρ. Since $x_1, \ldots, x_n$ are concentrations, we are concerned only with solutions of (1.49) which are in the first octant.

(a)   For $n = 3$, show that no solution of (1.49) escapes the first octant.

(b)   For $n = 3$, determine a closed bounded set E in the first octant such that every solution which passes through a point in the first octant ultimately enters and remains in the set E.

Chapter 2

Linear Systems

Existence Theorems for Linear Systems

Although useful information is obtained from the material in Chapter 1, it leaves us far from the specific knowledge of the solutions of a given differential equation. Now we begin to impose further hypotheses which can be used to obtain more detailed information about solutions.

Our next step is to study linear differential equations which are a special but very important class of differential equations. That is, we study systems of the form

$$x_1' = a_{11}(t)x_1 + \cdots + a_{1n}(t)x_n + u_1(t)$$
$$x_2' = a_{21}(t)x_1 + \cdots + a_{2n}(t)x_n + u_2(t)$$
$$\cdots$$
$$x_n' = a_{n1}(t)x_1 + \cdots + a_{nn}(t)x_n + u_n(t) \qquad (2.1)$$

in which the right-hand sides of the equations are linear in $x_1, \ldots, x_n$.

The study of such systems is crucially important for the following reasons. First, equations of this form arise in a wide variety of applications. Second, as happens in other parts of linear and nonlinear analysis, the theory for linear equations is the basis for much of the study of nonlinear equations. Finally, from the viewpoint of pure mathematics, the study is an elegant application of linear algebra. The canonical forms for matrices are used to calculate explicit solutions for systems which have constant coefficients and are homogeneous (i.e., $u_j(t) = 0$ for $j = 1, \ldots, n$). Also the theorems obtained for periodic systems of linear differential equations parallel closely theorems about solutions of systems of linear algebraic equations.

In a theoretical sense, this chapter contains all the topics on linear equations that are studied in an introductory course in differential equations. However, there are a number of useful computational techniques described in the introductory course which we have omitted. For example, a strategic transformation of variables can change a seemingly impossible equation into one which is easily solved. Consequently this chapter is a supplement to the introductory course. It does not supplant it.

All of the examples discussed at the end of Chapter 1 (Volterra equations, Hodgkin-Huxley equations, etc.) are nonlinear equations. Hence the theory in this chapter is not immediately applicable to the examples. However as we will see later, the linear theory will play an important part in the analysis of the examples, especially in studying the stability of the equilibrium points.

If $A(t)$ denotes the matrix

$$[a_{ij}(t)]$$

and if $u(t)$ denotes the vector

$$\begin{bmatrix} u_1(t) \\ \vdots \\ u_n(t) \end{bmatrix}$$

then the system (2.1) can be written in terms of vectors and matrices as

$$x' = A(t)x + u(t) \tag{2.2}$$

Because the properties of matrix $A(t)$ play a very important role in the study of the solutions of (2.1), we introduce some definitions used in the study of matrix $A(t)$.

Definition   If $A = [a_{ij}]$ is a constant matrix, the *norm* of A, denoted by $|A|$, is

$$|A| = \sum_{i,j=1}^{n} |a_{ij}|$$

if A and B are constant matrices and x is a constant vector, then it follows easily that:

$$|A + B| \leq |A| + |B|$$
$$|AB| \leq |A||B|$$
$$|Ax| \leq |A||x|$$

Definition   If $A(t) = [a_{ij}(t)]$, the *derivative of* $A(t)$, sometimes denoted by d/dt $A(t)$, is

$$\left[\frac{d}{dt} a_{ij}(t) \right]$$

The *integral of* $A(t)$ over $[a, b]$, sometimes denoted by

$$\int_a^b A(t)\, dt$$

is

$$\left[\int_a^b a_{ij}(t)\, dt \right]$$

and the *trace of* A, sometimes denoted by tr $A(t)$, is

$$\sum_{i=1}^{n} a_{ii}(t)$$

Existence Theorem 2.1 for Linear Systems. *If for $i, j = 1, \ldots, n$, each $a_{ij}(t)$ is continuous for all real t and if $u(t)$ is continuous for all real t, then if (t_0, x^0) is an arbitrary point in (t, x)-space, there is a unique solution $x(t, t_0, x^0)$ of equation (2.1) such that $x(t_0, t_0, x^0) = x^0$ and solution $x(t, t_0, x^0)$ has for its domain the real t-axis.*

Remark. Notice that this theorem shows that there is no extension problem for solutions of linear equations. That is, if the elements of $A(t)$ and $u(t)$ are sufficiently well-behaved (e.g., continuous for all real t), then the solution has for its domain the entire t-axis.

Instead of proving this theorem directly, we prove a somewhat more general theorem which is used less frequently, but is of sufficient interest to be presented for its own sake.

Existence Theorem 2.2 for Linear Equations. *Suppose $A(t)$ and $u(t)$ are Riemann integrable functions of t on (a, b), that is, the Riemann integrals over any interval $[c, d]$ contained in (a, b) of the elements of $A(t)$ and $u(t)$ exist, and suppose there exists a function $k(t)$ with domain (a, b) such that*

1. *$k(t)$ is continuous and bounded on (a, b)*

2. *if $t \in (a, b)$, then*

$$|A(t)| \le k(t)$$

and

$$|u(t)| \le k(t)$$

Let $t_0 \in (a, b)$ and suppose x^0 is a fixed n-vector. Then equation (2.1) or (2.2) has a unique solution $x(t)$ on (a, b) such that $x(t_0) = x^0$ in the following sense: if $t \in (a, b)$, then

$$x(t) = x^0 + \int_{t_0}^{t} A(s)x(s)\,ds + \int_{t_0}^{t} u(s)\,ds \qquad (2.3)$$

Proof The underlying idea of the proof is the same as the proof of Existence Theorem 1.1. For $t \in (a, b)$, we define

$$x_0(t) = x^0$$

$$\ldots$$

$$x_{n+1}(t) = x^0 + \int_{t_0}^{t} A(s)x_n(s)\,ds + \int_{t_0}^{t} u(s)\,ds \qquad (n = 0, 1, 2, \ldots)$$

If $x_n(t)$ is continuous on (a, b), then $A(s)x_n(s)$ is integrable over any interval $[c, d]$ contained in (a, b) and hence $x_{n+1}(t)$ is defined and continuous on (a, b). To show that the $x_n(t)$ converge uniformly, we proceed just as in the proof of the Existence

Theorem 1.1. First if $t \in (a, b)$,

$$|x_1(t) - x_0(t)| \leq \int_{t_0}^{t} \{|A(s)||x^0| + |u(s)|\} \, ds$$

$$\leq (1 + |x^0|) \int_{t_0}^{t} k(s) \, ds$$

Let $K(t) = \int_{t_0}^{t} k(s) \, ds$ and assume that for $t \in (a, b)$

$$|x_n(t) - x_{n-1}(t)| \leq (1 + |x^0|) \frac{(K(t))^n}{n!}$$

then

$$|x_{n+1}(t) - x_n(t)| \leq \int_{t_0}^{t} |A(s)x_n(s) - A(s)x_{n-1}(s)| \, ds$$

$$\leq (1 + |x^0|) \int_{t_0}^{t} k(s) \frac{(K(s))^n}{n!} \, ds$$

Since

$$\frac{d}{dt} K(t) = k(t)$$

and $K(t_0) = 0$, then

$$|x_{n+1}|(t) - x_n(t) \leq (1 + |x^0|) \frac{(K(t))^{n+1}}{(n+1)!}$$

Thus $\{x_n(t)\}$ converges uniformly on any closed interval $[c, d]$ in (a, b) to a continuous function $x(t)$. To complete the proof of existence of the solution, it is sufficient to show that

$$\lim_n \int_{t_0}^{t} A(s)x_n(s) \, ds = \int_{t_0}^{t} A(s)x(s) \, ds$$

Since $x(t)$ is continuous, $\int_{t_0}^{t} A(s)x(s) \, ds$ exists and

$$\left| \int_{t_0}^{t} A(s)[x_n(s) - x(s)] \, ds \right| \leq \int_{t_0}^{t} |A(s)| \, |x_n(s) - x(s)| \, ds \leq \varepsilon M$$

where M is a bound for $k(t)$ on (a, b).

The proof that $x(t)$ is a unique solution in any closed interval $[c, d]$ in (a, b) goes through just as for the proof of Lemma 1.5 in the proof of Existence Theorem 1.1. $\square$

The proof of Existence Theorem 2.1 for linear systems is obtained from Existence Theorem 2.2 for linear systems as follows. If the elements of $A(t)$ and $u(t)$ are continuous then since the solution $x(t)$ is continuous, equation (2.3) may be differentiated with respect to t and we obtain

$$\frac{d}{dt} x(t) = A(t)x(t) + u(t)$$

Homogeneous Linear Equations: General Theory

Definition   If $u(t)$ in equation (2.2) is identically 0, that is, if (2.2) has the form

$$x' = A(t)x \qquad (2.4)$$

then the equation is said to be a *homogeneous* linear equation.

Throughout this discussion we will study the n-dimensional equation (2.4) and it will be assumed that the elements of matrix $A(t)$ are continuous on an interval (a, b).

Theorem 2.1   *If $x(t)$ is a solution of (2.4) on (a, b) and if there exists $t_0 \in (a, b)$ such that $x(t_0) = 0$, then $x(t) = 0$ for all $t \in (a, b)$.*

Proof   Let $y(t) = 0$ for $t \in (a, b)$. It is clear that $y(t)$ is a solution of (2.4) on (a, b). But if $t = t_0$, $x(t) = y(t)$. Hence by the uniqueness condition in Existence Theorem 2.1 for linear systems, $x(t) = y(t)$ for all $t \in (a, b)$. ☐

Definition   Let $f_1(t), \ldots, f_q(t)$ be n-vector functions on (a, b). Then $f_1(t), \ldots, f_q(t)$ are *linearly dependent* if there exist constants $c_1, \ldots, c_q$ (not all zero) such that for all $t \in (a, b)$,

$$c_1 f_1(t) + \cdots + c_q f_q(t) = 0$$

If $f_1(t), \ldots, f_q(t)$ are not linearly dependent, they are *linearly independent*.

Theorem 2.2   *The collection of solutions of (2.4) is an n-dimensional linear space.*

Proof   The collection of solutions is clearly a linear space. Let $t_0 \in (a, b)$ and let

$$x_{(1)}^0 = (1, 0, \ldots, 0)$$
$$x_{(2)}^0 = (0, 1, 0, \ldots, 0)$$
$$\ldots$$
$$x_{(n)}^0 = (0, \ldots, 0, 1)$$

By the existence theorem, there are solutions $x^{(i)}(t)$ of (2.4) with $i = 1, \ldots, n$ such that

$$x^{(i)}(t_0) = x_{(i)}^0$$

But these solutions are linearly independent because suppose there are constants $c_1, \ldots, c_n$ such that for $t \in (a, b)$

$$c_1 x^{(1)}(t) + c_2 x^{(2)}(t) + \cdots + c_n x^{(n)}(t) = 0$$

Then

$$c_1 x^{(1)}(t_0) + c_2 x^{(2)}(t_0) + \cdots + c_n x^{(n)}(t_0) = 0$$

or

$$(c_1, c_2, \ldots, c_n) = 0$$

or

$$c_1 = c_2 = \cdots = c_n = 0$$

Thus, the collection of solutions is a linear space of dimension greater than or equal to n.

Now suppose $y^{(1)}(t), y^{(2)}(t), \ldots, y^{(n)}(t), y^{(n+1)}(t)$ are linearly independent solutions of (2.4). Let $t_0 \in (a, b)$. Then by Theorem 2.1, $y^{(1)}(t_0), y^{(2)}(t_0), \ldots, y^{(n)}(t_0), y^{(n+1)}(t_0)$ are linearly independent. That is, we have $(n + 1)$ linearly independent n-vectors. From linear algebra, this is impossible. □

Definition   A set of n linearly independent solutions of (2.4) is a *fundamental system* of (2.4).

Definition   An $n \times n$ matrix whose n columns are n linearly independent solutions of (2.4) is a *fundamental matrix* of (2.4).

Theorem 2.3   *Let X be an $n \times n$ matrix whose columns are solutions of (2.4). A necessary and sufficient condition that X be a fundamental matrix of (2.4) is that there exist $t_0 \in (a, b)$ such that* $\det X(t) \neq 0$ *at* $t = t_0$.

Proof   If $\det X(t) \neq 0$ at $t = t_0$, then the columns of $X(t_0)$ are linearly independent. Then by Theorem 2.1, the columns of $X(t)$ are linearly independent for each $t \in (a, b)$.

If the columns of $X(t)$ are linearly independent for each $t \in (a, b)$, then $\det X(t) \neq 0$ for each $t \in (a, b)$. This completes the proof of Theorem 2.3. □

Theorem 2.4   *If X is a fundamental matrix of (2.4) and C is a constant nonsingular matrix, then XC is a fundamental matrix. If X_1 is a second fundamental matrix, there exists a constant nonsingular matrix C_1 such that $X_1 = XC_1$.*

Proof   Since

$$X' = A(t)X$$

then

$$X'C = A(t)\{XC\}$$

or

$$(XC)' = A(t)\{XC\}$$

Also

$$\det XC = (\det X)(\det C) \neq 0$$

Thus XC is a fundamental matrix. For $t \in (a, b)$, let

$$(X(t))^{-1}(X_1(t)) = Y(t)$$

Then

$$X_1(t) = X(t)Y(t)$$

and

$$X_1' = XY' + X'Y$$

or

$$AX_1 = XY' + AXY = XY' + AX_1$$

Therefore

$$XY' = 0$$

Since $\det X(t) \neq 0$ for each $t \in (a, b)$, then for each $t \in (a, b)$

$$Y'(t) = 0$$

Therefore $Y(t)$ is a constant matrix. Since

$$\det Y(t) = \det\{[X(t)]^{-1}\} \det\{X_1(t)\} \neq 0$$

then $Y(t)$ is nonsingular. ▢

Theorem 2.5 *If $Y(t)$ is a matrix such that*

$$Y'(t) = A(t)Y(t) \tag{2.5}$$

then

$$\frac{d}{dt}[\det Y(t)] = [\operatorname{tr} A(t)][\det Y(t)] \tag{2.6}$$

Also if $t, t_0 \in (a, b)$, then

$$\det Y(t) = [\det Y(t_0)] \exp\left\{\int_{t_0}^{t} \operatorname{tr} A(s)\, ds\right\}$$

Proof   Let $Y(t) = (y_{ij}(t))$ and $A(t) = (a_{ij}(t))$. Then (by induction)

$$[\det Y(t)]'$$

$$= \begin{vmatrix} y'_{11} & y'_{12} & \cdots & y'_{1n} \\ & \vdots & & \\ y_{n1} & y_{n2} & \cdots & y_{nn} \end{vmatrix} + \cdots + \begin{vmatrix} y_{11} & y_{12} & \cdots & y_{1n} \\ & \vdots & & \\ y'_{n1} & y'_{n2} & \cdots & y'_{nn} \end{vmatrix} \qquad (2.7)$$

But by (2.5)

$$y'_{ij}(t) = \sum_{k=1}^{n} a_{ik}(t) y_{kj}(t)$$

Then (2.7) becomes

$$[\det Y(t)]' = \begin{vmatrix} \sum a_{1k}(t) y_{k1}(t) & \cdots & \sum a_{1k}(t) y_{kn}(t) \\ y_{21} & & y_{2n} \\ \vdots & & \vdots \\ y_{n1} & & y_{nn} \end{vmatrix} + \cdots$$

In the first determinant on the right, subtract from the first row the expression [a_{12} times second row $+a_{13}$ times third row $+ \cdots + a_{1n}$ times nth row]. Carrying out similar operations on the other determinants on the right, we obtain

$$[\det Y(t)]' = \begin{vmatrix} a_{11} y_{11} & \cdots & a_{11} y_{1n} \\ y_{21} & & y_{2n} \\ \vdots & & \vdots \\ y_{n1} & & y_{nn} \end{vmatrix} + \cdots$$

$$= a_{11} \det[Y(t)] + \cdots + a_{nn} \det[Y(t)]$$

$$= [\operatorname{tr} A(t)][\det Y(t)]$$

That is, equation (2.6) is satisfied.

A straightforward computation shows that

$$[\det Y(t_0)] \exp \left\{ \int_{t_0}^{t} \operatorname{tr} A(s) \, ds \right\}$$

is a solution of the scalar differential equation

$$x' = (\operatorname{tr} A(t)) x \qquad (2.8)$$

But (2.6) shows that $\det Y(t)$ is a solution of (2.8). Hence by the uniqueness of the solution we must have

$$\det Y(t) = [\det Y(t_0)] \exp \left\{ \int_{t_0}^{t} \operatorname{tr} A(s) \, ds \right\} \qquad \square$$

Corollary   *If the columns of X are solutions of* (2.4), *then X is a fundamental matrix if and only if* $\det X(t) \neq 0$ *for all* $t \in (a, b)$.

Now we obtain a more explicit representation for the solutions of some linear homogeneous equations by introducing the exponential of a matrix. If $B(t)$ is an $n \times n$ matrix such that the elements are functions of t consider the sums

$$\sum_{s=0}^{m} \frac{B^s}{s!}$$

where $B^0 = I$, the $n \times n$ identity matrix. For each t, these sums satisfy a Cauchy condition in the matrix norm because

$$\left| \sum_{s=0}^{p+q} \frac{B^s}{s!} - \sum_{s=0}^{p} \frac{B^s}{s!} \right| = \left| \sum_{s=p+1}^{p+q} \frac{B^s}{s!} \right| \leq \sum_{s=p+1}^{p+q} \frac{|B|^s}{s!}$$

But if p is sufficiently large, this last term is $< \varepsilon$ because it is part of the series expansion for $e^{|B|}$. From the definition of the matrix norm, it follows that the elements in the ith, jth position in the sums $\sum_{s=0}^{m} \frac{B^s}{s!} (m = 1, 2, 3, \ldots)$ satisfy a Cauchy condition and therefore converge to a function $c_{ij}(t)$ and the matrices $\sum_{s=0}^{m} \frac{B^s}{s!}$ converge in the matrix norm to the matrix $[c_{ij}(t)]$.

Definition   The matrix $[c_{ij}(t)]$ is called the *exponential* of $B(t)$ and is denoted by $e^{B(t)}$ or $\exp B(t)$.

Lemma 2.1   *If* $BD = DB$, *then* $e^{B+D} = e^B e^D$.

Proof   If b, d are real numbers, then

$$e^{b+d} = e^b e^d$$

or

$$1 + (b+d) + \frac{(b+d)^2}{2!} + \cdots = \left(1 + b + \frac{b^2}{2!} + \cdots\right)\left(1 + d + \frac{d^2}{2!} + \cdots\right) \quad (2.9)$$

Since $BD = DB$ then (2.9) is valid with b and d replaced by B and D respectively. $\square$

Theorem 2.6   *Given*

$$x' = A(t)x \quad (2.10)$$

let $B(t) = \int_{t_0}^{t} A(s)\,ds$ *for* $t_0, t \in (a, b)$ *and assume that for each* $t \in (a, b)$,

$$A(t)B(t) = B(t)A(t)$$

Then the solution $x(t)$ of (2.10) such that $x(t_0) = x^0$ is

$$x(t) = (\exp(B))x^0$$

Proof   We have already used the fact that the theorem for derivative of a product holds for the derivative of a product of matrices. That is, if $C = (c_{ij}(t))$, $D = (d_{ij}(t))$ and $c_{ij}(t)$, $d_{ij}(t)$ are differentiable functions $(i, j = 1, \ldots, n)$, then

$$\frac{d}{dt}(CD) = C\frac{dD}{dt} + \frac{dC}{dt}D \tag{2.11}$$

But we need to emphasize that since the multiplication of matrices is not commutative it is essential in using (2.11) to preserve the order in which the factors occur. From the definition of B, it follows that

$$\frac{d}{dt}B(t) = A(t)$$

Now assume that

$$\frac{d}{dt}[B(t)]^{m-1} = (m - 1)A(t)[B(t)]^{m-2}$$

Then using (2.11) and the hypothesis that A and B commute, we have

$$\begin{aligned}
\frac{d}{dt}[B(t)]^m &= \frac{d}{dt}\{(B(t))^{m-1}(B(t))\} \\
&= (B(t))^{m-1}A(t) + (m - 1)A(t)(B(t))^{m-2}B(t) \\
&= mA(t)(B(t))^{m-1} \tag{2.12}
\end{aligned}$$

Hence by induction, equation (2.12) holds for all positive integers m. By Existence Theorem 2.1 for linear equations, the desired solution $x(t)$ is the limit of the sequence

$$x_1(t) = x^0 + \int_{t_0}^t A(s)x^0 ds = (I + B)x^0$$

$$\cdots$$

$$x_m(t) = x^0 + \int_{t_0}^t A(s)x_{m-1}(s)\,ds$$

$$\cdots$$

Then

$$x'_m(t) = A(t)x_{m-1}(t) \tag{2.13}$$

Assume that

$$x_{m-1}(t) = \left(I + B + \frac{B^2}{2!} + \cdots + \frac{B^{m-1}}{(m-1)!}\right)x^0$$

Then by (2.13)

$$x'_m(t) = Ax^0 + ABx^0 + \frac{AB^2}{2!}x^0 + \cdots + \frac{AB^{m-1}}{(m-1)!}x^0$$

and by (2.12)

$$x_m(t) = K + \left(B + \frac{B^2}{2!} + \cdots + \frac{B^m}{m!} \right) x^0 \tag{2.14}$$

where K is an arbitrary constant vector. Since $B(t_0) = 0$, then if $t = t_0$, (2.14) yields

$$x_m(t_0) = K$$

But $x_m(t_0) = x^0$. Therefore (2.14) becomes

$$x_m(t) = x^0 + \left(B + \frac{B^2}{2!} + \cdots + \frac{B^m}{m!} \right) x^0$$

and we have

$$\lim_{m \to \infty} x_m(t) = e^B x^0$$

This completes the proof of Theorem 2.6. □

Homogeneous Linear Equations with Constant Coefficients

The condition in Theorem 2.6 that A and B commute is very strong, but there is one important class of equations for which this condition is satisfied, that is, the case in which A is a constant matrix. Our next objective is to obtain an explicit form for a fundamental matrix for (2.4) in the case where A is a constant.

Theorem 2.7 *If A is a constant matrix then e^{At} is a fundamental matrix for*

$$x' = Ax \tag{2.15}$$

Proof First, e^B is a fundamental matrix by Theorems 2.3 and 2.6. But by Lemma 2.1,

$$e^B = e^{A(t-t_0)} = e^{At} e^{-At_0}$$

Since e^{At_0} is nonsingular (its inverse is e^{At_0}) then by Theorem 2.4, e^{At} is a fundamental matrix. □

To study the form of e^{At}, we will use a canonical form of matrix A. So we first summarize briefly the results we shall need from the theory of canonical forms.

Definition If A is an $n \times n$ constant matrix and I is the identity $n \times n$ matrix, the roots of the equation $\det(A - \lambda I) = 0$ are the *eigenvalues* (or *characteristic roots* or *proper values*) of A. If λ is a root of multiplicity m of the equation $\det(A - \lambda I) = 0$, then λ is an *eigenvalue of A of algebraic multiplicity m*. The equation

$$\det(A - \lambda I) = 0$$

is the *characteristic equation* of A. If x is a nonzero vector such that $(A - \lambda I)x = 0$ or $Ax = \lambda x$, then x is an *eigenvector associated with eigenvalue λ*.

Definition   A real [complex] $n \times n$ matrix C is *similar over the real [complex] numbers* to a real [complex] $n \times n$ matrix D if there exists a nonsingular real [complex] matrix P such that $D = PCP^{-1}$.

(It is easy to verify that similarity is an equivalence relation.)

Lemma 2.2   *If C is similar to D, then the set of characteristic roots of C is the set of characteristic roots of D.*

Proof   If λ_1 is a characteristic root of C, then $\det(C - \lambda_1 I) = 0$. But

$$\begin{aligned}
\det(D - \lambda_1 I) &= \det(PCP^{-1} - \lambda_1 PP^{-1}) \\
&= \det(P(C - \lambda_1 I)P^{-1}) \\
&= \det P(\det(C - \lambda_1 I))\det P^{-1} \\
&= 0
\end{aligned}$$

$\quad\Box$

Jordan Canonical Form Theorem.   *If A is a complex $n \times n$ matrix, then there is a nonsingular complex matrix P such that*

$$P^{-1}AP = J$$

where the matrix J has the following form:

$$J = \begin{bmatrix} J_1 & & & \\ & J_2 & & \\ & & \ddots & \\ & & & J_s \end{bmatrix}$$

where all entries not written in are zero and J_i, $i = 1, \ldots, s$, is a square matrix that has the form

$$J_i = \begin{bmatrix} \lambda_i & 1 & & & \\ & \lambda_i & 1 & & \\ & & & \ddots & 1 \\ & & & & \ddots \\ & & & & \lambda_i \end{bmatrix}$$

where λ_i is an eigenvalue of A. Each eigenvalue of A appears in at least one J_i. Matrix J_i may be a 1×1 matrix. (In the case that each J_i is a 1×1 matrix, matrix J is a diagonal matrix.) The eigenvalues λ_i, $i = 1, \ldots, s$, are, in general, not distinct. Except for the order in which the matrices J_i ($i = 1, \ldots, s$) appear on the diagonal, the matrix J is unique. (The matrix J is called the Jordan *canonical form or the* rational canonical form *of matrix A. The matrices J_i will be termed blocks.)*

Proof The proof of this theorem is quite lengthy and can be found in any standard text on linear algebra. We describe the basic idea of the proof for the special case of an $n \times n$ matrix A which has just one eigenvalue λ. (Then λ has algebraic multiplicity n.) Let $v_1, \ldots, v_k$ be a basis for the linear space of solutions of the equation

$$(A - \lambda I)v = 0$$

That is, $v_1, \ldots, v_k$ is a basis for the space of eigenvectors associated with eigenvalue λ. Let $v_i^{(1)}$ denote a solution of the equation

$$(A - \lambda I)v = v_i$$

if it exists; and if j is an integer such that $j \geq 2$, let $v_i^{(j)}$ denote a solution of the equation

$$(A - \lambda I)v = v_i^{(j-1)}$$

if it exists. (These are called *generalized eigenvectors*.) It can be proved that there are n such vectors $v_1, \ldots, v_k, v_1^{(1)}, \ldots$ which are linearly independent. Suppose, for example, that the n vectors are

$$v_1, \ldots, v_k, v_1^{(1)}, v_\ell^{(1)}, v_k^{(1)}, v_1^{(2)} \tag{2.16}$$

where ℓ is an integer such that $1 < \ell < k$. Let P be the $n \times n$ matrix whose columns are the vectors

$$v_1, v_1^{(1)}, v_1^{(2)}, v_2, \ldots, v_\ell, v_\ell^{(1)}, v_{\ell+1}, \ldots, v_k, v_k^{(1)}$$

Since the vectors are linearly independent, P is nonsingular. Now

$$AP = \lambda v_1, \lambda v_1^{(1)} + v_1, \lambda v_1^{(2)} + v_1^{(1)}, \ldots, \lambda v_\ell, \lambda v_\ell^{(1)} + v_\ell, \ldots, \lambda v_k, \lambda v_k^{(1)} + v_k$$

and therefore

$$P^{-1}AP = P^{-1}\left[\lambda v_1, \lambda v_1^{(1)}, \lambda v_1^{(2)}, \ldots, \lambda v_\ell, \lambda v_\ell^{(1)}, \ldots, \lambda v_k, \lambda v_k^{(1)}\right]$$

$$+ P^{-1}\left[0, v_1, v_1^{(1)}, 0, \ldots, 0, v_\ell, \ldots, 0, v_k\right]$$

$$= P^{-1}\lambda P + P^{-1}\left[0, v_1, v_1^{(1)}, 0, \ldots, 0, v_\ell, \ldots, 0, v_k\right]$$

$$= \lambda I +
\begin{bmatrix}
0 & 1 & 0 & \cdots & 0 & \cdots & 0 \\
 & & 1 & & \cdot & & \cdot \\
 & \cdot & \cdot & \cdot & \cdots & 1 & \cdot \\
 & \cdot & \cdot & \cdot & \cdot & \cdot & \cdot \\
 & & & & \cdot & \cdot & 1 \\
0 & 0 & 0 & \cdots & 0 & & 0
\end{bmatrix}$$

$\uparrow\!\!\!\rule[0.5ex]{2em}{0.4pt}(\ell + 3)$th column

$$=
\begin{bmatrix}
\lambda & 1 & 0 & & & & & \\
 & \lambda & 1 & & & & & \\
 & & \lambda & & & & & \\
 & & & \cdot & & & & \\
 & & & & \lambda & 1 & & \\
 & & & & & \lambda & & \\
 & & & & & & \lambda & 1 \\
 & & & & & & & \lambda
\end{bmatrix}$$

$\uparrow\!\!\!\rule[0.5ex]{2em}{0.4pt}(\ell + 3)$th column

If the matrix A has more than one eigenvalue, the matrix P is constructed by finding sets of vectors (2.16) for each eigenvalue. It can be proved that the collection of all such sets of vectors consists of n linearly independent vectors.

The Jordan canonical form theorem says that A is similar over the complex numbers to the matrix J. In much of our work we will be interested in matrices A which are real. Since the eigenvalues of A need not be real, the matrix J may have complex entries, and a real canonical form is sometimes preferable.

Real Canonical Form Theorem. *If A is a real $n \times n$ matrix, then there is a real nonsingular matrix P such that $P^{-1}AP = J$ where matrix J is a real matrix which has the form*

$$
\begin{bmatrix}
J_1 & & & & \\
 & J_2 & & & \\
 & & \cdot & & \\
 & & & \cdot & \\
 & & & & J_s
\end{bmatrix}
$$

where all entries not written in are zero and J_j, $j = 1, \ldots, s$, is associated with eigenvalue λ_j and has one of the two following forms: If λ_j is real,

$$J_j = \begin{bmatrix} \lambda_j & 1 & & & & \\ & \lambda_j & \cdot & & & \\ & & \cdot & \cdot & & \\ & & & \cdot & 1 & \\ & & & & \cdot & \\ & & & & & \lambda_j \end{bmatrix}$$

If λ_j is a complex eigenvalue,

$$\lambda_j = \alpha_j + i\beta_j$$

where α_j, β_j are real and $\beta_j > 0$, then

$$J_j = \begin{bmatrix} \alpha_j & \beta_j & 1 & 0 & & & & \\ -\beta_j & \alpha_j & 0 & 1 & & & & \\ 0 & 0 & \cdot & & & & & \\ 0 & 0 & & \cdot & & & & \\ & & & & \cdot & & & \\ & & & & & \alpha_j & \beta_j & 1 & 0 \\ & & & & & -\beta_j & \alpha_j & 0 & 1 \\ & & & & & 0 & 0 & \alpha_j & \beta_j \\ & & & & & 0 & 0 & -\beta_j & \alpha_j \end{bmatrix}$$

If λ_j is real, matrix J_j may be a 1×1 matrix. If λ_j is complex, then J_j must be a $(2m \times 2m)$ matrix where $m \geq 1$. The $\lambda_1, \ldots, \lambda_s$ are not necessarily distinct. There is associated with each eigenvalue of A at least one J_j. (Note: If $\lambda_j = \alpha_j + i\beta_j$, where $\beta_j > 0$, is an eigenvalue then (since A is real) $\alpha_j - i\beta_j$ is also an eigenvalue. We say that $\alpha_j + i\beta_j$ and $\alpha_j - i\beta_j$ are both associated with J_j.) ⬚

Proof As in the Jordan canonical form, we simply describe the basic idea of the proof. First, suppose that A is a $(2m \times 2m)$ matrix which has $\alpha + i\beta$ as an eigenvalue of algebraic multiplicity m. Since A is real, then $\alpha - i\beta$ is also an eigenvalue of algebraic multiplicity m. It can be proved that the eigenvectors associated with $\alpha + i\beta$ have the form $u_j + iv_j$, $j = 1, \ldots, k$ where $u_1, \ldots, u_k, v_1, \ldots, v_k$ are real and linearly independent. Moreover $u_j - iv_j$ is an eigenvector associated with $\alpha - i\beta$. Let $u_j^{(1)} + iv_j^{(1)}$, where $u_j^{(1)}$ and $v_j^{(1)}$ are real, denote a solution of the equation

$$(A - \lambda I)w = u_j + iv_j$$

(if it exists) and if q is an integer such that $q \geq 2$ let $u_j^{(q)} + iv_j^{(q)}$ denote a solution of the equation

$$(A - \lambda I)w = u_j^{(q-1)} + iv_j^{(q-1)}$$

Linear Systems

(if it exists). It can be proved that there are exactly $2m$ vectors $u_1, \ldots, u_k, v_1, \ldots, v_k,$ $u_1^{(1)}, \ldots, v_1^{(1)}, \ldots$ which are linearly independent. Suppose the $2m$ vectors are:

$$u_1, u_1^{(1)}, u_1^{(2)}, u_2, u_2^{(1)}, u_3, \ldots, u_k, v_1, v_1^{(1)}, v_1^{(2)}, v_2, v_2^{(1)}, v_3, \ldots, v_k$$

Let P be the matrix

$$u_1, v_1, u_1^{(1)}, v_1^{(1)}, u_1^{(2)}, v_1^{(2)}, u_2, v_2, u_2^{(1)}, v_2^{(1)}, u_3, v_3, \ldots, u_k, v_k$$

and observe that

$$[A - (\alpha + i\beta)I]\left[u_1^{(2)} + iv_1^{(2)}\right] = u_1^{(1)} + iv_1^{(1)}$$

or

$$Au_1^{(2)} = \alpha u_1^{(2)} - \beta v_1^{(2)} + u_1^{(1)}$$
$$Av_1^{(2)} = \beta u_1^{(2)} + \alpha v_1^{(2)} + v_1^{(1)}$$

Matrix $A - (\alpha + i\beta)I$ acts similarly on the other vectors. Hence we obtain

$$\begin{aligned}
AP = \big[&\alpha u_1 \ -\beta v_1, \beta u_1 + \alpha v_1, \alpha u_1^{(1)} - \beta v_1^{(1)} + u_1, \beta u_1^{(1)} + \alpha v_1^{(1)} + v_1,\\
&\alpha u_1^{(2)} - \beta v_1^{(2)} + u_1^{(1)}, \beta u_1^{(2)} + \alpha v_1^{(2)} + v_1^{(1)},\\
&\alpha u_2 - \beta v_2, \beta u_2 + \alpha v_2,\\
&\alpha u_2^{(1)} - \beta v_2^{(1)} + u_2, \beta u_2^{(1)} + \alpha v_2^{(1)} + v_2,\\
&\alpha u_3 - \beta v_3, \beta u_3 + \alpha v_3, \ldots, \alpha u_k - \beta v_k, \beta u_k + \alpha v_k\big]
\end{aligned}$$

Hence

$$\begin{aligned}
P^{-1}AP = P^{-1}\big[&\alpha u_1, \alpha v_1, \alpha u_1^{(1)}, \alpha v_1^{(1)}, \alpha u_1^{(2)}, \alpha v_1^{(2)}, \alpha u_2, \alpha v_2\\
&\alpha u_2^{(1)}, \alpha v_2^{(1)}, \alpha u_3, \alpha v_3, \ldots, \alpha u_k, \alpha v_k\big]\\
+ P^{-1}\big[&-\beta v_1, \beta u_1, -\beta v_1^{(1)}, \beta u_1^{(1)}, -\beta v_1^{(2)}, \beta u_1^{(2)}, -\beta v_2, \beta u_2,\\
&-\beta v_2^{(1)}, \beta u_2^{(1)}, -\beta v_3, \beta u_3, \ldots, -\beta v_k, \beta u_k\big]\\
+ P^{-1}\big[&0, 0, u_1, v_1, u_1^{(1)}, v_1^{(1)}, 0, 0, u_2, v_2, 0, 0, \ldots, 0, 0\big]
\end{aligned}$$

From the definition of P and P^{-1}, it follows that

$$P^{-1}AP = \alpha I + \begin{bmatrix} 0 & \beta & 0 & 0 & \cdots & \\ -\beta & 0 & 0 & 0 & \cdots & \\ 0 & 0 & 0 & \beta & \cdots & \\ & & -\beta & 0 & \cdots & \\ & & & & & 0 & \beta \\ 0 & 0 & 0 & 0 & & -\beta & 0 \end{bmatrix}$$

$$+ \begin{bmatrix} 0 & 0 & 1 & 0 & 0 & 0 & \cdots & 0 \\ 0 & 0 & 0 & 1 & 0 & 0 & \cdots & 0 \\ & & & 0 & 1 & 0 & \cdots & 0 \\ & & & 0 & 0 & 1 & \cdots & 0 \\ & & & & & & \ddots & \vdots \\ & & & & & & \ddots & 0 \\ & & & & & & \ddots & 0 \\ 0 & 0 & & & & & & 0 \end{bmatrix}$$

For more general kinds of matrices A, extensions of the above idea can be made as in the proof of the Jordan canonical form theorem.

Now let us see what the fundamental matrix e^{tA} looks like. First we apply the Jordan canonical form theorem. Since $A = PJP^{-1}$, then

$$e^{tA} = e^{tPJP^{-1}}$$

$$= I + tPJP^{-1} + \frac{t^2}{2!}(PJP^{-1}) + \cdots + \frac{t^n}{n!}(PJP^{-1})^n + \cdots$$

$$= PP^{-1} + PtJP^{-1} + P\frac{t^2}{2!}JP^{-1} + \cdots + P\frac{t^n}{n!}J^nP^{-1} + \cdots$$

$$= Pe^{tJ}P^{-1}$$

Since $e^{tA} = Pe^{tJ}P^{-1}$ is a fundamental matrix, then by Theorem 2.4 $Pe^{tJ}P^{-1}(P) = Pe^{tJ}$ is also a fundamental matrix. To study the form of Pe^{tJ}, let us first see what e^{tJ} looks like. It follows from a simple observation about matrix multiplication that

$$e^{tJ} = \exp \begin{bmatrix} tJ_1 & & \\ & tJ_2 & \\ & & tJ_s \end{bmatrix} = \begin{bmatrix} e^{tJ_1} & & \\ & e^{tJ_2} & \\ & & e^{tJ_s} \end{bmatrix}$$

where, as usual, entries which are not written are zeros. If J_i is a 1×1 matrix, then e^{tJ_i} is just the scalar $e^{t\lambda_i}$. If J_i is an $r \times r$ matrix where $r > 1$, then

$$J_i = \lambda_i I + D$$

where I is the $r \times r$ identity matrix and

$$D = \begin{bmatrix} 0 & 1 & 0 & \cdots & 0 \\ 0 & 0 & 1 & \cdots & 0 \\ & & \ldots\ldots & & \\ 0 & 0 & 0 & \cdots & 1 \\ 0 & 0 & 0 & \cdots & 0 \end{bmatrix}$$

Since $ID = DI$, then by Lemma 2.1,

$$e^{tJ_i} = e^{t(\lambda_i I + D)} = e^{t\lambda_i I} e^{tD}$$

but

$$e^{tD} = \begin{bmatrix} 1 & 0 & \cdots & 0 \\ 0 & 1 & \cdots & 0 \\ & \cdots & & \\ 0 & & & 1 \end{bmatrix} + \begin{bmatrix} 0 & t & 0 & \cdots & 0 \\ 0 & 0 & t & \cdots & 0 \\ & & \cdots & & \\ 0 & 0 & 0 & \cdots & t \\ 0 & 0 & 0 & \cdots & 0 \end{bmatrix}$$

$$+ \frac{1}{2!} \begin{bmatrix} 0 & t & 0 & \cdots & 0 \\ 0 & 0 & t & \cdots & 0 \\ & & \cdots & & \\ 0 & 0 & 0 & \cdots & t \\ 0 & 0 & 0 & \cdots & 0 \end{bmatrix}^2 \qquad (2.17)$$

$$+ \cdots + \frac{1}{n!} \begin{bmatrix} 0 & t & 0 & \cdots & 0 \\ 0 & 0 & t & \cdots & 0 \\ & & \cdots & & \\ 0 & 0 & 0 & \cdots & t \\ 0 & 0 & 0 & \cdots & 0 \end{bmatrix}^n + \cdots$$

and a simple computation shows that if J_i is an $r \times r$ matrix, then if $h \geq r$, $D^h = 0$.

Thus (2.17) is an easily computed finite sum and we have

$$
e^{tJ_i} = \begin{bmatrix} e^{t\lambda_i} & & & & \\ & \cdot & & & \\ & & \cdot & & \\ & & & \cdot & \\ & & & & e^{t\lambda_i} \end{bmatrix} \begin{bmatrix} 1 & t & \frac{t^2}{2!} & \cdots & & \frac{t^{r-1}}{(r-1)!} \\ & 1 & t & \frac{t^2}{2!} & & \\ & & & \cdot & & \vdots \\ & & & & & \frac{t^2}{2!} \\ & & & & & t \\ & & & & & 1 \end{bmatrix}
$$

$$
= \begin{bmatrix} e^{t\lambda_i} & te^{t\lambda_i} & \cdot & & \frac{t^{r-1}e^{t\lambda_i}}{(r-1)!} \\ & e^{t\lambda_i} & te^{t\lambda_i} & \cdot & \\ & & \cdot & \cdot & \\ & & & \cdot & te^{\lambda_i} \\ & & & \cdot & e^{t\lambda_i} \end{bmatrix} \qquad (2.18)
$$

If $P = [p_{ij}]$ the fundamental matrix Pe^{tJ} has the form

$$
\begin{bmatrix} p_{11} & p_{12} & \cdots & p_{1n} \\ p_{21} & p_{22} & & p_{2n} \\ & & & \\ p_{n1} & p_{n2} & & p_{nn} \end{bmatrix} \begin{bmatrix} e^{tJ_1} & & \\ & e^{tJ_2} & \\ & & e^{tJ_s} \end{bmatrix}
$$

where each e^{tJ_i} ($i = 1, \ldots, s$) has the form given in (2.18). For definiteness, assume that the 1×1 matrices among the J_i are $J_1, \ldots, J_k$. Then if J_{k+1} is an $r \times r$ matrix, the columns of Pe^{tJ} are

First Column	Second Column	$\cdots$	k-th Column	$(k + 1)$-th Column
$p_{11}e^{t\lambda_1}$	$p_{12}e^{t\lambda_2}$		$p_{1k}e^{t\lambda_k}$	$p_{1,k+1}e^{t\lambda_{k+1}}$
$p_{21}e^{t\lambda_1}$	$p_{22}e^{t\lambda_2}$		$p_{2k}e^{t\lambda_k}$	$p_{2,k+1}e^{t\lambda_{k+1}}$
$\vdots$	$\vdots$		$\vdots$	$\vdots$
$p_{n1}e^{t\lambda_1}$	$p_{n2}e^{t\lambda_2}$		$p_{nk}e^{t\lambda_k}$	$p_{n,k+1}e^{t\lambda_{k+1}}$

$(k+2)$-th Column $(k+r)$-th Column

$(tp_{1,k+1} + p_{1,k+2})e^{t\lambda_{k+1}}$ $\left[\dfrac{t^{r-1}}{(r-1)!}p_{1,k+1} + \dfrac{t^{r-2}}{(r-2)!}p_{1,k+2}\right.$

$(tp_{2,k+1} + p_{2,k+2})e^{t\lambda_{k+1}}$ $\left. + \cdots + tp_{1,k+r-1} + p_{1,k+r}\right]e^{t\lambda_{k+1}}$

$\vdots$

$(tp_{n,k+1} + p_{2,k+2})e^{t\lambda_{k+1}}$

(Because of the bulkiness of the expressions, we have written only the first element of the $(k+r)$-th column.)

In applications of differential equations the case in which A is real is of particular importance, and we want to apply the Real Canonical Form Theorem to obtain real solutions. It is sufficient to study in detail the case in which there exists a real nonsingular matrix P such that

$$P^{-1}AP = \begin{bmatrix} \alpha & \beta & 1 & 0 & & & & & & \\ -\beta & \alpha & 0 & 1 & & & & & & \\ & & \alpha & \beta & 1 & 0 & & & & \\ & & -\beta & \alpha & 0 & 1 & & & & \\ & & & & \cdot & & & & & \\ & & & & & \cdot & & & & \\ & & & & & & \alpha & \beta & 1 & 0 \\ & & & & & & -\beta & \alpha & 0 & 1 \\ & & & & & & & & \alpha & \beta \\ & & & & & & & & -\beta & \alpha \end{bmatrix}$$

Let

$$B = \begin{bmatrix} \alpha & & & & 0 \\ & \alpha & & & \\ & & \cdot & & \\ & & & \cdot & \\ & & & & \cdot \\ 0 & & & & \alpha \end{bmatrix}$$

$$C = \begin{bmatrix} 0 & \beta & & & \\ -\beta & 0 & & & \\ & & \ddots & & \\ & & & 0 & \beta \\ & & & -\beta & 0 \end{bmatrix}$$

$$D = \begin{bmatrix} 0 & 0 & 1 & & 0 \\ & & & \ddots & \\ & & & & 1 \\ & & & & 0 \\ 0 & & & & 0 \end{bmatrix}$$

Matrix B commutes with matrix $C + D$ and $CD = DC$. Hence

$$e^{tPAP^{-1}} = e^{tB}e^{tC}e^{tD}$$

From the definition of the exponential of a matrix, it follows that

$$e^{tB} = \begin{bmatrix} e^{\alpha t} & & & \\ & e^{\alpha t} & & \\ & & \ddots & \\ 0 & & & e^{\alpha t} \end{bmatrix}$$

and

$$e^{tD} = \begin{bmatrix} 1 & 0 & t & 0 & \frac{t^2}{2!} & \cdots & \\ & 1 & 0 & t & 0 & \frac{t^2}{2!} & \cdots \\ & & & \ddots & & & \\ & & & & & t & \\ & & & & & 0 & \\ & & & & & & 1 \end{bmatrix}$$

where the entry in the upper right hand corner is 0 since A is an $n \times n$ matrix where n is even. In order to compute e^{tC} we remark first that

$$\begin{bmatrix} 0 & 1 \\ -1 & 0 \end{bmatrix}^2 = \begin{bmatrix} -1 & 0 \\ 0 & -1 \end{bmatrix}$$

$$\begin{bmatrix} 0 & 1 \\ -1 & 0 \end{bmatrix}^3 = \begin{bmatrix} 0 & -1 \\ 1 & 0 \end{bmatrix}$$

$$\begin{bmatrix} 0 & 1 \\ -1 & 0 \end{bmatrix}^4 = \begin{bmatrix} 1 & 0 \\ 0 & 1 \end{bmatrix}$$

From the series expansions for the sine and cosine functions, it follows that

$$e^{tC} = \begin{bmatrix} \cos \beta t & \sin \beta t & & & & & \\ -\sin \beta t & \cos \beta t & & & & & \\ & & \cos \beta t & \sin \beta t & & & \\ & & -\sin \beta t & \cos \beta t & & & \\ & & & & \ddots & & \\ & & & & & \cos \beta t & \sin \beta t \\ & & & & & -\sin \beta t & \cos \beta t \end{bmatrix}$$

Hence

$$
e^{tP^{-1}AP} = \begin{bmatrix}
\begin{matrix} e^{\alpha t}\cos\beta t & e^{\alpha t}\sin\beta t \\ -e^{\alpha t}\sin\beta t & e^{\alpha t}\cos\beta t \end{matrix} & & \\
& \ddots & \\
& & \begin{matrix} e^{\alpha t}\cos\beta t & e^{\alpha t}\sin\beta t \\ -e^{\alpha t}\sin\beta t & e^{\alpha t}\cos\beta t \end{matrix}
\end{bmatrix}
$$

$$
\times \begin{bmatrix}
1 & 0 & t & & 0 & \frac{t^2}{2!} & \\
\cdot & \cdot & & \cdot & & & \ddots \\
& & & & \cdot & & \frac{t^2}{2!} \\
& & & t & & \cdot & 0 \\
& & & & \cdot & \cdot & t \\
& & & & & & 0 \\
& & & & & & 1
\end{bmatrix}
$$

or

$$
e^{tP^{-1}AP} = \begin{bmatrix}
e^{\alpha t}\cos\beta t & e^{\alpha t}\sin\beta t & te^{\alpha t}\cos\beta t & te^{\alpha t}\sin\beta t & \\
-e^{\alpha t}\sin\beta t & e^{\alpha t}\cos\beta t & -te^{\alpha t}\sin\beta t & te^{\alpha t}\cos\beta t & \cdots \\
\cdot & \cdot & e^{\alpha t}\cos\beta t & e^{\alpha t}\cos\beta t & \\
\cdot & \cdot & -e^{\alpha t}\sin\beta t & e^{\alpha t}\cos\beta t & \\
\cdot & \cdot & \cdot & & \\
\cdot & \cdot & \cdot & &
\end{bmatrix}
$$

As pointed out before $Pe^{tP^{-1}AP} = Pe^{tJ}$ is a fundamental matrix. The treatment of arbitrary real A requires direct extensions of the procedure described above.

Finally, we point out an important case. Suppose that the equation

$$x = Ax \tag{2.19}$$

corresponds to a single n-th order equation

$$x^{(n)} + a_1 x^{(n-1)} + a_2 x^{(n-2)} + \cdots + a_{n-1}x' + a_n x = 0$$

where $a_1, a_2, \ldots, a_n$ are constants. Then (2.19) has the form

$$
\begin{bmatrix} x_1' \\ x_2' \\ \vdots \\ x_n' \end{bmatrix} = \begin{bmatrix}
0 & 1 & 0 & \cdots & 0 \\
0 & 0 & 1 & \cdots & 0 \\
& & \cdot & \cdot & \cdot \\
0 & 0 & 0 & \cdots & 1 \\
-a_n & -a_{n-1} & & \cdots & -a_1
\end{bmatrix} \begin{bmatrix} x_1 \\ x_2 \\ \vdots \\ x_n \end{bmatrix}
$$

First it is easy to prove (see Exercise 8) that the characteristic equation of

$$A = \begin{bmatrix} 0 & 1 & 0 & \cdots & 0 \\ 0 & 0 & 1 & \cdots & 0 \\ & & \cdot & \cdot & \\ 0 & 0 & 0 & \cdots & 1 \\ -a_n & -a_{n-1} & & \cdots & -a_1 \end{bmatrix}$$

is $\lambda^n + a_1\lambda^{n-1} + a_2\lambda^{n-2} + \cdots + a_{n-1}\lambda + a_n = 0$. If λ is an eigenvalue of A and the Jordan canonical form of A is

$$\begin{bmatrix} J_1 & & & \\ & J_2 & & \\ & & \ddots & \\ & & & J_s \end{bmatrix}$$

then λ appears in just one of the matrices J_i. The reason for this is that the linear space of eigenvectors associated with λ is 1-dimensional. In order to prove this, suppose that

$$\begin{bmatrix} x_1 \\ x_2 \\ \vdots \\ x_n \end{bmatrix}$$

is an eigenvector associated with an eigenvalue λ of A. Then

$$\begin{bmatrix} \lambda & -1 & 0 & \cdot & \cdot & 0 \\ 0 & \lambda & -1 & \cdot & \cdot & 0 \\ & \cdot & \cdot & & & \\ 0 & 0 & \cdot & \cdot & \lambda & -1 \\ a_n & a_{n-1} & \cdot & \cdot & \cdot & \lambda + a_1 \end{bmatrix} \begin{bmatrix} x_1 \\ x_2 \\ \vdots \\ x_n \end{bmatrix} = 0$$

Then

$$\lambda x_1 - x_2 = 0$$
$$\lambda x_2 - x_3 = 0$$
$$\cdots$$
$$\lambda x_{n-1} - x_n = 0$$
$$a_n x_1 + a_{n-1} x_2 + \cdots + (\lambda + a_1)x_n = 0$$

or

$$\lambda x_1 = x_2$$
$$\lambda x_2 = x_3$$
$$\cdots$$
$$\lambda x_{n-1} = x_n$$
$$a_n x_1 + a_{n-1}\lambda x_1 + \cdots + (\lambda + a_1)\lambda^{n-1} x_1 = 0$$

Since λ is an eigenvalue, the last equation is satisfied for all real x_1 and hence each eigenvector has the form

$$\begin{bmatrix} x_1 \\ \lambda x_1 \\ \lambda^2 x_1 \\ \vdots \\ \lambda^{n-1} x_1 \end{bmatrix}$$

Thus the linear space is one-dimensional.

The fundamental matrices which we have just computed give at once important information about the behavior of the solutions. That is, we have

Theorem 2.8   *Let $Re(\lambda)$ denote the real part of λ. Then each solution of*

$$x' = Ax \tag{2.20}$$

approaches zero as $t \to \infty$ $[t \to -\infty]$ iff $Re(\lambda) < 0[Re(\lambda) > 0]$ for all the eigenvalues λ of A. Each solution of (2.20) is bounded on the set (a, ∞) where a is any fixed real number (the set $(-\infty, a)$ where a is any fixed real number) iff

1) *$Re(\lambda) \leq 0[Re(\lambda) \geq 0]$ for all the eigenvalues λ of A;*

2) *$Re(\lambda) = 0$ implies that in the Jordan canonical form, the eigenvalue λ appears only in matrices J_i such that J_i is a 1×1 matrix.*

Proof   The proof follows from inspection of fundamental matrix Pe^{tJ}, where J is the Jordan canonical form, and application of L'Hospital's Rule.   ⬜

Theorem 2.9   *Suppose $\lambda = i\mu \neq 0$ is a pure imaginary eigenvalue of A. Suppose J is the real canonical form of A and*

$$J = \begin{bmatrix} J_1 & & & & \\ & \ddots & & & \\ & & J_m & & \\ & & & \ddots & \\ & & & & J_s \end{bmatrix}$$

where $J_1, \ldots, J_m$ are the blocks with which $i\mu$ and $-i\mu$ are associated. Suppose that for $q - 1, \ldots, m$, the J_q is a $(k_q \times k_q)$ matrix. Suppose L is the linear space of solutions of (minimal) period $\frac{2\pi}{\mu}$ of (2.20) then the dimension of L is 2m.

Proof   Suppose $x(t)$ is a solution of period $\frac{2\pi}{\mu}$. Since Pe^{tJ} is a fundamental matrix, there exists a vector c such that

$$Pe^{tJ}c = x(t)$$

Then $P^{-1}x(t)$ has period $\frac{2\pi}{\mu}$ and hence $e^{tJ}c$ has period $\frac{2\pi}{\mu}$. But the vector

$$e^{tJ_1}\begin{bmatrix} c_1 \\ \vdots \\ c_{k_1}-1 \\ c_{k_1} \end{bmatrix}$$

is periodic if and only if

$$c_1 = c_2 = \cdots = c_{k_1-2} = 0$$

A similar remark holds J_q with $q = 2, \ldots, m$, and the conclusion of the theorem follows. □

Homogeneous Linear Equations with Periodic Coefficients: Floquet Theory

Next we obtain some results for homogeneous systems with periodic coefficients which parallel the results already obtained for homogeneous systems with constant coefficients. As might be expected, the results are not as explicit as the theorems for systems with constant coefficients. The theory to be described is called Floquet theory. For this work, we need a lemma about matrices which says roughly that a log function for nonsingular matrices can be defined.

Lemma 2.3 *If C is a constant nonsingular matrix, there exists a constant matrix K such that*

$$C = e^K$$

Remark As would be expected, the matrix K is not unique. For example, if

$$K_1 = \begin{bmatrix} 2n\pi i & & 0 \\ & \ddots & \\ 0 & & 2n\pi i \end{bmatrix}$$

then $e^{K_1} = I$, the identity. Since $K_1 K = K K_1$ then

$$e^{K+K_1} = e^K e^{K_1} = e^K = C$$

Proof Since C is nonsingular,

$$\det C \neq 0$$

and it follows that all eigenvalues $\lambda_1, \ldots, \lambda_n$ are nonzero. By the Jordan canonical form theorem, there exists a constant nonsingular matrix P such that

$$P^{-1}CP = J$$

where J is the Jordan canonical form. We prove the lemma first for J. We have

$$J = \begin{bmatrix} J_1 & & & \\ & \cdot & & \\ & & \cdot & \\ & & & \cdot \\ & & & & J_s \end{bmatrix}$$

where, for $i = 1, \ldots, s$, either J_i is a 1×1 matrix or

$$J_i = \lambda_k \left(I + \frac{D}{\lambda_i} \right)$$

and

$$D = \begin{bmatrix} 0 & 1 & & & \\ & \cdot & \cdot & & \\ & & \cdot & \cdot & \\ & & & \cdot & 1 \\ & & & & 0 \end{bmatrix}$$

In analogy with the familiar infinite series from calculus,

$$\log(1 + x) = x - \frac{x^2}{2} + \frac{x^3}{3} - \frac{x^4}{4} + \cdots \quad (x \text{ real}, |x| < 1)$$

we define

$$\log \left(I + \frac{D}{\lambda_i} \right) = \sum_{k=1}^{\infty} \frac{(-1)^{k+1}}{k} \left(\frac{D}{\lambda_i} \right)^k \tag{2.21}$$

If J_i is an $r_i \times r_i$ matrix, then if $k \geq r_i$,

$$D^k = 0 \tag{2.22}$$

Hence the series on the right-hand side of equation (2.21) has only a finite number of nonzero terms and thus always converges. Moreover, it follows, again from (2.22), that

$$\exp \left[\log \left(I + \frac{D}{\lambda_i} \right) \right]$$

is also a polynomial in D/λ_i. Next we show that

$$\exp \left[\log \left(I + \frac{D}{\lambda_i} \right) \right] = I + \frac{D}{\lambda_i}$$

Using the infinite series for $\log(1 + x)$ and e^x, we have if x is real and $|x| < 1$, then

$$1 + x = \exp[\log(1 + x)] \tag{2.23}$$

$$= 1 + [\log(1 + x)] + \frac{[\log(1 + x)]^2}{2!} + \frac{[\log(1 + x)]^3}{3!} + \cdots$$

$$= 1 + \left(x - \frac{x^2}{2} + \frac{x^3}{3} - \cdots \right) + \frac{1}{2!} \left(x - \frac{x^2}{2!} + \frac{x^3}{3} - \cdots \right)^2$$

$$+ \frac{1}{3!} \left(x - \frac{x^2}{2} + \frac{x^3}{3} - \cdots \right)^3 + \cdots$$

By standard theorems from calculus, it follows that the terms of the right-hand side of (2.23) can be rearranged to obtain a power series $\sum_{n=0}^{\infty} a_n x^n$ which has the same value as the original expression. Thus we have for all real x such that $|x| < 1$,

$$1 + x = \sum_{n=0}^{\infty} a_n x^n$$

Hence by the Identity Theorem for Power Series it follows that $a_0 = 1$, $a_1 = 1$, and $a_k = 0$ for $k > 1$. Since $\log(I + D/\lambda_i)$ is defined in terms of the same formal expansion as the series expansion for $\log(1 + x)$, it follows that the corresponding coefficients in $\exp[\log(I + D/\lambda_i)]$ must have the same values, that is, we must have

$$\exp\left[\log\left(I + \frac{D}{\lambda_i} \right) \right] = I + \frac{D}{\lambda_i}$$

Now if J_i is a 1×1 matrix, let

$$K_i = \log \lambda_i$$

(As pointed out earlier, each λ_i is nonzero. Hence $\log \lambda_i$ is defined. Note, however, that λ_i may be negative or complex, and hence $\log \lambda_i$ is, in general, complex-valued.) If J_i is an $r_i \times r_i$ matrix where $r_i > 1$, let

$$K_i = (\log \lambda_i)I + \log\left(I + \frac{D}{\lambda_i} \right)$$

where I is the $r_i \times r_i$ identity matrix. Then if $r_i > 1$,

$$\exp K_i = \exp[(\log \lambda_i)I] \exp\left[\log\left(I + \frac{D}{\lambda_i} \right) \right]$$

$$= (e^{\log \lambda_i} I)\left(I + \frac{D}{\lambda_i} \right)$$

$$= \lambda_i \left(I + \frac{D}{\lambda_i} \right) = J_i$$

Then if K is the matrix

$$K = \begin{bmatrix} K_1 & & \\ & \ddots & \\ & & K_s \end{bmatrix}$$

we have

$$\exp(K) = J$$

In the general case

$$C = PJP^{-1} = P(\exp\ K)P^{-1} = \exp(PKP^{-1})$$

This completes the proof of Lemma 2.3.   ☐

Using Lemma 2.3, we prove a theorem which makes possible some useful definitions.

Theorem 2.10   *Suppose that $X(t)$ is a fundamental matrix of*

$$x' = A(t)x$$

where $A(t)$ is continuous for all real t and A has period T, that is, for all t,

$$A(t + T) = A(t)$$

Then $Y(t) = X(t+T)$ is a fundamental matrix. Also there exists a continuous matrix $P(t)$ of period T such that $P(t)$ is nonsingular for all t and a constant matrix R such that

$$X(t) = P(t)e^{tR}$$

Proof   First since $X(t)$ is a fundamental matrix and A has period T, we have

$$Y'(t) = X'(t + T) = A(t + T)X(t + T) = A(t)Y(t)$$

and for all real t,

$$\det\ Y(t) = \det\ X(t + T) \neq 0.$$

Thus $Y(t)$ is a fundamental matrix. Hence by Theorem 2.4, there exists a nonsingular constant matrix C such that

$$X(t + T) = Y(t) = X(t)C$$

By Lemma 2.3, there exists a constant matrix R such that $C = e^{TR}$. Let

$$P(t) = X(t)e^{-tR} \tag{2.24}$$

Then for all t, matrix $P(t)$ is nonsingular because $X(t)$, being a fundamental matrix, is nonsingular for all t and e^{-tR} has the inverse e^{tR}. Multiplying (2.24) on the right by e^{tR}, we obtain:

$$P(t)e^{tR} = X(t)$$

Also $P(t)$ has period T because

$$P(t + T) = X(t + T)e^{-(t+T)R}$$
$$= X(t)e^{TR}e^{-(t+T)R}$$

$$= X(t)e^{-tR}$$
$$= P(t)$$

This completes the proof of Theorem 2.10. $\quad\Box$

Definition The characteristic roots or eigenvalues of $C = e^{TR}$ are the *characteristic multipliers* of $A(t)$. The characteristic roots or eigenvalues of R are the *characteristic exponents* of $A(t)$.

(Note that since $X(t + T) = X(t)C$, then $X(t)$ is "multiplied by" the eigenvalues of C to get $X(t + T)$. Hence the name "characteristic multipliers.")

Theorem 2.11 *The characteristic multipliers $\lambda_1, \ldots, \lambda_n$ are uniquely determined by $A(t)$ and all the characteristic multipliers are nonzero.*

Proof If $X_1(t)$ is a fundamental matrix of

$$x' = A(t)x$$

then by Theorem 2.4 there exists a matrix C_1, constant and nonsingular, such that

$$X = X_1 C_1$$

Hence

$$X_1(t)C_1 e^{TR} = X(t)e^{TR} = X(t + T) = X_1(t + T)C_1,$$

and therefore

$$X_1(t + T) = X_1(t)C_1 e^{TR} C_1^{-1}$$

Since e^{TR} and $C_1 e^{TR} C_1^{-1}$ are similar, then by Lemma 2.2 they have the same characteristic roots. If $\lambda_1, \ldots, \lambda_n$ are the characteristic multipliers, then since C is nonsingular,

$$\left| \prod_{i=1}^{n} \lambda_i \right| = |\det C| \neq 0$$

This completes the proof of Theorem 2.11. $\quad\Box$

Theorem 2.12 *If $\rho_1, \ldots, \rho_n$ are the characteristic exponents of $A(t)$, then the characteristic multipliers of $A(t)$ are $e^{T\rho_1}, \ldots, e^{T\rho_n}$. Also if ρ_i appears in blocks of dimensions $i_1, \ldots, i_q$ in the Jordan canonical form of R, then $e^{T\rho_i}$ appears in blocks of dimensions $i_1, \ldots, i_q$ in the Jordan canonical form of C; and conversely.*

Proof There exists a constant nonsingular matrix P such that

$$R = P^{-1}JP$$

where J denotes the Jordan canonical form of R. Then

$$e^{TR} = e^{TP^{-1}JP} = P^{-1}e^{TJ}P$$

and hence by Lemma 2.2 the eigenvalues of e^{TR} are the eigenvalues of e^{TJ}.

We indicate the remainder of the argument by considering one block of J, say the $q \times q$ block

$$J_q = \begin{bmatrix} \rho & 1 & & & \\ & \rho & 1 & & \\ & & \ddots & & 1 \\ & & & \ddots & \\ & & & & \rho \end{bmatrix}$$

where ρ is an eigenvalue of R, that is, a characteristic exponent of $A(t)$. Then as in (2.18)

$$e^{TJ_q} = \exp \begin{bmatrix} T\rho & T & & \\ & T\rho & \ddots & \\ & & \ddots & T \\ & & & T\rho \end{bmatrix}$$

$$= \begin{bmatrix} e^{T\rho} & & \\ & \ddots & \\ & & e^{T\rho} \end{bmatrix} \begin{bmatrix} 1 & T & \cdots & \frac{T^{q-1}}{(q-1)!} \\ & 1 & T & \\ & & \cdot & \cdot \\ & & & T \\ & & & 1 \end{bmatrix}$$

and $e^{T\rho}$ is an eigenvalue of e^{TJ_q}. Finally, we must prove that the block associated with eigenvalue $e^{T\rho}$ is a $q \times q$ block. In order to prove this, let

$$B = \begin{bmatrix} 1 & T & \cdots & \frac{T^{q-1}}{(q-1)!} \\ & 1 & T & \\ & & \cdot & \cdot \\ & & & T \\ & & & 1 \end{bmatrix}$$

A straightforward calculation shows that

$$(B - I)^q = 0$$

but if r is an integer such that $1 \le r < q$, then

$$(B - I)^r \ne 0$$

Hence since

$$e^{TJ_q} - e^{T\rho}I = e^{T\rho}[B - I]$$

it follows that the block associated with eigenvalue $e^{T\rho}$ is a $q \times q$ block. This completes the proof of Theorem 2.12. □

Theorem 2.13   *Each solution of $x' = A(t)x$ approaches zero as $t \to \infty$ if and only if $|\lambda_i| < 1$ for each characteristic multiplier λ_i {if and only if $Re(\rho_i) < 0$ for each characteristic exponent ρ_i}. Each solution of $x' = A(t)x$ is bounded if and only if*

1. *$|\lambda_i| \leq 1$ for each characteristic multiplier λ_i {$Re(\rho_i) \leq 0$ for each characteristic exponent ρ_i} and*

2. *If $|\lambda_i| = 1$, each J_i in the canonical form of*

$$C = [X(t)]^{-1}X(t + T)$$

in which λ_i appears is a 1×1 matrix.

Proof   The theorem is proved by showing that there exists a change of variables under which the equation $x = A(t)x$ becomes a linear homogeneous system with constant coefficients. Using the notation of Theorem 2.10, let

$$Z(t) = e^{tR}[X(t)]^{-1} = [P(t)]^{-1}$$

and define

$$v(t) = Z(t)x(t)$$

where $x(t)$ is a solution, that is,

$$\frac{d}{dx}x(t) = A(t)x(t)$$

Then $x(t) = [Z(t)]^{-1}v(t)$ and

$$\frac{d}{dt}(Z^{-1}v) = A(t)Z^{-1}v$$

and we have

$$\left(\frac{d}{dt}Z^{-1}\right)v + Z^{-1}\frac{dv}{dt} = AZ^{-1}v$$

$$Z^{-1}\frac{dv}{dt} = AZ^{-1}v - \left(\frac{d}{dt}Z^{-1}\right)v$$

$$\frac{dv}{dt} = (ZAZ^{-1})v - Z\left(\frac{d}{dt}Z^{-1}\right)v$$

$$= [e^{tR}X^{-1}AXe^{-tR}]v - e^{tR}X^{-1}\left[\frac{dX}{dt}e^{-tR} + X\frac{d}{dt}e^{-tR}\right]v$$

$$= e^{tR}X^{-1}[AX - X']e^{-tR}v - e^{tR}X^{-1}X(-R)e^{-tR}v \qquad (2.25)$$

Since X is a fundamental matrix, then

$$X' = AX$$

Thus (2.25) becomes

$$\frac{dv}{dt} = Rv$$

(Note: The reduction of $\frac{dx}{dt} = A(t)x$ to the equation $\frac{dv}{dt} = Rv$ is called *reduction in the sense of Lyapunov.* It must be emphasized that this reduction procedure is generally not computationally practical because in order to find R, it is necessary to find the characteristic multipliers, that is, the eigenvalues of $[X(t)]^{-1}[X(t+T)]$. Thus one must first determine a fundamental matrix. Observe also that we have shown that

$$ZAZ^{-1} - Z\left(\frac{d}{dt}Z^{-1}\right) = R$$

Thus since A is real, it follows that R is real.)

The proof of the theorem follows from Theorem 2.12, the application of Theorem 2.8 to the equation

$$\frac{dv}{dt} = Rv$$

and the fact (from Theorem 2.10) that $|P(t)|$ is bounded. ▢

Definition   The *geometric multiplicity* of an eigenvalue λ of matrix A is the dimension of the linear subspace

$$L = \{x/Ax = \lambda x\}$$

Theorem 2.14   *If no characteristic multiplier equals one, the equation*

$$x' = A(t)x \tag{2.26}$$

has no nontrivial periodic solutions of period T. (That is, the only periodic solution is the identically zero solution.) If one is a characteristic multiplier of geometric multiplicity m, then the dimension of the linear space L of solutions with period T of (2.26) is m.

Proof   Returning to Theorem 2.10, we have

$$X(t) = P(t)e^{tR} = P(t)e^{tSJS^{-1}} = P(t)Se^{tJ}S^{-1}$$

where S is a nonsingular matrix and J is the Jordan canonical form of R. Denoting $P(t)S$ by $P_1(t)$, we conclude by Theorem 2.4 that $P_1(t)e^{tJ}$ is a fundamental matrix. If no characteristic multiplier equals one, then no characteristic root of J equals 0 or

$\frac{2p\pi i}{T}$ ($p = \pm 1, \pm 2, \dots$). Hence no column of e^{tJ} has period T or period $\frac{T}{p}$. But $P_1(t)$ has period T. Hence no column of $P_1(t)e^{tJ}$ has period T. Finally, we need to show that no linear combination of columns of $P_1(t)e^{tJ}$ has period T. The existence of such a linear combination means that there exists a constant vector c such that

$$P_1(t)e^{tJ}c = x(t)$$

where $x(t)$ has period T. Then

$$e^{tJ}c = [P_1(t)]^{-1}x(t)$$

Thus $e^{tJ}c$ has period T. But because of the form of e^{tJ}, this is impossible.

Now suppose one is a characteristic multiplier of geometric multiplicity m. It follows that e^{tJ} has m columns of period T because if

$$\lambda_i = e^{T\rho_i} = 1$$

then

$$T\rho_i = 2p\pi i \qquad (p = 0, \pm 1, \pm 2, \dots)$$

and

$$e^{t\rho_i} = e^{t\frac{2p\pi i}{T}}$$

Hence the fundamental matrix $P_1(t)e^{tJ}$ has m columns of period T, and thus the dimension of $\mathcal{L}$ is greater than or equal to m.

Finally we show that the dimension of $\mathcal{L}$ is less than or equal to m. If $x(t)$ is a solution of period T of (2.26), there is a constant vector c such that

$$P_1(t)e^{tJ}c = x(t)$$

Hence

$$e^{tJ}c = [P_1(t)]^{-1}x(t)$$

has period T. The components of $e^{tJ}c$ have typically the form

$$e^{t\rho_k}\left[c_k + c_{k+1}t + \cdots + c_{k+w}\frac{t^w}{w!}\right] \qquad (2.27)$$

where $c_k, c_{k+1}, \dots, c_{k+w}$ are the kth, $(k+1)$th, $\dots$, $(k+w)$th components of vector c. If expression (2.27) is not identically zero, then it has period T if and only if

$$\rho_k = \frac{2\pi ni}{T}$$

where $n = 0, \pm 1, \dots$, and

$$c_{k+1} = \cdots = c_{k+w} = 0$$

Therefore $[P_1(t)]^{-1}x(t)$ is a linear combination of columns of e^{tJ} which have period T. But there are m such columns of e^{tJ}. Thus the dimension of $\mathcal{L}$ is less than or equal to m. This completes the proof of Theorem 2.14. $\qquad\square$

A further examination of the fundamental matrix $P_1(t)e^{tJ}$ shows that equation (2.26) may have other periodic solutions. For example, if there is a characteristic exponent ρ_j such that

$$\rho_j = \frac{2\pi i}{mT}$$

where m is a positive integer, then there is a column of $P_1(t)e^{tJ}$ which has period mT.

Inhomogeneous Linear Equations

Next we obtain a useful and important formula for the solution of linear inhomogeneous equations: the variation of constants formula.

Theorem (Variation of Constants Formula) *Suppose the $n \times n$ matrix $A(t)$ and the n-vector $u(t)$ are continuous on (a, b). Let $t_0 \in (a, b)$. If $X(t)$ is a fundamental matrix of*

$$x' = A(t)x$$

then the solution $x(t)$ of

$$x' = A(t)x + u(t) \tag{2.28}$$

which satisfies the initial condition

$$x(t_0) = 0$$

is, for $t \in (a, b)$,

$$x(t) = X(t) \int_{t_0}^{t} \{X(s)\}^{-1} u(s) \, ds$$

Proof   The proof is straightforward verification by computation as follows:

$$\frac{d}{dt}[x(t)] = X'(t) \int_{t_0}^{t} [X(s)]^{-1} u(s) \, ds + X(t)\{X(t)\}^{-1} u(t)$$

$$= A(t)X(t) \int_{t_0}^{t} [X(s)]^{-1} u(s) \, ds + u(t)$$

$$= A(t)x(t) + u(t) \qquad\qquad \square$$

Corollary 2.15   *The solution $x(t)$ of (2.28) which satisfies the initial condition*

$$x(t_0) = x_0$$

is

$$x(t) = y(t) + X(t) \int_{t_0}^{t} [X(s)]^{-1} u(s) \, ds$$

where y(t) is the solution of

$$x' = A(t)x$$

which satisfies the initial condition: $y(t_0) = x_0$.

The idea behind the variation of constants formula is this: If c is a constant vector then

$$x(t) = X(t)c$$

is a solution of

$$x' = A(t)x$$

One tries to obtain a solution of (2.28) by replacing the constant vector c by a vector function $c(t)$. Suppose that

$$x(t) = X(t)c(t)$$

is a solution of (2.28). Then

$$x' = Ax + u$$
$$x' = [X']c + [X]c' = AXc + Xc' = Ax + Xc'$$

Thus

$$Xc' = u$$

or

$$c' = X^{-1}u$$

Hence if $c(t_0) = 0$

$$c(t) = \int_{t_0}^{t} [X(s)]^{-1} u(s) \, ds$$

We have described the variation of constants formula as being useful and important. But anyone who has actually applied the formula to calculate a solution of an inhomogeneous equation might take issue with that claim. Applying the formula requires calculating a fundamental matrix of the associated homogeneous equation, finding a particular solution of that homogeneous equation, calculating the inverse of the fundamental matrix, and, finally, performing a number of more or less tedious integrations. If $A(t)$ is a constant matrix, then we have a procedure for carrying out these steps. But if $n > 2$, this procedure becomes quite laborious. Moreover, if $A(t)$ is not constant, we have no explicit procedure for carrying out these calculations.

The variation of constants formula has, of course, the saving grace that it is applicable in the general case (unlike other techniques such as the method of undetermined coefficients described in, for example, Edwards and Penney [1989, p. 149ff]).

But the variation of constants formula has far greater importance in developing the theory of other classes of equations. This will be illustrated in the next section of this chapter. Also the variation of constants formula plays a fundamental role in the study of nonlinear problems as we will see in Chapter 7. Finally, the formula can be used to define a Green's function for ordinary differential equations as we shall illustrate a little later in the description of the Sturm-Liouville theory.

Periodic Solutions of Linear Systems with Periodic Coefficients

We consider n-dimensional systems of the form

$$\frac{dx}{dt} = A(t)x + f(t) \tag{2.29}$$

where the matrix $A(t)$ has period T, the vector $f(t)$ has period T, and $A(t)$ and $f(t)$ both have continuous first derivatives for all t. We investigate the existence of solutions of period T of equation (2.29). The results obtained have intrinsic value; and they are crucial in the study of a large class of nonlinear problems, as will be seen in Chapter 7.

First, we obtain a result which is foreshadowed by Theorem 2.14. Let $X(t)$ be a fundamental matrix of the corresponding homogeneous equation

$$\frac{dx}{dt} = A(t)x \tag{2.30}$$

such that $X(0) = I$, the identity matrix. Let $x(t)$ denote solution of (2.29) such that

$$x(0) = x_0$$

Then by the variation of constants formula,

$$x(t) = X(t)x_0 + \int_0^t X(t)[X(s)^{-1} f(s) \, ds \tag{2.31}$$

Lemma 2.4   *A nasc that $x(t)$ have period T is*

$$x(T) = x(0) \tag{2.32}$$

Proof   The condition is clearly necessary. In order to prove the sufficiency, let

$$y(t) = x(t + T)$$

Then

$$y(0) = x(T) = x(0)$$

Also $y(t)$ is a solution of (2.29) because

$$\frac{dy(t)}{dt} = \frac{d}{dt}x(t + T) = A(t + T)x(t + T) + f(t + T)$$
$$= A(t)x(t + T) + f(t)$$
$$= A(y)y(t) + f(t)$$

Since $y(t)$ is a solution and $y(0) = x(0)$ then by the uniqueness of solution, for all t

$$y(t) = x(t)$$

but by definition

$$y(t) = x(t + T)$$ ∎

Using (2.31), we may rewrite (2.32) as

$$X(T)x_0 + \int_0^T X(T)[X(s)]^{-1} f(s)\,ds = x_0$$

or

$$[X(T) - I]x_0 + \int_0^T X(t)[X(s)]^{-1} f(s)\,ds = 0 \tag{2.33}$$

By Lemma 2.4 and the derivation of (2.33), it follows that a nasc that a solution $x(t)$ of (2.29) with $x(0) = x_0$ have period T is that x_0 satisfies (2.33). Thus, to find the periodic solutions of (2.29), it is sufficient to find the solutions x_0 of the linear system (2.33).

Since $X(0) = I$, then the eigenvalues of $X(T)$ are the characteristic multipliers of $A(t)$. Now suppose that $A(t)$ has no characteristic multiplier equal to one. Then

$$\det[X(T) - I] \neq 0 \tag{2.34}$$

and (2.33) has a unique solution, say $\bar{x}_0$ and the desired period solution $\bar{x}(t)$ is the solution of (2.29) such that

$$\bar{x}(0) = \bar{x}_0$$

(If $x_0 = 0$, then $\bar{x}(t)$ is identically zero.) Thus we have

Theorem 2.15   *If $A(t)$ has no characteristic multiplier equal to one, then equation (2.29) has a unique solution of period T (which may be identically zero).*

The stability properties of this periodic solution will be discussed in Chapter 4.

Theorem 2.15 is analogous to and follows from the familiar theorem about solutions of a system of linear algebraic equations. Now we must investigate the case in which $A(t)$ may have characteristic multipliers equal to one. That is, the case in which equation (2.30) has solutions of period T.

We seek results which are analogous to the corresponding results for linear algebraic systems. As might be expected, the situation is more complicated: there may be no periodic solutions or there may be infinite sets of periodic solutions. However, the approach to the problem remains the same: we investigate the solutions of (2.33). Only this time, we start with the assumption that equation (2.30) has nontrivial solutions of period T.

First we introduce two definitions.

Definition   The *adjoint matrix* of the matrix

$$A = [a_{ij}]$$

is the matrix

$$A^* = [a_{ij}^*]$$

where

$$a_{ij}^* = \bar{a}_{ji}$$

and $\bar{a}_{ji}$ denotes the conjugate value of a_{ji}.

It is straightforward to show by calculation that

$$(A^*)^* = A$$
$$(A^{-1})^* = (A^*)^{-1}$$
$$(A_1 A_2)^* = A_2^* A_1^*$$

Definition   The *adjoint equation* of the homogeneous equation

$$\frac{dx}{dt} = A(t)$$

is the equation

$$\frac{dx}{dt} + A^*(t)x = 0$$

It is straightforward to verify that the set of solutions of period T of equation (2.30) is a linear space of dimension not exceeding n. (Cf. Theorem 2.2). The same result holds for the set of solutions of period T of the adjoint equation

$$\frac{dx}{dt} + A^*(t)x = 0 \tag{2.35}$$

In the following discussion we will be dealing with these two linear spaces, and it will be shown that the two linear spaces have the same dimension. However, it is convenient to start with the hypothesis that the dimension of the linear space of solutions of period T of the adjoint equation (2.35) is given as m where $1 \le m \le n$. The fact that the dimension of the linear space of solutions of period T of equation (2.30) is also m will follow from our discussion.

By the hypothesis that the dimension is m, there exists a fundamental matrix $U(t)$ of equation (2.35) such that the first m columns of $U(t)$ each have period T. That is, if

$$U(t) = \begin{bmatrix} u_{11}(t) & u_{12}(t) & \cdot & \cdot & \cdot & u_{1n}(t) \\ u_{21}(t) & u_{23}(t) & & & & \\ & \cdot & & & & \\ & \cdot & & & & \\ & \cdot & & & & \\ u_{n1}(t) & u_{n2}(t) & \cdot & \cdot & \cdot & u_{nn}(t) \end{bmatrix}$$

then the column

$$u^{(j)}(t) = \begin{bmatrix} u_{1j}(t) \\ \cdot \\ \cdot \\ \cdot \\ u_{nj}(t) \end{bmatrix}$$

is a solution of period T of (2.35) if $j = 1, \ldots, m$; and if $j = m + 1, \ldots, n$, then $u^{(j)}(t)$ does not have period T.

Now we make the following change of variables in equation (2.29):

$$y = U^*(0)x$$

Then (2.29) becomes

$$[U^*(0)]^{-1}\frac{dy}{dt} = A(t)[U^*(0)]^{-1}y + f(t)$$

or

$$\frac{dy}{dt} = U^*(0)A(t)[U^*(0)]^{-1}y + U^*(0)f(t) \qquad (2.36)$$

Let

$$B(t) = U^*(0)A(t)[U^*(0)]^{-1}$$

and

$$g(t) = U^*(0)f(t)$$

Then (2.36) becomes

$$\frac{dy}{dt} = B(t)y + g(t) \qquad (2.37)$$

It is clear that a solution $x(t)$ of (2.29) has period T if and only if the solution

$$y(t) = [U^*(0)]x(t)$$

of (2.37) has period T. Before proceeding to the study of (2.37), we make some preliminary remarks. If we let

$$V(t) = U^{-1}(0)U(t)$$

then since

$$\frac{dU}{dt} + A^*(t)U = 0$$

we obtain

$$U(0)\frac{dV}{dt} + A^*(t)U(0)V = 0$$

or

$$\frac{dV}{dt} + U^{-1}(0)A^*(t)U(0)V = 0$$

or

$$\frac{dV}{dt} + [U^*(0)A(t)\{U^*(0)\}^{-1}]^*V = 0$$

or

$$\frac{dV}{dt} + B^*(t)V = 0 \tag{2.38}$$

Thus $V(t)$ plays the same role with equation (2.37) that $U(t)$ plays with equation (2.29). That is, V is a fundamental matrix of the equation

$$\frac{dy}{dt} + B^*(y) = 0 \tag{2.39}$$

Lemma 2.5   *If $Y(t)$ is a fundamental matrix of*

$$\frac{dy}{dt} = B(t)y$$

such that $Y(0) = I$, then

$$[Y(t)]^{-1} = V^*(t)$$

Proof   First

$$\frac{d(Y^*)}{dt} = \left[\frac{dY}{dt}\right]^* = [B(t)Y]^* = Y^*B^*(t)$$

Then

$$\frac{d(Y^*V)}{dt} = \frac{dY^*}{dt}V + Y^*\frac{dV}{dt}$$
$$= [Y^*B^*(t)]V + Y^*[-B^*V]$$
$$= 0$$

Hence Y^*V is a constant matrix. Since $Y^*(0) = I$ and $V(0) = I$, it follows that for all t,

$$Y^*(t)V(t) = I$$

Then

$$I = [Y^*V]^* = V^*(Y^*)^* = V^*Y$$

or, multiplying on the right by Y^{-1}, we have

$$Y^{-1} = V^* \qquad\qquad\qquad\qquad \Box$$

Now we are ready to investigate the periodic solutions of (2.37). By the variation of constants formula, if we impose the condition that $Y(0) = I$, then a solution $y(t)$ of (2.37) is given by

$$y(t) = Y(t)y(0) + \int_0^t Y(t)[Y(s)]^{-1}g(s)\,ds \tag{2.40}$$

By Lemma 2.4, solution $y(t)$ has period T if and only if

$$y(0) = y(T) \tag{2.41}$$

Rewriting (2.41) using (2.40), we obtain

$$[I - Y(T)]y(0) - \int_0^T Y(T)[Y(s)]^{-1} g(s)\, ds = 0 \tag{2.42}$$

Multiplying (2.42) by $[Y(T)]^{-1}$ and observing that, by Lemma 2.5, $[Y(s)]^{-1} = V^*(s)$ we have

$$[V^*(T) - I]y(0) - \int_0^T V^*(s)g(s)\, ds = 0 \tag{2.43}$$

But (2.43) is a system of n linear equations in the n components of $y(0)$. To investigate the solutions of (2.43), we first calculate $V^*(T) - I$.

As defined earlier,

$$V(t) = U^{-1}(0)U(t)$$

That is,

$$V(t) = U^{-1}(0) \begin{bmatrix} u_{11} & \cdots & u_{1m} & \cdots & u_{1n} \\ \vdots & & \cdots & & \\ u_{n1} & \cdots & u_{nm} & & u_{nn} \end{bmatrix}$$

and we may choose the columns

$$\begin{bmatrix} u_{1j} \\ \vdots \\ u_{nj} \end{bmatrix} \tag{2.44}$$

so that if $j = 1, \ldots, m$, the columns have period T and if $j = m + 1, \ldots, n$ the columns do not have period T. Let

$$V(t) = \begin{bmatrix} v_{11} & \cdots & v_{1n} \\ \vdots & & \vdots \\ v_{n1} & \cdots & v_{nn} \end{bmatrix}$$

and let

$$v^{(j)}(t) = U^{-1}(0) \begin{bmatrix} u_{1j} \\ \vdots \\ u_{nj} \end{bmatrix} = \begin{bmatrix} v_{1j} \\ \vdots \\ v_{nj} \end{bmatrix}$$

Then if $j = 1, \ldots, m$, the $v^{(j)}(t)$ has period T. But if $j = m + 1, \ldots, n$, then $v^{(j)}(t)$ does not have period T because if $m + 1 \le j \le n$, then since

$$U(0)v^{(j)} = u^{(j)}$$

it would follow that if $v^{(j)}$ were periodic, $u^{(j)}$ would be periodic.

Since

$$V(0) = I$$

then by the periodicity (with period T) of the first m columns of $V(t)$, it follows that

$$V(T) = \begin{bmatrix} 1 & 0 & \cdots & 0 & v_{1m+1} & \cdots & v_{1n} \\ 0 & 1 & \cdots & 0 & & & \\ & \vdots & & & & & \\ 0 & 0 & \cdots & 1 & & & \\ 0 & 0 & \cdots & 0 & v_{m+1\,m+1} & & \\ & \cdot & \cdot & \cdot & \cdot & & \\ 0 & 0 & \cdots & 0 & v_{nm+1} & \cdots & v_{nn} \end{bmatrix} \tag{2.45}$$

and

$$V^*(T) = \begin{bmatrix} 1 & 0 & \cdots & 0 & 0 & \cdots & 0 \\ 0 & 1 & \cdots & 0 & 0 & \cdots & 0 \\ & & \cdot & \cdot & \cdot & & \\ 0 & \cdots & & 1 & 0 & \cdots & 0 \\ v_{1m+1} & \cdots & v_{m+1m+1} & \cdots & v_{nm+1} & & \\ & & \cdot & \cdot & \cdot & & \\ v_{1n} & & & v_{m+1\,n} & & v_{nn} \end{bmatrix} \tag{2.46}$$

where each entry in (2.45) and (2.46) is evaluated at T. Thus

$$V^*(T) - I = \begin{bmatrix} 0 & \cdots & 0 & 0 & \cdots & 0 \\ & \vdots & & & & \\ 0 & \cdots & 0 & & & \\ v_{1m+1}(T) & \cdots & v_{m+1m+1}(T) - 1 & \cdots & v_{nm+1}(T) \\ & \vdots & & & & \\ v_{1n}(T) & \cdots & v_{m+1n}(T) & \cdots & v_{nn}(T) - 1 \end{bmatrix} \tag{2.47}$$

Lemma 2.6   *The rank of $V^*(T) - I$ is $n - m$.*

Proof   Since the first m rows of $V^*(T) - I$ consist of 0's, then

$$[\text{rank of } V^*(T) - I] \leq n - m$$

Suppose the rank is less than $n - m$, that is, suppose there is a linear combination (with not all coefficients equal to zero) of the last $(n - m)$ rows of $V^*(T) - I$ such that this combination is zero (more precisely, the zero n-vector). Let the coefficients in this linear combination be: $a_{m+1}, a_{m+2}, \ldots, a_n$. That is, for $k = 1, \ldots, n - m$, the a_{m+k} is the coefficient of the $(m + k)$th row in $V^*(T) - I$. Let

$$S(t) = a_{m+1} v^{m+1}(t) + \cdots + a_n v^{(n)}(t)$$

As a linear combination of columns of $V(t)$, the function $S(t)$ is a solution of (2.39). Also from the definition of $S(t)$ and the fact that $V(t)$ is a fundamental matrix of (2.39) it follows that $S(t), v^{(1)}(t), \ldots, v^{(m)}(t)$ are $(m + 1)$ linearly independent solutions of (2.39). Next we show that $S(t)$ has period T. By Lemma 2.4, it is sufficient to show that

$$S(0) = S(T)$$

Since $V(0) = I$, then

$$S(0) = a_{m+1} v^{m+1}(0) + \cdots + a_n \, v^{(n)}(0)$$

$$= \begin{bmatrix} 0 \\ \vdots \\ 0 \\ a_{m+1} \\ \vdots \\ a_n \end{bmatrix}$$

To calculate $S(T)$, we note that by the properties of $a_{m+1}, \ldots, a_n$, we have

$$a_{m+1}[v_{m+1\,m+1}(T) - 1] + a_{m+2} v_{m+1\,m+2}(T)$$
$$+ \cdots + a_n v_{m+1\,n}(T) = 0$$
$$a_{m+1}[v_{m+2\,m+1}(T)] + a_{m+2}[v_{m+2\,m+2}(T) - 1]$$
$$+ \cdots + a_n \, v_{m+2\,n}(T) = 0$$
$$\cdots$$
$$a_{m+1} v_{n\,m+1}(T) + a_{m+2} v_{n\,m+2}(T)$$
$$+ \cdots + a_n[v_{n\,n}(T) - 1] = 0$$

or

$$a_{m+1} = a_{m+1} v_{m+1\,m+1}(T) + a_{m+2} v_{m+1\,m+2}(T) + \cdots + a_n v_{m+1\,n}(T)$$
$$a_{m+2} = a_{m+1} v_{m+2\,m+1}(T) + a_{m+2} v_{m+2\,m+2}(T) + \cdots + a_n v_{m+2\,n}(T)$$
$$\cdots$$
$$a_n = a_{m+1} v_{n\,m+1}(T) + a_{m+2} v_{n\,m+2}(T) + \cdots + a_n v_{n\,n}(T) \qquad (2.48)$$

System (2.48) shows that the last $(n - m)$ components of $S(T)$ are $a_{m+1}, a_{m+2}, \ldots, a_n$. From inspection of $V^*(T)$, that is, equation (2.46), and the properties of $a_{m+1}, a_{m+2}, \ldots, a_n$, it follows that the first m components of $S(T)$ are all zero. Thus $S(0) = S(T)$

and hence $S(t)$ is a periodic solution of (2.39). Thus $S(t), v_1(t), \ldots, v_m(t)$ are $(m+1)$ linearly independent solutions of (2.39). But

$$
\begin{aligned}
S(t) &= a_{m+1}v^{m+1}(t) + \cdots + a_n v^{(n)}(t) \\
&= a_{m+1}U^{-1}(0)u^{m+1}(t) + \cdots + a_n U^{-1}(0)u^{(n)}(t)
\end{aligned}
$$

or

$$
U(0)S(t) = a_{m+1}u^{m+1}(t) + \cdots + a_n u^{(n)}(t)
$$

Since $U(0)S(t)$ has period T, this yields a contradiction to our original assumption that the linear space of solutions of period T of equation (2.35) has dimension m. This completes the proof of Lemma 2.6   □

Now we are ready to study the solutions

$$
y(0) = (\bar{y}_1, \ldots, \bar{y}_n) \tag{2.49}
$$

of the system (2.43). Consider first the case in which equation (2.29) is homogeneous, that is, $f(t) \equiv 0$. Then (2.43) becomes the homogeneous system

$$
[V^*(T) - I]y(0) = 0 \tag{2.50}
$$

From the properties of $V^*(T) - I$ and the fact that $g(t) \equiv 0$, it follows that a solution of (2.50) is a vector

$$
(\bar{y}_1, \ldots, \bar{y}_m, \bar{y}_{m+1}, \ldots, \bar{y}_n)
$$

where the values $\bar{y}_1, \ldots, \bar{y}_m$ are arbitrarily chosen and

$$
\bar{y}_{m+j} = 0 \qquad (j = 1, \ldots, n - m)
$$

That is, the set of periodic solutions of (2.37) is the set of solutions with initial values

$$
(\bar{y}_1, \ldots, \bar{y}_m, 0, \ldots, 0)
$$

where $\bar{y}_1, \ldots, \bar{y}_m$ are arbitrarily chosen. Thus there is a $1-1$ correspondence between the collection of periodic solutions and the set of points in R^m. Also the periodic solutions corresponding to the finite set of m-vectors

$$
\bar{y}^{(1)}, \ldots, \bar{y}^{(q)}
$$

are linearly independent if and only if the m-vectors $\bar{y}^{(1)}, \ldots, \bar{y}^{(q)}$ are linearly independent. (This follows by the same kind of argument used in the proof of Theorem 2.2.)

Thus, as promised earlier, we have proved that the linear space of solutions of period T of (2.32) has dimension m.

Finally we consider the unhomogeneous case, that is, the case in which $f(t) \not\equiv 0$. Applying (2.47) and letting $g_1, \ldots, g_n$ denote components of g, we may write (2.43) as

$$\sum_{j=1}^{n} \int_{0}^{T} v_{jk}(s)g_j(s)\,ds = 0 \qquad (k = 1, \ldots, m) \tag{2.51}$$

$$\sum_{j=1}^{n} [v_{jk}(T) - \delta_{jk}]y_j(0) = \sum_{j=1}^{n} \int_{0}^{T} v_{jk}(s)g_j(s)\,ds \quad (k = m+1, \ldots, n) \tag{2.52}$$

and we have: a necessary and sufficient condition that equation (2.37) have a solution of period T is that the equations (2.51) with $k = 1, \ldots, n$ are true where $(y_1(0), \ldots, y_n(0))$ is the initial condition of the periodic solution.

If $g_1, \ldots, g_n$ are such that the equations (2.51), $k = 1, \ldots, m$, are satisfied, then (2.52), $k = m + 1, \ldots, n$, becomes a system of $(n - m)$ equations and from the properties of $V^*(T) - I$ it follows that this system can be solved uniquely for $y_{m+1}(0), \ldots, y_n(0)$.

Summarizing, we have

Theorem 2.16 *Suppose that the equations (2.51), $k = 1, \ldots, m$ are satisfied and suppose $y_{m+1}(0), \ldots, y_n(0)$ satisfy equations (2.51), $k = m + 1, \ldots, n$. Then the set of solutions of period T of equation (2.37) consists of the solutions of (2.37) with initial value at $t = 0$*

$$(\bar{y}_1, \ldots, \bar{y}_m, y_{m+1}(0), \ldots, y_n(0))$$

where $(\bar{y}_1, \ldots, \bar{y}_m)$ is an arbitrary point in R^m.

Sturm-Liouville Theory

So far in this book, we have been studying the initial value problem for differential equations. The initial value problem is the fundamental problem for differential equations, but there is another problem, the boundary value problem, which is also of great importance. Boundary value problems have a long history: they arose very early in the study of partial differential equations and have played an important role ever since. At this juncture, we will give a brief account of the boundary value problem for an important class of linear equations, that is, the Sturm-Liouville theory. Although the results we will describe are very old (the original papers by Sturm and Liouville appeared in the 1830's) they remain important today, both in themselves and because they form the prototype for an important segment of linear functional analysis. Our account will be brief and uneven: brief because we will consider a special case and will omit proofs of serious theorems; uneven because we will go into detail at certain points where we want to explain connections between various results and, on the other hand, we will omit careful considerations of important topics.

It is natural to ask why these acknowledged omissions occur. A complete self-contained account would seem more useful. However, as will be seen from our summary of the Sturm-Liouville theory, the theory of boundary value problems is intimately dependent on functional analysis. As will be described, the Sturm-Liouville theory can be regarded as an infinite-dimensional analog of some matrix theory (linear algebra). This analog takes place in the Hilbert space $\mathcal{L}^2$. A complete exposition of the Sturm-Liouville theory should have functional analysis for its background. Since such a background requirement is contrary to the spirit of this book, we will settle for a lesser discussion. For an extensive account of boundary value problems, see Coddington and Levinson [1955].

First we describe how boundary value problems arise by looking at the heat equation. Consider a bar of uniform cross-section, homogeneous material and with its sides insulated so that no heat passes through them. Assume that the cross-section is small enough so that the temperature on any cross-section is constant. Let ℓ be the length of the bar. Then the temperature u at any point on the bar depends only on the position x on the bar and on the time t. That is, $u = u(x, t)$. The temperature $u(x, t)$ is prescribed by the heat equation

$$a^2 u_{xx} = u_t \tag{2.53}$$

where a is a positive constant depending on the material of the bar, and $0 < x < \ell$ and $t > 0$. (The endpoints $x = 0$, $x = \ell$, and $t = 0$ are not included because, for example, $u(x, t)$ is not defined for $x < 0$ and thus, strictly speaking, the derivative $u_x(0, t)$ is not defined. Only the right-hand derivative is defined at $x = 0$. As it turns out, this is not a serious complication in this work.) We assume that an initial temperature distribution is given in the bar, that is,

$$u(x, 0) = f(x) \tag{2.54}$$

where $0 \le x \le \ell$ and $f(x)$ is a given function and we assume that the temperature at the ends of the bar is 0. That is,

$$u(0, t) = 0, \quad u(\ell, t) = 0 \tag{2.55}$$

for $t > 0$.

The objective is to solve equation (2.53) with initial condition (2.54) and boundary condition (2.55). We use the conventional approach of separation of variables and assume that

$$u(x, t) = X(x)T(t) \tag{2.56}$$

Substituting from (2.56) into (2.53), we obtain

$$\alpha^2 X'' T = X T'$$

or

$$\frac{X''}{X} = \frac{1}{\alpha^2} \frac{T'}{T} \tag{2.57}$$

Since the left-hand side of (2.57) is a function of x only and the right-hand side is a function of t only it follows that both sides of (2.57) are equal to a constant k and thus solving (2.53) is reduced to solving the two ordinary differential equations

$$\frac{X''}{X} = k, \quad \frac{1}{\alpha^2}\frac{T'}{T} = k \tag{2.58}$$

The constant k remains unspecified. Applying the boundary conditions (2.55), we have

$$X(0)T(t) = 0, \quad X(\ell)T(t) = 0 \tag{2.59}$$

Conditions (2.59) imply

$$X(0) = 0, \quad X(\ell) = 0 \tag{2.60}$$

unless $T(t) = 0$ for all t. But we exclude the possibility that $T(t) = 0$ for all t because this would imply that

$$u(x,t) = 0$$

for all x and t. Then condition (2.54) could not be fulfilled except for the case in which $f(x) = 0$ for $0 \leq x \leq \ell$.

Thus we must solve the equation

$$X'' - kX = 0 \tag{2.61}$$

subject to the boundary conditions (2.60). This problem clearly has the trivial solution $X(x) = 0$ for all x. *The essential point to be made here is that the problem may or may not have nontrivial solutions. What happens depends on the value of k.* (Notice the radical difference between this problem and the initial value problem. The initial value problem has, under reasonable hypothesis on the differential equation, a unique solution. The boundary value problem may or may not have a solution.)

Suppose that k is a positive number, say $k = m^2$ where $m > 0$. Then the general solution of (2.61) is

$$X(x) = e_1 e^{mx} + c_2 e^{-mx}$$

and it follows that the problem described by (2.60) and (2.61) has no nontrivial solutions. Further calculations show that if $k = 0$ or if k is a complex number then the problem has no nontrivial solutions. (See, e.g., Boyce and DiPrima, Chapter 10.) However if k is a negative number, say $k = -q^2$ where $q > 0$, then the general solution of (2.61) is

$$X(x) = c_1 \cos qt + c_2 \sin qt$$

and from (2.60) it follows that $c_1 = 0$ and that

$$q = \frac{n\pi}{\ell} \quad (n = \pm 1, \pm 2, \ldots)$$

Hence

$$X(x) = c \sin \frac{n\pi}{\ell} x$$

where c is any nonzero constant. (To complete the solution of our original problem, described by equations (2.53), (2.54) and (2.55), we would then return to the equation

$$\frac{1}{\alpha^2} \frac{T'}{T} = -q^2$$

which is easily solved.) With considerably more work, we can establish the remarkable fact that the solutions

$$\sin \frac{n\pi}{\ell} x \qquad (n = 1, 2, \ldots)$$

of our problem can be used to describe the initial temperature distribution $f(x)$ if $f(x)$ is a sufficiently well-behaved function. (For example, if f is continuous on $[0, \ell]$, $f(0) = f(\ell)$ and f has a continuous first derivative on $(0, \ell)$.) In fact, if

$$b_n = \frac{2}{\ell} \int_0^\ell f(x) \sin \frac{n\pi x}{\ell} \, dx \qquad n = 1, 2, \ldots$$

then the infinite series

$$\sum_{j-1}^{\infty} b_n \sin \frac{n\pi x}{\ell} \qquad (2.62)$$

converges uniformly on $[0, \ell]$ to the function $f(x)$. This "remarkable fact" is, of course, a special case of the basic theorem of Fourier series. (We get a sine series because our function $f(x)$ is defined only on $[0, \ell]$. Extending $f(x)$ to an odd function on $[-\ell, \ell]$ and applying the basic Fourier theorem to that odd function yields the expansion (2.62).) For an introduction to Fourier series, see Churchill and Brown [1987].

The Sturm-Liouville theory which we will describe may be regarded as a generalization of the material summarized above. We deal with a class of ordinary differential equations which arise from separation of variables applied to certain partial differential equations. (The equation (2.61) with boundary conditions (2.60) is a prototype of this class of ordinary differential equations.)

The theory to be obtained can be regarded as an infinite-dimensional version of the eigenvalue theory in linear algebra. So, we will first list some theorems from linear algebra. These will then serve as a kind of guide through the Sturm-Liouville theory.

Let $A = [a_{ij}]$ be an $n \times n$ matrix with real constant entries. We will assume that A is a symmetric matrix, that is,

$$A = A^T$$

where A^T is the transpose of A, that is, $A^T = [b_{ij}]$ where $b_{ij} = a_{ji}$. It is easy to show that if A is symmetric then

$$(Ax, y) = (x, Ay)$$

where x, y are n-vectors and (x, y) is the usual inner product, that is, if $x = (x_1, \ldots, x_n)$ and $y = (y_1, \ldots, y_n)$, then $(x, y) = \sum_{i=1}^{n} x_i y_i$. We have the following results concerning matrix A:

Theorem 1A. *If the eigenvlaues of A are counted with their algebraic multiplicities, then A has n eigenvalues. Each eigenvalue is real, and if λ_0 is an eigenvalue of multiplicity q, then corresponding to λ_0 there are q linearly independent eigenvectors.*

Theorem 2A. *Suppose that $\lambda_1, \ldots, \lambda_m$ are the distinct eigenvalues of A (thus $m \leq n$) and λ_j has multiplicity r_j $(j = 1, \ldots,)$ and suppose that $x_{(j)}^{(1)}, \ldots, x_{(j)}^{(r_j)}$ is a set of linearly independent eigenvectors of λ_j $(j = 1, \ldots, m)$. Then the set of all the eigenvectors, that is, the set*

$$S = \{x_{(1)}^{(1)}, \ldots, x_{(1)}^{(r_1)}, x_{(2)}^{(1)}, \ldots, x_{(2)}^{(r_2)}, \ldots, x_{(m)}^{(1)}, \ldots, x_{(m)}^{(r_m)}\}$$

is a set of n linearly independent eigenvectors. (Thus S is a basis for real Euclidean n-space. That is, if y is a real n-vector, there is a unique linear combination of vectors in S which is equal to y.)

Theorem 3A. *If λ is not an eigenvalue of A, then the vector equation*

$$(A - \lambda I)x = b$$

where b is a given n-vector, has a unique solution x. (This is just a special case of Cramer's Rule since $\det(A - \lambda I) \neq 0$.)

Theorem 4A. *If λ is an eigenvalue of A, then the vector equation*

$$(A - \lambda I)x = b \tag{2.63}$$

where b is given, has a solution if and only if for each vector y which is an eigenvector of A corresponding to eigenvalue λ it is true that

$$(b, y) = 0$$

Theorem 5A. *If (2.63) has a solution, say $\tilde{x}$, then (2.63) has an infinite set of solutions of the form*

$$\tilde{x} + y$$

where y is any element in the linear space of eigenvectors corresponding to eigenvalue λ.

In describing the Sturm-Liouville theory, we work with real-valued functions on the interval $[0, 1]$ and the inner product

$$(f, g) = \int_0^1 f(x)\bar{g}(x)\,dx$$

where $\bar{g}(x)$ denotes the complex conjugate of $g(x)$. In a rigorous description this integral is a Lebesgue integral and function f, g are measurable functions which are elements of $\mathcal{L}^2[0, 1]$, that is, $\int_0^1 |f(x)|^2 \, dx < \infty$ and $\int_0^1 |g(x)|^2 \, dx < \infty$.

We study the differential equation

$$[p(x)y']' - q(x)y + \lambda y = 0 \tag{2.64}$$

where $p(x)$, $q(x)$ are continuous on $[0, 1]$, and $p(x)$ is positive and differentiable on $[0, 1]$ and λ is a number. We seek solutions $y(x)$ which satisfy the boundary conditions

$$ay(0) + by'(0) = 0$$
$$cy(1) + dy'(1) = 0 \tag{2.65}$$

where a, b, c, d are constants such that $a^2 + b^2 > 0$ and $c^2 + d^2 > 0$. The problem of finding numbers λ and corresponding nontrivial solutions of (2.64) which satisfy condition (2.65) is called a *Sturm-Liouville problem*. We shall refer to it as the S-$\mathcal{L}$ problem. Let L denote the operator which takes the function u into

$$L(u) = -[p(x)u']' + q(x)u$$

It is clear, from the definition, that L is linear, that is, if α, β are constants, then

$$L(\alpha u + \beta v) = \alpha L(u) + \beta L(v)$$

Simple calculations (see Boyce and DiPrima, Chapter 11 [1986]) using integration by parts show that if u and v satisfy the boundary conditions (2.65) then

$$(L(u), v) - (u, L(v)) = 0 \tag{2.66}$$

If L is a linear operator which satisfies (2.66), then L is said to be *self-adjoint*. (The self-adjointness of L is a generalization or extension of the symmetry property of matrix A, and it plays a crucial role in the development of the Sturm-Liouville theory.) If equation (2.64) has, for a fixed value of λ, a nontrivial solution $y(x)$ which satisfies boundary conditions (2.65), then λ is called an *eigenvalue of the S-$\mathcal{L}$ problem*. Since equation (2.64) can be written as

$$L(y) = \lambda y$$

then we say also that λ is an *eigenvalue* of *the operator* L. The nontrivial solution $y(x)$ is called an *eigenfunction corresponding to eigenvalue* λ.

To anyone who encounters Sturm-Liouville theory for the first time, the form of equation (2.64) may seem strange and arbitrarily given. Actually the equation has neither of these properties. First, it includes equations which arise in a large number of physical problems. Second, writing the equation in this form simplifies proofs that must be carried out, for example, the proof of (2.66).

It is straightforward to show (see Boyce and DiPrima [1986]) that the eigenvalues (and eigenfunctions) of the S-$\mathcal{L}$ problem are real, that each eigenvalue is simple (each eigenvalue has exactly one linearly independent eigenfunction) (see Exercise 11) and that linearly independent eigenfunctions (which must correspond to distinct eigenvalues) are orthogonal, that is, if $\phi_1(x)$, $\phi_2(x)$ are two such eigenvalues then

$$\int_0^1 \phi_1(x)\phi_2(x)\,dx = 0$$

A little more "machinery" mostly from complex variable theory can be used to prove that the set of eigenvalues is finite or is a denumerable set with no cluster points (see Coddington and Levinson, Chapter 7 [1955]). Notice that we have not considered the question of whether there exist *any* eigenvalues. We have simply stated that *if* there exist eigenvalues and eigenfunctions they have the properties described above.

Now we indicate how to prove that the S-$\mathcal{L}$ problem has an infinite set of eigenvalues. (We note first that our prototype S-$\mathcal{L}$ problem, described by (2.60) and (2.61), has an infinite set of eigenvalues $\frac{n^2\pi^2}{\ell^2}$ ($n = 1, 2, \ldots$).) To prove that the $\mathcal{S}$-$\mathcal{L}$ problem has an infinite set of eigenvalues we introduce Green's function which we will define by using the variation of constants formula. Suppose that λ is not an eigenvalue of the $\mathcal{S}$-$\mathcal{L}$ problem. Let $\phi_1(x)$, $\phi_2(x)$ be the (linearly independent) solutions of

$$L(u) - \lambda u = 0 \tag{2.67}$$

which satisfy the conditions

$$\phi_1(0) = 1, \qquad \phi_1'(0) = 0$$

$$\phi_2(0) = 0, \qquad \phi_2'(0) = 1$$

(Such solutions exist by Theorem 2.2.)

If $f(x)$ is a continuous function on $[0, 1]$, then by the variation of constants formula, a solution of

$$L(u) - \lambda u = f(x) \tag{2.68}$$

is

$$u(x) = \int_0^x \left\{ \frac{\phi_1(x)\phi_2(s) - \phi_2(x)\phi_1(s)}{p(s)[W(\phi_1, \phi_2)(s)]} \right\} f(s)\, ds \tag{2.69}$$

where $W(\phi_1, \phi_2)(s)$ is the Wronskian of ϕ_1 and ϕ_2 evaluated at s. We define the function $K_\lambda(t, s)$ as follows: if $s > x$, then

$$K_\lambda(x, s) = 0$$

if $s \leq x$, then

$$K_\lambda(x, s) = \frac{\phi_1(x)\phi_2(s) - \phi_2(x)\phi_1(s)}{p(s)[W(\phi_1, \phi_2)(s)]}$$

Then we may rewrite (2.69) as

$$u(x) = \int_0^1 \{K_\lambda(x, s)\}\, f(s)\, ds \tag{2.70}$$

Now let

$$w(x) = \int_0^1 \{K_\lambda(x, s) + k_1\phi_1(x) + k_2\phi_2(x)\}\, f(s)\, ds \tag{2.71}$$

where k_1, k_2 are functions of s. Since $u(x)$, defined in (2.69) and (2.70), is a solution of (2.68) and ϕ_1, ϕ_2 are solutions of the corresponding homogeneous equation (2.67), then it follows that $w(x)$ is a solution of (2.68). Also it can be proved that continuous functions $k_1(s)$, $k_2(s)$ can be chosen so that, independent of f, $w(x)$ satisfies the boundary conditions (2.65). (See Coddington and Levinson, Chapter 7. For this proof, the condition that λ is not an eigenvalue of the $\mathcal{S}$-$\mathcal{L}$ problem is required.) So $w(x)$ is the desired solution.

Equation (2.71) yields an analog of Theorem 3A. We have already shown that $w(x)$ is a solution of equation (2.68) with boundary conditions (2.65). The proof of uniqueness runs as follows: if $w_1(x)$ and $w_2(x)$ are solutions of (2.68) which satisfy boundary conditions (2.65), then

$$L(w_1 - w_2) + \lambda(w_1 - w_2) = 0$$

Also $w_1 - w_2$ satisfies boundary conditions (2.65). Since λ is not an eigenvalue of the $\mathcal{S}$-$\mathcal{L}$ problem, then for $t \in [0, 1]$,

$$w_1(x) - w_2(x) = 0$$

or

$$w_1(x) = w_2(x)$$

The function

$$G_\lambda(x, s) = K_\lambda(x, s) + k_1(s)\phi_1(x) + k_2(s)\phi_2(x)$$

is called the *Green's function* for the $\mathcal{S}$-$\mathcal{L}$ problem. We obtained Green's function by mathematical considerations starting from the variation of constants formula. For a different and interesting approach to Green's function in which one starts from a physical interpretation of equation (2.68), see Courant-Hilbert [1953], p. 351 ff.

Once we have Green's function, the set of eigenvalues is obtained from the following procedure. Without loss of generality, we may assume that $\lambda = 0$ is not an eigenvalue of the $\mathcal{S}$-$\mathcal{L}$ problem. (See Coddington and Levinson, p. 193.) Then $G_0(x, s)$ is defined. Let $\mathfrak{G}$ be the linear integral operator on $C[0, 1]$ defined by

$$\mathfrak{G}[f(x)] = \int_0^1 G_0(x, s)f(s)\,ds$$

First, from (2.71) and (2.68) we have: if $f(x)$ is continuous on $[0, 1]$, then

$$L\mathfrak{G}[f(x)] = L[w(x)] = f(x) \tag{2.72}$$

(We note that if u has a second derivative and satisfies the boundary conditions (2.65), then applying (2.71) with $f(x) = Lu$ and using the fact that $w(x) = u(x)$, we obtain:

$$u = \mathfrak{G}[L(u)] \tag{2.73}$$

Equations (2.72) and (2.73) show that $\mathfrak{G}$ is a kind of inverse of L.)

Now suppose λ is an eigenvalue of the $\mathcal{S}$-$\mathcal{L}$ problem and ϕ is an eigenfunction corresponding to λ. Then

$$L\phi = \lambda\phi$$

and

$$\phi = \int_0^1 G_0(s, t)\lambda\phi(s)\,ds$$

or

$$\phi = \lambda\mathfrak{G}(\phi) \tag{2.74}$$

That is, the operator $\mathfrak{G}$ has eigenvalue $\frac{1}{\lambda}$ and ϕ is an eigenfunction corresponding to that eigenvalue. Note that by assumption, $\lambda \neq 0$.

Conversely, suppose (2.74) holds. By (2.72), we have

$$\phi = L\mathfrak{G}(\phi)$$

and from (2.74)

$$L\mathfrak{G}(\phi) = L\left[\mathfrak{G}(\phi)\right] = L\left(\frac{\phi}{\lambda}\right)$$

Therefore

$$\phi = L\left(\frac{\phi}{\lambda}\right)$$

or

$$L(\phi) = \lambda\phi$$

Also it follows from (2.74) that since $\mathfrak{G}(\phi)$ satisfies the boundary conditions (2.65) then so does ϕ. Consequently λ is an eigenvalue and ϕ is a corresponding eigenfunction of the $\mathcal{S}\text{-}\mathcal{L}$ problem. Thus we have shown that the set of eigenvalues of the $\mathcal{S}\text{-}\mathcal{L}$ problem is the set of reciprocals of the eigenvalues of $\mathfrak{G}$. Hence it is sufficient to investigate the eigenvalues of $\mathfrak{G}$. It can be shown that the operator $\mathfrak{G}$ is self-adjoint; that $\mathfrak{G}$ is a completely continuous or compact operator from $C[0, 1]$ into $C[0, 1]$; and that $\mathfrak{G}$ has a denumerably infinite set of eigenvalues which have zero as a cluster point. (See Coddington and Levinson, Chapter 7.)

The arguments sketched above yield an infinite-dimensional result which is analogous to Theorem 1A stated earlier. The contrast between the two results is worth noting. To obtain the existence of eigenvalues of matrix A, we need only apply the fundamental theorem of algebra. Studying the eigenvalues of the $\mathcal{S}\text{-}\mathcal{L}$ problem requires considerably more effort.

Now we turn to an infinite-dimensional analog of Theorem 2A for the $\mathcal{S}\text{-}\mathcal{L}$ problem. Let $\{\lambda_m\}$ be the sequence of eigenvalues and let $\{\Psi_m(t)\}$ be a sequence of corresponding orthonormal eigenvectors. The analogous result (Coddington and Levinson, Chapter 7) is

Theorem *If $f \in \mathcal{L}^2[0, 1]$, then*

$$f = \sum_{m=0}^{\infty}(f, \Psi_m)\Psi_m$$

that is,

$$\lim_{k \to \infty}\left\| f - \sum_{m=0}^{k}(f, \Psi_m)\Psi_m \right\| = 0$$

(The series $\sum_{m=1}^{\infty}(f, \Psi_m)\Psi_m$ is a generalized Fourier series.)

Finally we consider equation (2.68) with boundary conditions (2.65) and assume that λ is an eigenvalue of our original $\mathcal{S}\text{-}\mathcal{L}$ problem (defined by (2.64) and (2.65)). We indicate how to obtain analogs of Theorems 4A and 5A by using some formal computations. (We emphasize that our argument will be only formal. To make the argument rigorous, we would need to justify each of the formal steps.)

By the theorem stated immediately above, we have

$$f = \sum_{j=1}^{\infty} (f, \Psi_j) \Psi_j$$

Assume that the desired solution u has the form

$$u = \sum_{j=1}^{\infty} \alpha_j \Psi_j$$

Then we must determine the α_j's.

Since

$$Lu = L \left\{ \sum_{j=1}^{\infty} \alpha_j \Psi_j \right\} = \sum_{j=1}^{\infty} \alpha_j L(\Psi_j) = \sum_{j=1}^{\infty} \alpha_j \lambda_j \Psi_j$$

then substituting in (2.68), we have

$$\sum_{j=1}^{\infty} \alpha_j \lambda_j \Psi_j - \sum_{j=1}^{\infty} \lambda \alpha_j \Psi_j = \sum_{j=1}^{\infty} (f, \Psi_j) \Psi_j$$

or

$$\sum_{j=1}^{\infty} \{\alpha_j \lambda_j - \lambda \alpha_j - (f, \Psi_j)\} \Psi_j = 0 \tag{2.75}$$

Since the Ψ_j's are orthogonal, it follows that each coefficient in (2.75) is zero. Thus

$$\alpha_j \lambda_j - \lambda \alpha_j = (f, \Psi_j) \qquad (j = 1, 2, \dots) \tag{2.76}$$

By hypothesis, λ is an eigenvalue, say $\lambda = \lambda_m$. Then if $j \neq m$, we have

$$\alpha_j (\lambda_j - \lambda) = (f, \Psi_j)$$

and

$$\alpha_j = \frac{(f, \Psi_j)}{\lambda_j - \lambda}$$

But if $j = m$, then (2.76) becomes:

$$0 = (f, \Psi_m)$$

Thus if the problem defined by (2.68) and (2.65) has a solution, we must have

$$(f, \Psi_m) = 0 \tag{2.77}$$

Conversely if (2.77) holds, then the problem has the solution

$$u = \sum_{j=1}^{\infty} \alpha_j \Psi_j$$

where, if $j \neq m$, we have

$$\alpha_j = \frac{(f, \Psi_j)}{\lambda_j - \lambda}$$

If $j = m$, we have

$$\alpha_m(\lambda_m - \lambda_m) = (f, \Psi_m)$$

Since $\lambda_m - \lambda_m = 0$ and $(f, \Psi_m) = 0$, there is no condition on α_m and hence equation (2.68) with boundary conditions (2.65) has an infinite set of solutions of the form

$$\sum_{j=1}^{m-1} \alpha_j \Psi_j + \sum_{j=m+1}^{\infty} \alpha_j \Psi_j + c\Psi_m$$

where c is any constant.

The Sturm-Liouville problem we have discussed is an especially simple case, and we want to indicate how this problem should be generalized in order to obtain a more complete treatment of Sturm-Liouville theory. Among the extensions of the theory which can be made are the following. First in equation (2.64), the term λy can be replaced by $\lambda r(x)y$ where $r(x)$ is continuous and positive on $[0, 1]$. Then the inner product becomes

$$(f, g) = \int_0^1 f(x)g(x)r(x)\,dx$$

For simplicity we have restricted ourselves to initial conditions (2.65) which are called *separated conditions*, that is, the first condition refers to values at $x = 0$ and the second equation to values at $x = 1$. (Cf. Exercise 12.) The theory may be extended to cases where $p(x)$ is 0 at one or both of the endpoints of $[0, 1]$. Bessel's equation is an important example in which $p(x)$ is 0 at both endpoints. See Courant-Hilbert, [1953]. The theory can be generalized to treatment of an nth-order differential equation (with appropriate boundary conditions). See Coddington and Levinson [Chapter 7, 1955]. Finally the Sturm-Liouville theory can be used to initiate studies of nonlinear boundary condition problems.

Exercises

1. Find the eigenvalues and corresponding eigenvectors and generalized eigenvectors of the matrix

$$A = \begin{bmatrix}
1 & 6 & -5 & 6 & -6 & 1 & -1 \\
-3 & 7 & -3 & 7 & -6 & 1 & -1 \\
0 & 0 & 1 & 6 & -5 & 1 & -1 \\
0 & 0 & -3 & 7 & -2 & 1 & -1 \\
0 & 0 & 0 & 0 & 2 & 1 & -1 \\
0 & 0 & 0 & 0 & 0 & 2 & 0 \\
0 & 0 & 0 & 0 & 0 & 0 & 2
\end{bmatrix}$$

2. Find the Jordan canonical form and the real canonical form of the matrix A in Exercise 1.

3. Find a real fundamental matrix for the equation

$$\frac{dx}{dt} = Ax$$

where A is the matrix in Exercise 1.

4. Given the nth order differential equation

$$\frac{d^n x}{dt^n} + a_1(t)\frac{d^{n-1} x}{dt^{n-1}} + \cdots + a_{n-1}(t)\frac{dx}{dt} + a_n(t)x = 0 \qquad (2.78)$$

where $a_i(t)$ is continuous on (t_1, t_2) for $i = 1, \ldots, n$; let $\phi_1(t), \ldots,$ $\phi_n(t)$ be solutions of (2.78) on $(t_1 t_2)$.

Definition   The determinant

$$W(\phi_1, \ldots, \phi_n) = \det \begin{bmatrix} \phi_1 & \cdots & \phi_n \\ \frac{d\phi_1}{dt} & \cdots & \frac{d\phi_n}{dt} \\ \cdot & \cdot & \cdot \\ \frac{d^{n-1}\phi_1}{dt^{n-1}} & \cdots & \frac{d^{n-1}\phi_n}{dt^{n-1}} \end{bmatrix}$$

is the *Wronskian* of (2.78) with respect to $\phi_1, \ldots, \phi_n$.

Prove that a necessary and sufficient condition that $\phi_1, \ldots, \phi_n$ are linearly independent on (t_1, t_2) is that $W(\phi_1, \ldots, \phi_n) \neq 0$ for all $t \in (t_1, t_2)$. If t_0 and t are two points in (t_1, t_2), find a relationship between the values of the Wronskian at t_0 and t.

5. Suppose

$$P(t) = a_0 t^m + a_1 t^{m-1} + \cdots + a_{m-1}t + a_m$$
$$Q(t) = b_0 t^n + b_1 t^{n-1} + \cdots + b_{n-1}t + b_n$$

where $a_0 \neq 0$, $b_0 \neq 0$, and $m > n$. Prove that $P(t)$ and $Q(t)$, as functions on the real line, are linearly independent.

6. Let $f(t) = t^3$, $g(t) = t^4$. By Exercise 5, functions f and g are linearly independent as functions on the real line. Prove that f and g cannot be solutions of a differential equation of the form

$$\frac{d^2 x}{dt^2} + a(t)\frac{dx}{dt} + b(t)x = 0 \qquad (2.79)$$

where $a(t)$, $b(t)$ are continuous for all real t.

7. Suppose $p(t)$ is a continuous real-valued function of period T. The equation

$$\frac{d^2x}{dt^2} + p(t)x = 0 \tag{H}$$

is called Hill's equation.

Write Hill's equation as a two-dimensional first-order system and let

$$X(t) = \begin{bmatrix} x_{11}(t) & x_{12}(t) \\ x_{21}(t) & x_{22}(t) \end{bmatrix}$$

be a fundamental matrix of the system such that $X(0) = I$, the identity matrix. If

$$X(t + T) = X(t)C$$

where $C = [c_{ij}]$ is a constant nonsingular matrix, prove that det $C = 1$. If

$$2\mathcal{A} = c_{11} + c_{22}$$

show that if $\mathcal{A}^2 < 1$, all the solutions of equation (H) are bounded. Show that if $\mathcal{A}^2 > 1$, equation (H) has solutions which are not bounded.

8. Prove (by induction) that if

$$A = \begin{bmatrix} 0 & 1 & 0 & \cdots & 0 \\ 0 & 0 & 1 & \cdots & 0 \\ & & \cdots & & \\ 0 & 0 & 0 & \cdots & 1 \\ -a_n & -a_{n-1} & & \cdots & -a_1 \end{bmatrix}$$

then

$$\det[A - \lambda I] = (-1)^n(\lambda^n + a_1\lambda^{n-1} + a_2\lambda^{n-2} + \cdots + a_{n-1}\lambda + a_n).$$

9. Given the single n-th order equation

$$x^{(n)} + a_1x^{(n-1)} + a_2x^{(n-2)} + \cdots + a_{n-1}x' + a_nx = 0 \tag{2.80}$$

show that if the characteristic equation is

$$\lambda^n + a_1\lambda^{n-1} + \cdots + a_{n-1}\lambda + a_n = (\lambda - \lambda_1)^r(\lambda - \lambda_2)^s = 0$$

where $r + s = n$, then all the solutions of (2.80) are of the form

$$A_1e^{\lambda_1 t} + A_2te^{\lambda_1 t} + \ldots A_r t^{r-1}e^{\lambda_1 t}$$
$$+ B_1e^{\lambda_2 t} + B_2te^{\lambda_2 t} + \cdots + B_s t^{s-1}e^{\lambda_2 t} \tag{2.81}$$

where $A_1, \ldots, A_r, B_1, \ldots, B_s$ are constants.

10. Derive a variation of constants formula for

$$x^{(n)} + a_1(t)x^{(n-1)} + \cdots + a_{n-1}(t)x^{(1)} + a_n^{(t)}x = b(t) \qquad (2.82)$$

where $a_1(t), a_2(t), \ldots, a_n(t)$ and $b(t)$ are continuous on (α, β) and $\alpha < 0 < \beta$, which yields the solution $x(t)$ of (2.82) that satisfies the initial condition $x(0) = \bar{x}_0, x'(0) = \bar{x}_1, \ldots x^{(n-1)}(0) = \bar{x}_{n-1}$.

11. Prove that each eigenvalue of the $\mathcal{S}\text{-}\mathcal{L}$ problem is simple.

12. It can be proved that the eigenvalues of the Sturm-Liouville problem

$$y'' + \lambda y = 0$$
$$y(-1) = y(1)$$
$$y'(-1) = y'(1)$$

are real. Show that the eigenvalues are a denumerable set and that corresponding to each nonzero eigenvalue, there are two linearly independent eigenfunctions. (This latter statement does not contradict the result in Exercise 11. The boundary conditions in this problem are not separated. So the result in Exercise 11 is not applicable.)

Chapter 3

Autonomous Systems

Introduction

Although we obtained some explicit formulas for solutions of linear systems in Chapter 2, the results obtained suggest that there is very little chance of obtaining such formulas for more extensive classes of equations. For example, we obtained the explicit formula Pe^{tJ} for a fundamental matrix of a linear homogeneous system with constant coefficients $x' = Ax$. However, even if we consider the slightly more general system $x' = A(t)x$ where $A(t)$ has period τ, the results obtained are no longer explicit formulas. Consequently, in studying larger classes of equations which include certain nonlinear equations, we must resign ourselves to obtaining limited information about the solutions. We will look for nonnumerical or "qualitative" properties of solutions. In this chapter we take the first steps in these qualitative studies by studying the qualitative properties of solutions of autonomous systems, which arise in many applications.

First we study some general properties of solutions of autonomous systems which are very useful in the analysis of autonomous systems. After that, we restrict ourselves to the study of two-dimensional systems, where much more detailed analysis is possible.

We consider a system

$$\frac{dx}{dt} = f(x, y)$$
$$\frac{dy}{dt} = g(x, y)$$

and study the behavior of the solution curves near an isolated equilibrium point, that is, a point (x_0, y_0) such that

$$f(x_0, y_0) = g(x_0, y_0) = 0$$

First we consider the linear case

$$\frac{dx}{dt} = ax + by$$
$$\frac{dy}{dt} = cx + dy$$

and then proceed to analysis of the nonlinear case. We obtain a general description, due to Lefschetz [1962], of the solution curves in the neighborhood of an equilibrium point: (i) the case in which the neighborhood consists of elliptic, hyperbolic, and parabolic sectors; (ii) the case in which solution curves are closed curves or spirals.

Then, we introduce the concept of the index of an equilibrium point and obtain the Bendixson theorem which concerns a system of the form

$$\frac{dx}{dt} = ax + by + F(x, y)$$

$$\frac{dy}{dt} = cx + dy + G(x, y)$$

where F and G are higher order terms and the eigenvalues of the matrix

$$\begin{bmatrix} a & b \\ c & d \end{bmatrix}$$

are $\lambda(\neq 0)$ and 0. Lastly we prove the famous Poincaré-Bendixson theorem which plays an important role in both pure and applied mathematics.

General Properties of Solutions of Autonomous Systems

Definition   A system of differential equations in which the independent variable does not appear explicitly, that is, a system of the form

$$x' = f(x)$$

is an *autonomous* system.

Autonomous systems arise naturally in the study of conservative mechanical systems. For example, the equations arising in celestial mechanics are autonomous.

We consider the n-dimensional autonomous system

$$x'_i = f_i(x_1, \ldots, x_n) \qquad (i = 1, \ldots, n) \tag{3.1}$$

where for $i = 1, \ldots, n$, the domain of f_i is an open set D in R^n and f_i satisfies a local Lipschitz condition, that is, for each point $(x_1, \ldots, x_n) \in D$, there is a neighborhood N of $(x_1, \ldots, x_n)$ such that in N, the functions $f_1, \ldots, f_n$ satisfy a Lipschitz condition. (As usual, the Lipschitz condition is invoked to ensure uniqueness of solution.) We assume that all the solutions of (3.1) to be considered are maximal.

Definition   Let $(x_1(t), \ldots, x_n(t))$ be a solution of (3.1) such that not all the functions $x_1(t), \ldots, x_n(t)$ are constant functions. Let I be the domain of $(x_1(t), \ldots, x_n(t))$. The underlying point set of the solution, that is, the set of points

$$C = \{(x_1(t), \ldots, x_n(t)) \mid t \in I\}$$

which is a curve in the intuitive sense, is an *orbit* of (3.1). If $n = 2$, the orbit is often called a *path* or *characteristic*.

(Notice that we do not use the term "curve" in a precise sense. We could approach the concept of orbit in this way, but it would be unnecessarily elaborate.)

Our first step is to obtain a fundamental result concerning orbits, a result which can be described intuitively as follows: If the open set D is the domain of f_i ($i = 1, \ldots, n$) in (3.1), then each point of D is contained in exactly one orbit or one constant solution of (3.1). That is, the orbits and constant solutions cover D but they do not intersect one another. Also no orbit crosses itself. (If an orbit intersects itself, it is a simple closed curve.) These properties of orbits are intuitively reasonable, but rigorous proofs require some rather fussy steps which we carry out in Lemmas 3.1, 3.2, 3.3, 3.4 and Theorems 3.1 and 3.2. We emphasize that these results are *not* valid for nonautonomous systems. (See Exercises 1 and 2.) After obtaining this basic result we introduce some other concepts which will be useful later in the study of autonomous systems: Ω-limit sets, invariant sets, and minimal sets. Also we will discuss systems of the form

$$x' = ax + by$$
$$y' = cx + dy$$

and this gives us an introduction to the concept of stability.

Lemma 3.1 *If* $(x_1(t), \ldots, x_n(t))$ *is a solution of* (3.1) *with domain I and if h is a real number, then* $(x_1(t + h), \ldots, x_n(t + h))$ *is a solution of* (3.1) *with domain*

$$I_h = \{t - h / t \in I\}$$

Proof Let

$$\bar{x}_i(t) = x_i(t + h) \qquad (i = 1, \ldots, n)$$

Then letting $s = t + h$, we have, for $i = 1, \ldots, n$,

$$\frac{d\bar{x}_i(t)}{dt} = \frac{dx_i}{ds}(s)\frac{ds}{dt} = \frac{dx_i}{ds}(s) = f_i[x_1(s), \ldots, x_n(s)]$$
$$= f_i[x_1(t + h), \ldots, x_n(t + h)]$$
$$= f_i[\bar{x}_1(t), \ldots, \bar{x}_n(t)]$$

This completes the proof of Lemma 3.1. ∎

Notice that the argument used in the proof of Lemma 3.1 breaks down if (3.1) is nonautonomous, that is, if

$$f_i = f_i(t, x_1, \ldots, x_n)$$

because then

$$\frac{d}{dt}\bar{x}_i(t) = \frac{d}{dt}x_i(t+h) = f_i[t+h, x_1(t+h), \ldots, x_n(t+h)]$$
$$= f_i[t+h, \bar{x}_1(t), \ldots, \bar{x}_n(t)]$$

Lemma 3.1 shows that there are many more solutions than orbits because if C is an orbit which is the underlying point set of the solution $(x_1(t), \ldots, x_n(t))$ with domain I, then C is also the underlying point set of the solution $(x_1(t+h), \ldots, x_n(t+h))$ with domain I_h.

Lemma 3.2   *If $(x_1(t), \ldots, x_n(t))$ is a solution of (3.1) with domain I and $(\bar{x}_1(t), \ldots, \bar{x}_n(t))$ is a solution of (3.1) with domain $\bar{I}$ and there exist $t_o \in I$, $\bar{t}_0 \in \bar{I}$ such that*

$$(x_1(t_0), \ldots, x_n(t_0)) = (\bar{x}_1(\bar{t}_0), \ldots, x_n(\bar{t}_0))$$

then for all $t \in \bar{I}$,

$$(x_1(t+t_0-\bar{t}_0), \ldots, x_n(t+t_0-\bar{t}_0)) = (\bar{x}_1(t), \ldots, \bar{x}_n(t)) \tag{3.2}$$

Proof   By Lemma 3.1

$$(x_1(t+t_0-\bar{t}_0), \ldots, x_n(t+t_0-\bar{t}_0))$$

is a solution of (3.1) with domain

$$I_{t_0-\bar{t}_0} = \{t-(t_0-\bar{t}_0)/t \in I\}$$

But if $t = \bar{t}_0$

$$(x_1(t+t_0-\bar{t}_0), \ldots, x_n(t+t_0-\bar{t}_0))$$
$$= (x_1(t_0), \ldots, x_n(t_0)) = (\bar{x}_1(\bar{t}_0), \ldots, \bar{x}_n(\bar{t}_0))$$

By the uniqueness of solution property of (3.1) and the fact that the solutions are maximal, it follows that (3.2) holds for all $t \in \bar{I}$. This completes the proof of Lemma 3.2. □

Lemma 3.2 shows that if orbit C is the underlying point set of a solution $(x_1(t), \ldots, x_n(t))$ then every solution of which C is the underlying point set has the form $(x_1(t+k), \ldots, x_n(t+k))$ where k is a real constant. Thus for every solution of which C is the underlying point set the direction of increasing t on C is the same. Hence C may be regarded as a directed curve.

Definition   If $(x_1^0, \ldots, x_n^0) \in D$ is such that for $i = 1, \ldots, n$

$$f_i(x_1^0, \ldots, x_n^0) = 0$$

then $(x_1^0, \ldots, x_n^0)$ is an *equilibrium point* (or *rest point or critical point* or *singular point*) of (3.1). Note that if $(x_1^0, \ldots, x_n^0)$ is an equilibrium point, and if for all real t,

$$x_i(t) = x_i^0 \qquad (i = 1, \ldots, n)$$

then $(x_1(t), \ldots, x_n(t))$ is a solution of (3.1).

Theorem 3.1 *If* $(x_1^0, \ldots, x_n^0) \in D$, *then either* $(x_1^0, \ldots, x_n^0)$ *is an equilibrium point of* (3.1) *or* $(x_1^0, \ldots, x_n^0)$ *is contained in exactly one orbit of* (3.1).

Proof Suppose that $(x_1^0, \ldots, x_n^0)$ is not an equilibrium point of (3.1). By Existence Theorem 1.1, there exists a solution $(x_1(t), \ldots, x_n(t))$ of (3.1) which satisfies the initial condition

$$(x_1(t_0), \ldots, x_n(t_0)) = \left(x_1^0, \ldots, x_n^0 \right)$$

for some t_0 in the domain of $(x_1(t), \ldots, x_n(t))$. Thus $(x_1^0, \ldots, x_n^0)$ is contained in at least one orbit of (3.1). Now suppose that $(x_1^0, \ldots, x_n^0)$ is contained in orbits C and $\bar{C}$ which are the underlying point set of solutions $(x_1(t), \ldots, x_n(t))$ and $(\bar{x}_1(t), \ldots, \bar{x}_n(t))$ with domains I and $\bar{I}$, respectively. Then there exist $t_0 \in I, \bar{t}_0 \in \bar{I}$ such that

$$\left(x_1^0, \ldots, x_n^0 \right) = (x_1(t_0), \ldots, x_n(t_0)) = (\bar{x}_1(\bar{t}_0), \ldots, \bar{x}_n(\bar{t}_0))$$

Hence by Lemma 3.2

$$C = \bar{C}$$

This completes the proof of Theorem 3.1. ∎

Theorem 3.1 can be stated intuitively as: Orbits do not intersect.

Lemma 3.3 *If* $(x_1(t), \ldots, x_n(t))$ *is a solution of* (3.1) *with domain I such that there exist* $t_1, t_2 \in I$, *where* $t_1 \neq t_2$, *and such that*

$$(x_1(t_1), \ldots, x_n(t_1)) = (x_1(t_2), \ldots, x_n(t_2)) \qquad (3.3\text{a})$$

then I is the real t-axis and for all real t,

$$x_i(t_1 + t) = x_i(t_2 + t) \qquad i = 1, \ldots, n \qquad (3.3\text{b})$$

or

$$x_i(t + t_2 - t_1) = x_i(t) \qquad i = 1, \ldots, n$$

Proof For definiteness assume that $t_1 < t_2$. Then

$$[t_1, t_2] \subset I$$

Let

$$E = \{s/s \geq 0 \text{ and } (3.3\text{b}) \text{ holds for } 0 \leq t \leq s\}$$

Set E is nonempty because, by hypothesis, $0 \in E$. Let t_0 be the lub of E. Suppose first that $t_0 = 0$. By Lemma 3.1, $(x_1(t_1 + t), \ldots, x_n(t_1 + t))$ and $(x_1(t_2 + t), \ldots, x_n(t_2 + t))$ are solutions of (3.1) and by (3.3a), these two solutions have the same value at $t = 0$. Hence by the uniqueness of solution, there exists $\delta > 0$ such that if $|t| < \delta$,

$$x_i(t_1 + t) = x_i(t_2 + t) \qquad i = 1, \ldots, n$$

This contradicts the condition that $t_0 = 0$.

Next suppose that t_0 is a positive number. Then $[t_1, t_2 + t_0] \subset I$ and hence $t_1 + t_0 \in I$. If $0 \le t < t_0$

$$x_i(t_1 + t) = x_i(t_2 + t) \qquad i = 1, \ldots, n$$

Hence for $i = 1, \ldots, n$,

$$x_i(t_1 + t_0) = \lim_{t \to t_0} x_i(t_1 + t) = \lim_{t \to t_0} x_i(t_2 + t)$$

Hence the domain of $(x_1(t_2 + t), \ldots, x_n(t_2 + t))$ contains t_0. For if it did not contain t_0, then the domain could be extended to include t_0 by the same argument as used in the proof of Extension Theorem 1.3, and this would contradict the maximality which is assumed for all solutions discussed. Now $(x_1(t_1 + t_0 + t), \ldots, x_n(t_1 + t_0 + t))$ and $(x_1(t_2 + t_0 + t), \ldots, x_n(t_2 + t_0 + t))$ are both solutions of (3.1) by Lemma 3.1 and have the same value at $t = 0$. Hence by the uniqueness of solutions, there exists $\delta > 0$ such that if $|t| < \delta$,

$$x_i(t_1 + t_0 + t) = x_i(t_2 + t_0 + t) \qquad (i = 1, \ldots, n)$$

This contradicts the definition of t_0. Thus E is the set of nonnegative reals. A similar argument shows that (3.3b) holds for $t \le 0$. $\qquad \Box$

Lemma 3.4   *If there exists a set of pairs $\{(t_\nu, h_\nu)\}$ where each h_ν is positive and such that*

$$x_i(t_\nu + h_\nu) = x_i(t_\nu) \qquad (i = 1, \ldots, n)$$

and if $(x_1(t), \ldots, x_n(t))$ is a nonconstant solution (i.e., a solution which is not an equilibrium point), then there exists a minimal positive number h such that for all real t,

$$x_i(t + h) = x_i(t) \qquad (i = 1, \ldots, n)$$

Proof   By Lemma 3.3

$$x_i(t + h_\nu) = x_i(t) \qquad (i = 1, \ldots, n) \tag{3.4}$$

for each ν and for all real t. Suppose there is a monotonic sequence h_{ν_n} which converges to zero. Let τ_1 and τ_2 be arbitrary numbers such that $\tau_1 < \tau_2$. For each h_{ν_n}, there is an integer k_n such that

$$\tau_1 + (k_n - 1)h_{\nu_n} \le \tau_2 \le \tau_1 + k_n h_{\nu_n}$$

By (3.4),

$$x_i(\tau_1) = x_i(\tau_1 + mh_{v_n}) \qquad (i = 1, \ldots, n)$$

for all integers m. Hence since

$$\lim_n (\tau_1 + k_n h_{v_n}) = \tau_2$$

we have

$$x_i(\tau_1) = x_i(\tau_2)$$

Since τ_1, τ_2 are arbitrary, then $x_1(t), \ldots, x_n(t)$ is an equilibrium point. Thus the $\text{glb}_v\{h_v\}$ is positive. It follows at once that $h = \text{glb}_v\{h_v\}$. $\quad\square$

From Lemmas 3.3 and 3.4, we have at once:

Theorem 3.2 *If $(x_1(t), \ldots, x_n(t))$ is a solution of (3.1) with domain I which is not an equilibrium point and if there exist $t_1, t_2 \in I$, where $t_1 \neq t_2$, and such that*

$$(x_1(t_1), \ldots, x_n(t_1)) = (x_1(t_2), \ldots, x_n(t_2))$$

then I is the positive t-axis and there is a minimal positive number h such that for all real t,

$$x_i(t + h) = x_i(t) \qquad (i = 1, \ldots, n)$$

Also the orbit underlying solution $x_1(t), \ldots, x_n(t)$ is a simple closed curve.

Theorem 3.2 can be stated intuitively as: If an orbit intersects itself, it is a simple closed curve.

Definition The number h in the conclusion of Theorem 3.2 is the *period* of the solution $(x_1(t), \ldots, x_n(t))$ and $(x_1(t), \ldots, x_n(t))$ is said to be *periodic*.

Definition Equilibrium point $(x_1^0, \ldots, x_n^0)$ of (3.1) is *isolated* if there exists a neighborhood N of $(x_1^0, \ldots, x_n^0)$ such that the only equilibrium point of (3.1) in N is $(x_1^0, \ldots, x_n^0)$.

Theorem 3.3 *Suppose $(x_1^0, \ldots, x_n^0) \in D$ is an equilibrium point of (3.1) and suppose that for $i = 1, \ldots, n$, the function f_i has continuous first derivatives in a neighborhood of $(x_1^0, \ldots, x_n^0)$. Let*

$$f_{ij} = \frac{\partial f_i}{\partial x_j} (x_1^0, \ldots, x_n^0).$$

Then if $\det[f_{ij}] \neq 0$, the point $(x_1^0, \ldots, x_n^0)$ is an isolated equilibrium point.

Proof Since the mapping from R^n into R^n given by

$$x \to f(x)$$

is differentiable, there is a neighborhood N of $(x_1^0, \ldots, x_n^0)$ such that if $x = (x_1, \ldots, x_n) \in N$ and $x^0 = (x_1^0, \ldots, x_n^0)$, then

$$f(x) = [f_{ij}](x - x^0) + R(x)$$

where

$$\lim_{|x-x^0| \to 0} \frac{|R(x)|}{|x - x^0|} = 0 \tag{3.5}$$

Hence if $x \in N$ and $x - x^0 \neq 0$,

$$[f_{ij}]^{-1} f(x) = x - x^0 + [f_{ij}]^{-1} R(x) \tag{3.6}$$

But if $|x - x^0|$ is sufficiently small, then by (3.5),

$$|[f_{ij}]^{-1}| \, |R(x)| < \frac{1}{2} |x - x^0|$$

Thus if $|x - x^0|$ is sufficiently small by nonzero,

$$f(x) \neq 0$$

This completes the proof of Theorem 3.3. $\square$

Notation   If $S = x(t)$ denotes a solution of a system

$$x' = f(x) \tag{3.7}$$

the underlying point set of solution S, that is, the orbit of S, will be denoted by $0(S)$ or $0[x(t)]$.

Definition   If $x(t)$ is a solution of the system

$$x' = f(x)$$

and if there exists a number t_0 such that $x(t)$ is defined for all $t \geq t_0$, then an *ω-limit point* of solution $x(t)$ is a point x^0 such that there exists a sequence of real numbers $\{t_n\}$ with

$$\lim_n t_n = \infty$$

and

$$\lim_n x(t_n) = x^0$$

If S denotes the solution $x(t)$, then the set of ω-limit points of S will be denoted by $\Omega(S)$ or $\Omega[x(t)]$.

Definition   Suppose that f is continuous on an open set $D \subset R^n$. A set $E \subset D$ is *invariant* if and only if for each point $x^0 \in E$, it is true that, if $x(t)$ is a solution of (3.7) such that for some t, $x(t) = x^0$, then $0[x(t)] \subset E$.

Notation   If A, B are subsets of R^n, let $d(A, B)$ denote the number

$$\operatorname*{glb}_{\substack{p \in A \\ q \in B}} |p - q|$$

Theorem 3.4   *Suppose that f is continuous on an open set $D \subset R^n$ and suppose that $x(t)$ is a solution of (3.7) which is not an equilibrium point and is such that there exists a number t_0 so that $x(t)$ is defined for all $t \geq t_0$ and there exists a number $B > 0$ such that for all $t \geq t_0$,*

$$|x(t)| < B \tag{3.8}$$

Then $\Omega[x(t)]$ is nonempty, closed, connected, and invariant. Also

$$\lim_{t \to \infty} d[x(t), \Omega[x(t)]] = 0$$

Proof   That $\Omega[x(t)]$ is nonempty follows from (3.8) and the Weierstrass-Bolzano theorem.

To show that $\Omega[x(t)]$ is closed, suppose that q is a limit point of $\Omega[x(t)]$. Then there exists a sequence $\{q_n\} \subset \Omega[x(t)]$ such that

$$\lim_n |q_n - q| = 0 \tag{3.9}$$

Since $q_n \in \Omega[x(t)]$, there exists $\{t_n\}$ such that $t_n \to \infty$ and

$$|x(t_n) - q_n| < \frac{1}{n} \tag{3.10}$$

From (3.9) and (3.10), it follows that

$$\lim_n x(t_n) = q$$

Thus $q \in \Omega[x(t)]$ and hence $\Omega[x(t)]$ is closed.

Now suppose $\Omega[x(t)]$ is not connected. Then there exist disjoint nonempty open sets U_1 and U_2 such that

$$\Omega[x(t)] \subset U_1 \cup U_2$$
$$\Omega[x(t)] \cap U_1 \neq \phi$$
$$\Omega[x(t)] \cap U_2 \neq \phi$$

Since $\Omega[x(t)]$ is closed and bounded, the sets

$$M = \Omega[x(t)] \cap U_1$$

and

$$N = \Omega[x(t)] \cap U_2$$

are closed and bounded. Hence since $M \cap N = \phi$, then $d(M, N) = \delta > 0$. Since $M \cup N = \Omega[x(t)]$, there exists a monotonic increasing sequence $\{t_n\}$ such that

$$\lim_n t_n = \infty$$

and

$$\left.\begin{array}{ll} d[x(t_n), M] < \frac{\delta}{4} & \text{for } n \text{ odd} \\ d[x(t_n), N] < \frac{\delta}{4} & \text{for } n \text{ even} \end{array}\right\} \tag{3.11}$$

Since $x(t)$ is a continuous function of t, then the functions

$$\left.\begin{array}{l} g(t) = d[x(t), M] \\ h(t) = d[x(t), N] \end{array}\right\} \tag{3.12}$$

are continuous. From (3.11), it follows that if $t = t_{2m-1}$,

$$g(t) - h(t) < 0$$

and if $t = t_{2m}$, then

$$g(t) - h(t) > 0$$

Hence there exists $t_m^{(1)}$ such that

$$t_m^{(1)} \in (t_{2m-1}, t_{2m})$$

and

$$g\left(t_m^{(1)}\right) - h\left(t_m^{(1)}\right) = 0 \tag{3.13}$$

But there is a point $\bar{x} \in \Omega[x(t)]$ such that a subsequence of $\{x(t_m^{(1)})\}$ converges to $\bar{x}$. Hence from (3.12) and (3.13), it follows that

$$d[\bar{x}, M] = d[\bar{x}, N] \tag{3.14}$$

Since $\bar{x} \in \Omega[x(t)]$, then $\bar{x} \in M \cup N$. If $\bar{x} \in M$, then

$$d[\bar{x}, M] = 0$$

and

$$d[\bar{x}, N] \geq \delta$$

This contradicts (3.14). The assumption that $\bar{x} \in N$ leads to a similar contradiction. Hence $\Omega[x(t)]$ is connected.

Finally we prove that $\Omega[x(t)]$ is invariant. Suppose $\bar{x}(t)$ is a solution of (3.7) which is not an equilibrium point and which is such that for $t = \bar{t}$,

$$\bar{x}(t) \in \Omega[x(t)]$$

Let t_0 be a point in the domain of $\bar{x}(t)$. Since

$$\bar{x}(\bar{t}) \in \Omega[x(t)]$$

then there exists $\{t_n\}$ with $t_n \to \infty$ such that

$$\lim_n x(t_n) = \bar{x}(\bar{t})$$

Then if $\tau = t_0 - \bar{t}$, it follows from the continuity of the solution as a function of the initial value (Corollary 1.1) that

$$\lim_n x(t_n + \tau) = \bar{x}(\bar{t} + \tau) = \bar{x}(t_0)$$

This completes the proof of Theorem 3.4. $\Box$

Definition A *minimal* set relative to equation (3.7) is a set E which is: (1) nonempty, (2) closed, (3) invariant; and is such that no proper subset of E has these three properties.

Theorem 3.5 *If E is a nonempty invariant compact set, then E contains a minimal set.*

Proof Let $\{E_v\}$ be the collection of nonempty closed invariant subsets of E. Since $E \in \{E_v\}$, the collection is nonempty, and $\{E_v\}$ is partially ordered by inclusion. Also each linearly ordered subset of $\{E_v\}$ has an upper bound, that is, the set E. Hence by Zorn's lemma, there is a maximal linearly ordered subset $\{E_{v_n}\}$ of $\{E_v\}$. Let $\tilde{E} = \cap_n E_{v_n}$. Then $\tilde{E} \neq \phi$ because $\{E_{v_n}\}$ has the finite intersection property (if $\{E_{v_n}^k\}$ is a finite subset of $\{E_{v_n}\}$ then $\cap_k E_{v_n}^k \neq \phi$) and E is compact. Since $\tilde{E}$ is the intersection of sets which are closed and invariant, then $\tilde{E}$ is closed and invariant.

Now suppose $\tilde{E}$ is not minimal. Then there exists a proper subset G of $\tilde{E}$ such that G is nonempty, closed, and invariant. This contradicts the condition that $\{E_{v_n}\}$ is a maximal linearly ordered subset of $\{E_v\}$. This completes the proof of Theorem 3.5. $\Box$

The material in this section lies at the basis of an important contemporary subject: the theory of dynamical systems. To sketch roughly how a dynamical system is defined, we suppose that M is an invariant set of an autonomous system

$$\frac{dx}{dt} = f(x) \tag{3.15}$$

and suppose $\bar{x} \in M$. There exists a solution $x(t, \bar{x})$ of (3.15) such that the orbit of $x(t, \bar{x})$ contains $\bar{x}$. That is, there exists t_0 such that

$$x(t_0, \bar{x}) = \bar{x}$$

Let the transformation T_t be such that

$$T_t: \bar{x} \to x(t_0 + t, \bar{x})$$

Note that T_0 is the identity mapping. Also

$$T_s T_t \bar{x} = T_s [x(t_0 + t), \bar{x}]$$
$$= x(t_0 + t + s, \bar{x})$$
$$= T_{s+t} \bar{x}$$

Thus the collection of mappings $\{T_t\}$ is a 1-parameter group of transformations of M into itself. Such a group of transformations is called a *flow*. It is an example of a dynamical system.

A dynamical system is defined by specifying a set M which has a given structure (very frequently M is a manifold) and a group of transformations T_t where $t \in R$ or R^+ (the nonnegative reals) in the case of a flow. A second kind of dynamical system is defined by considering transformations T_t where t is an integer. Such a dynamical system is called a *cascade*. Various requirements are imposed on the transformations T_t. For example, for a flow, it may be required that $T_t(\bar{x})$ be continuous in $(t, \bar{x})$. For a detailed discussion and further references, see Anasov and Arnold [1988].

The term, "dynamical system," stems from historical considerations and the concept of a dynamical system is not formally related to a mechanical system.

Orbits Near an Equilibrium Point:
The Two-Dimensional Case

In the remainder of this chapter, we restrict outselves to orbits of two-dimensional systems, that is, orbits which are curves in the xy-plane. Since this is a very strong restriction and involves some of the oldest results in qualitative theory, it is natural to ask why bother with such aged specal cases, why not work directly with the n-dimensional case. There are a number of important answers to this question. From the point of view of pure mathematics, the two-dimensional case can be expected to yield more results because in that case the orbit (a one-dimensional entity) is restricted to the plane. (We will realize the depth of this observation when we get to the Poincaré-Bendixson theorem later in this chapter.) Second, there are many important applications of two-dimensional ordinary, differential equations. These include classical applications in physics and electrical engineering (see Andronov and Chaikin [1949]) and more recent applications in biology (see Beuter, Glass, Mackey, and Titcombe [2003]). Also this material plays a role in the analysis of higher-dimensional singularly perturbed systems.

We first describe in detail the behavior of orbits near an isolated equilibrium point of a two-dimensional linear system. This study, originated by Poincaré, is the beginning of the qualitative theory of ordinary differential equations. Part of Poincaré's genius lay in his remarkable ability to perceive the right directions for study, that is, directions in which significant and extensive progress could be made. The study of orbits near equilibrium points is such a direction.

Linear Homogeneous Systems

We study first the two-dimensional linear homogeneous system

$$x' = ax + by$$
$$y' = cx + dy \tag{3.16}$$

where a, b, c, d are real constants. We assume that

$$\det \begin{bmatrix} a & b \\ c & d \end{bmatrix} \neq 0$$

and hence that $(0, 0)$ is the only equilibrium point of (3.16). Let A denote the matrix

$$\begin{bmatrix} a & b \\ c & d \end{bmatrix}$$

If P is a real nonsingular matrix such that $P^{-1}AP = J$ where J is the real canonical form, then a fundamental matrix of (3.16) is Pe^{tJ}. Hence, except for the distortion introduced by multiplying by P, the orbits of (3.16) are the underlying point sets of curves described by linear combinations of the columns of e^{tJ}. We disregard this distortion, that is, we assume that A is in canonical form. Let λ_1, λ_2 be the eigenvalues of A, that is, suppose that λ_1, λ_2 are the roots of the characteristic equation of A:

$$\lambda^2 - (a + d)\lambda + (ad - bc) = 0$$

Case I: λ_1, λ_2 are real, unequal and have the same sign. Then

$$A = \begin{bmatrix} \lambda_1 & 0 \\ 0 & \lambda_2 \end{bmatrix}$$

and

$$e^{tA} = \begin{bmatrix} e^{t\lambda_1} & 0 \\ 0 & e^{t\lambda_2} \end{bmatrix}$$

and an arbitrary solution is

$$\begin{bmatrix} c_1 e^{t\lambda_1} \\ c_2 e^{t\lambda_2} \end{bmatrix}$$

where c_1, c_2 are real constants. Suppose first that $\lambda_1 < \lambda_2 < 0$. Then if $c_1 = 0, c_2 > 0$, the corresponding orbit is the positive y-axis and is directed toward the origin. Similar orbits are obtained if $c_1 = 0, c_2 < 0$ or if $c_1 < 0, c_2 = 0$ or if $c_1 > 0, c_2 = 0$. If $c_1 \neq 0, c_2 \neq 0$

$$\lim_{t \to \infty} \frac{c_2 e^{\lambda_2 t}}{c_1 e^{\lambda_1 t}} = \lim_{t \to \infty} \frac{c_2}{c_1} e^{(\lambda_2 - \lambda_1)t} = \left(\text{sign} \frac{c_2}{c_1} \right) \infty$$

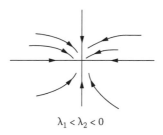

$$\lambda_1 < \lambda_2 < 0$$

Figure 3.1

and

$$\lim_{t \to -\infty} \frac{c_2 e^{\lambda_2 t}}{c_1 e^{\lambda_1 t}} = 0$$

Typical orbits are sketched (with arrows indicating their direction) in Figure 3.1.

Definition   If λ_1, λ_2 are unequal and negative, the equilibrium point $(0, 0)$ is a *stable node*.

If $\lambda_1 > \lambda_2 > 0$, then the typical orbits look the same as the orbits for the case $\lambda_1 < \lambda_2 < 0$ except that they are oppositely directed. See Figure 3.2.

Definition   If $\lambda_1 > \lambda_2 > 0$, the equilibrium point is an *unstable node*.

Case II. $\lambda_1 = \lambda_2 = \lambda$ and the matrix A is

$$A = \begin{bmatrix} \lambda & 0 \\ 0 & \lambda \end{bmatrix}$$

Then an arbitrary solution is

$$\begin{bmatrix} c_1 e^{t\lambda} \\ c_2 e^{t\lambda} \end{bmatrix}$$

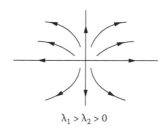

$$\lambda_1 > \lambda_2 > 0$$

Figure 3.2

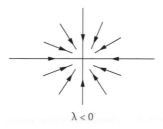

$\lambda < 0$

Figure 3.3

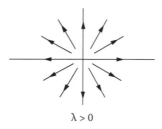

$\lambda > 0$

Figure 3.4

If $c_1 \neq 0$, $c_2 \neq 0$, the corresponding orbit is the intersection of the line, that passes through the origin and has slope c_2/c_1, with the interior of the quadrant which contains the point (c_1, c_2). If $c_1 > 0$ and $c_2 = 0$, the corresponding orbit is the positive x-axis. Similar orbits are obtained if $c_1 < 0$, $c_2 = 0$, or $c_1 = 0$, $c_2 > 0$ or $c_1 = 0$, $c_2 < 0$. If $\lambda < 0$, all the orbits approach $(0, 0)$ as indicated in Figure 3.3. If $\lambda > 0$, the orbits are oppositely directed as indicated in Figure 3.4.

Definition   If $\lambda_1 = \lambda_2 = \lambda$ and the canonical form is diagonal, the equilibrium point is a *stable node* if $\lambda < 0$ and is an *unstable node* if $\lambda > 0$.

Case III. $\lambda_1 = \lambda_2 = \lambda$ and the matrix A is

$$A = \begin{bmatrix} \lambda & 1 \\ 0 & \lambda \end{bmatrix}$$

Then a fundamental matrix is

$$\begin{bmatrix} e^{t\lambda} & te^{t\lambda} \\ 0 & e^{t\lambda} \end{bmatrix}$$

Autonomous Systems

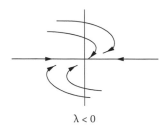

$$\lambda < 0$$

Figure 3.5

and an arbitrary solution is

$$\begin{bmatrix} c_1 e^{t\lambda} + c_2 t e^{t\lambda} \\ c_2 e^{t\lambda} \end{bmatrix}$$

If $c_1 > 0$, $c_2 = 0$ and $c_1 < 0$, $c_2 = 0$, the corresponding orbits are the positive and negative x-axes. If $c_2 \neq 0$ then

$$\lim_{t \to \infty} \frac{c_2 e^{t\lambda}}{c_1 e^{t\lambda} + c_2 t e^{t\lambda}} = \lim_{t \to \infty} \frac{c_2}{c_1 + c_2 t} = 0$$

and

$$\lim_{t \to -\infty} \frac{c_2 e^{t\lambda}}{c_1 e^{t\lambda} + c_2 t e^{t\lambda}} = 0$$

If $c_2 > 0$ [< 0], the orbit is in the upper [lower] half-plane. If $\lambda < 0$, the orbits approach $(0, 0)$ and are sketched in Figure 3.5. If $\lambda > 0$, the orbits are oppositely directed as indicated in Figure 3.6.

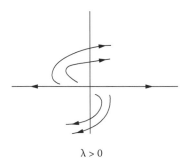

$$\lambda > 0$$

Figure 3.6

Definition   If $\lambda_1 = \lambda_2 = \lambda$ and the canonical form is

$$\begin{bmatrix} \lambda & 1 \\ 0 & \lambda \end{bmatrix}$$

the critical point is a *stable node* if $\lambda < 0$ and is an *unstable node* if $\lambda > 0$.

Case IV. $\lambda_2 < 0 < \lambda_1$. Then

$$A = \begin{bmatrix} \lambda_1 & 0 \\ 0 & \lambda_2 \end{bmatrix}$$

and an arbitrary solution is

$$\begin{bmatrix} c_1 e^{t\lambda_1} \\ c_2 e^{t\lambda_2} \end{bmatrix}$$

where c_1, c_2 are constants. If $c_1 > 0$, $c_2 > 0$, then both components are positive for all t. As $t \to \infty$, then $c_1 e^{t\lambda_1} \to \infty$ and $c_2 e^{t\lambda_2} \to 0$; as $t \to -\infty$, then $c_1 e^{t\lambda_1} \to 0$ and $c_2 e^{t\lambda_2} \to \infty$. The corresponding orbit is as sketched in the first quadrant in Figure 3.7. If c_1, c_2 are both negative or if one is positive and the other negative, then the corresponding orbits are in other quadrants but have similar properties and are also sketched in Figure 3.7. If $\lambda_2 = -\lambda_1$, the orbits are actually branches of hyperbolas. If $\lambda_2 \neq -\lambda_1$, then the orbit is a little more complicated: its equation is:

$$x \left(y^{-\frac{\lambda_1}{\lambda_2}} \right) = c_1 (c_2)^{-\frac{\lambda_1}{\lambda_2}}$$

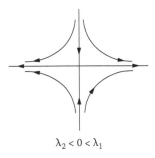

$$\lambda_2 < 0 < \lambda_1$$

Figure 3.7

Definition If $\lambda_2 < 0 < \lambda_1$, the equilibrium point is a *saddle point*.

Case V. λ_1 and λ_2 are complex conjugate numbers, that is, $\lambda_1 = \alpha + i\beta$, $\lambda_2 = \alpha - i\beta$.

First if $\alpha = 0$ then an arbitrary solution is

$$\begin{bmatrix} c_1 \cos \beta t + c_2 \sin \beta t \\ -c_1 \sin \beta t + c_2 \cos \beta t \end{bmatrix} \tag{3.17}$$

where c_1, c_2 are real constants. Let ϕ be such that

$$\cos \phi = \frac{c_1}{\sqrt{c_1^2 + c_2^2}}$$

$$\sin \phi = \frac{c_2}{\sqrt{c_1^2 + c_2^2}}$$

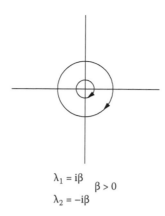

$$\lambda_1 = i\beta$$
$$\beta > 0$$
$$\lambda_2 = -i\beta$$

Figure 3.8

and let

$$K = \sqrt{c_1^2 + c_2^2}$$

Then (3.17) can be written:

$$\begin{bmatrix} K\ \cos(\beta t - \phi) \\ -K\ \sin(\beta t - \phi) \end{bmatrix}$$

and the corresponding orbit is a circle with center $(0, 0)$ and radius K. The orbits are sketched in Figure 3.8.

Definition   If $\lambda_1 = i\beta$, $\lambda_2 = -i\beta$, the equilibrium point is a *center*.

If $\alpha \neq 0$, then an arbitrary solution is

$$\begin{bmatrix} c_1 e^{t\alpha} \cos \beta t + c_2 e^{t\alpha} \sin \beta t \\ -c_1 e^{t\alpha} \sin \beta t + c_2 e^{t\alpha} \cos \beta t \end{bmatrix}$$

or

$$\begin{bmatrix} K e^{t\alpha} \cos(\beta t - \phi) \\ -K e^{t\alpha} \sin(\beta t - \phi) \end{bmatrix}$$

and the corresponding orbit is a spiral which spirals outward [inward] if $\alpha > 0$ [< 0]. The orbits are sketched in Figures 3.9 and 3.10.

Definition   If $\lambda_1 = \alpha + i\beta$, $\lambda_2 = \alpha - i\beta$, where $\alpha \neq 0$, then the equilibrium point is a *spiral point* or *focus*. If $\alpha < 0$ [> 0], the equilibrium point is a *stable* [*unstable*] spiral point.

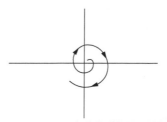

$$\lambda_1 = \alpha + i\beta$$
$$\lambda_2 = \alpha - i\beta \qquad \alpha > 0, \beta > 0$$

Figure 3.9

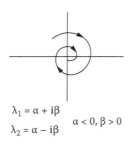

$$\lambda_1 = \alpha + i\beta$$
$$\lambda_2 = \alpha - i\beta \qquad \alpha < 0, \beta > 0$$

Figure 3.10

Stability of an Equilibrium Point

Now that the study of orbits near the equilibrium point $(0, 0)$ of (3.16) is complete, it is rather natural to raise the following question. Suppose we consider the system

$$x' = ax + by + f(x, y)$$
$$y' = cx + dy + g(x, y) \qquad (3.18)$$

where

$$\det \begin{bmatrix} a & b \\ c & d \end{bmatrix} \neq 0$$

and f and g are continuous, satisfy a local Lipschitz condition at each point in some neighborhood of $(0, 0)$, and are higher-order terms, that is,

$$\lim_{|x|+|y| \to 0} \frac{|f(x, y)| + |g(x, y)|}{|x| + |y|} = 0$$

Then $(0, 0)$ is an equilibrium point of (3.18) and the question is whether the behavior of the orbits of (3.18) in a small enough neighborhood of $(0, 0)$ is determined by the linear terms on the right-hand side of (3.18). Later we shall consider this question in a more general, n-dimensional context. For the present, we merely point out a result which is a special case of a theorem to be proved later. To state this result, we introduce the following definition.

Definition   The origin is an *asymptotically stable equilibrium point* of (3.16) [(3.18)] if given $\varepsilon > 0$ there exists $\delta > 0$ such that if $(x(t), y(t))$ is a solution of (3.16) [(3.18)] for which there is a number t_0 with

$$|x(t_0)| + |y(t_0)| < \delta$$

then it is true that the solution $(x(t), y(t))$ is defined for all $t \geq t_0$, and

$$\mathop{\text{lub}}_{t \geq t_0} |x(t)| + |y(t)| < \varepsilon$$

and

$$\lim_{t \to \infty} x(t), y(t) = (0, 0)$$

(If $(0, 0)$ is a stable node or a stable spiral point of (3.16), then $(0, 0)$ is clearly an asymptotically stable equilibrium point of (3.16).)

Later (Chapter 4) we will prove a more general version of the following theorem.

Theorem 3.6   *If* $(0, 0)$ *is an asymptotically stable equilibrium point of* (3.16), *then* $(0, 0)$ *is an asymptotically stable equilibrium point of* (3.18).

This theorem can be paraphrased roughly as: "If the behavior of the orbits of (3.16) near $(0, 0)$ is not seriously affected by small changes, then the orbits near $(0, 0)$ of (3.16) and (3.18) are about the same." But we are left then with the question of what happens if the orbits are seriously affected by small changes. For example, if $(0, 0)$ is a center of (3.16) (and hence is not asymptotically stable), then a small change can seriously affect the orbits near $(0, 0)$. It is easy to show that if $(0, 0)$ is a center of (3.16), the addition of higher-order terms (so that we have equation (3.18)) may make all the orbits move away from $(0, 0)$ with increasing t or may make all the orbits move toward $(0, 0)$ with increasing t. More precisely, motion with increasing t along all orbits is away from $(0, 0)$ or motion with increasing t along all orbits is toward $(0, 0)$.

For example, consider the system

$$x' = y + x^3$$
$$y' = -x + y^3$$

Let

$$r^2 = x^2 + y^2$$

Then

$$rr' = xx' + yy'$$
$$= xy + x^4 - xy + y^4$$
$$r' = \frac{x^4 + y^4}{r}$$

Thus for each solution other than the equilibrium point, r' is always positive. So each orbit moves away from $(0, 0)$ as t increases. Thus $(0, 0)$ is clearly not asymptotically stable.

On the other hand, for the system

$$x' = y - x^3$$
$$y' = -x - y^3$$

we have

$$r' = \frac{-x^4 - y^4}{r}$$

Thus r' is always negative, each orbit moves toward $(0, 0)$ as t increases, and $(0, 0)$ is asymptotically stable. One might say that the behavior of the orbits of the linear system

$$x' = y$$
$$y' = x$$

is not decisive and is strongly influenced by small or higher-order terms.

These examples show that, with the addition of certain nonlinear terms, the orbits all move away from $(0, 0)$ or all move toward $(0, 0)$. The behavior of the orbits may be more complicated, as the following example shows.

$$x' = y + xr^2 \sin \frac{\pi}{r}$$

$$y' = -x + yr^2 \sin \frac{\pi}{r} \tag{3.19}$$

where $r^2 = x^2 + y^2$. Then

$$rr' = xx' + yy'$$
$$= xy + x^2 r^2 \sin \frac{\pi}{r} - xy + y^2 r^2 \sin \frac{\pi}{r} = r^4 \sin \frac{\pi}{r}$$

and

$$r' = r^3 \sin \frac{\pi}{r}$$

Letting

$$\theta = \arctan \frac{y}{x}$$

we obtain

$$\theta' = \frac{xy' - yx'}{x^2 + y^2}$$

$$= \frac{-x^2 + xyr^2 \sin \frac{\pi}{r} - y^2 - xyr^2 \sin \frac{\pi}{r}}{x^2 + y^2}$$

$$= \frac{-(x^2 + y^2)}{x^2 + y^2}$$

$$= -1$$

Thus

$$r = \frac{1}{n}$$

$$\theta = -t$$

describes an orbit which is a simple closed curve, that is, a circle with center 0 and radius $1/n$. If $r > 1$, then $1/r < 1$ and hence $r' > 0$. Thus any orbit which passes through a point in the xy-plane that is outside the circle $r = 1$ moves away from the circle $r = 1$ with increasing t. Since $\theta' = -1$, the orbit spirals outward in the clockwise direction. Also, if

$$\frac{1}{2q} < r < \frac{1}{2q - 1} \qquad (q = 1, 2, 3, \dots)$$

then

$$2q > \frac{1}{r} > 2q - 1$$

and hence

$$r' = r^3 \sin \frac{\pi}{r} < 0$$

If

$$\frac{1}{2q + 1} < r < \frac{1}{2q} \qquad (q = 1, 2, 3, \dots ,)$$

then

$$2q + 1 > \frac{1}{r} > 2q$$

and hence

$$r' > 0$$

Thus the orbits of (3.19) appear as sketched in Figure 3.11.

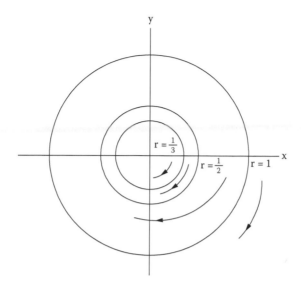

Figure 3.11

Orbits Near an Equilibrium Point of a Nonlinear System

Now we consider a somewhat more general question concerning orbits near an equilibrium point. Suppose that $X(x, y)$ and $Y(x, y)$ are power series in x and y which converge in a neighborhood N of $(0, 0)$ and suppose that $(0, 0)$ is an isolated equilibrium point of the system

$$x' = X(x, y)$$
$$y' = Y(x, y) \tag{E_2}$$

We say that $(0, 0)$ is a *general equilibrium point* of (E_2). Then what do the orbits of (E_2) in the neighborhood N look like? Theorem 3.6 gives a partial answer to this question because it says that under certain conditions, the linear terms in X and Y determine the behavior of the orbits. But considerably further analysis is needed. For example, so far we have no information about the behavior of the orbits if X or Y is a series which starts with terms of degree greater than one.

It might be thought that the answer to this question would be very complicated. There is, however, a remarkably simple answer to this question which can be described roughly as follows. If $(0, 0)$ is an isolated equilibrium point of (E_2), then either $(0, 0)$ is a spiral, that is, the orbits move toward $(0, 0)$ as t increases [decreases], or a center (there is a neighborhood N of $(0, 0)$ such that every orbit which has a nonempty intersection with N is a simple closed curve) or a neighborhood of $(0, 0)$ can be divided into a finite number of "sectors" of the types illustrated in Figure 3.12.

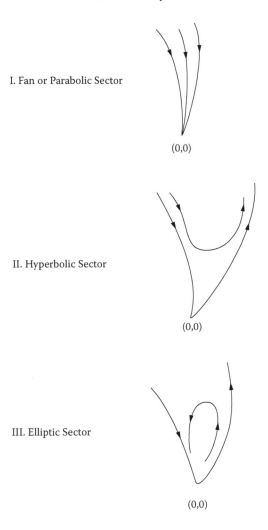

I. Fan or Parabolic Sector

(0,0)

II. Hyperbolic Sector

(0,0)

III. Elliptic Sector

(0,0)

Figure 3.12

In each of these sketches, the arrows indicate the direction of increasing or decreasing time. Note that the only essentially new configuration that arises when nonlinear equilibrium points are studied is the elliptic sector.

A. The general case

We consider the system

$$\frac{dx}{dt} = X(x, y)$$
$$\frac{dy}{dt} = Y(x, y) \tag{3.20}$$

where $(0, 0)$ is an isolated equilibrium point of (3.20), that is,

$$X(0, 0) = 0$$
$$Y(0, 0) = 0$$

and there is a neighborhood of $(0, 0)$ such that no other point in the neighborhood is an equilibrium point of (3.20), that is, $(0, 0)$ is an isolated equilibrium point of (3.20).

We follow the analysis of Lefschetz [1962, Chapter X] and prove the results under hypotheses which include as a special case the condition that X and Y are power series in x and y (are analytic in x and y). An essential step in the procedure is determining the curves described by the equations

$$\frac{dr}{dt} = 0$$
$$\frac{d\theta}{dt} = 0$$

or, equivalently,

$$r\frac{dr}{dt} = xX(x, y) + yY(x, y) = 0 \tag{3.21}$$

and

$$r^2\frac{d\theta}{dt} = xY(x, y) - yX(x, y) = 0 \tag{3.22}$$

Now if X and Y are analytic, then by the Weierstrass preparation theorem (see Appendix), there exists an integer n such that

$$xX(x, y) + yY = [y^n + a_{n-1}(x)y^{n-1} + \cdots + a_1(x)y + a_0(x)]E(x, y)$$

where $a_j(x)$ is an analytic function such that $a_j(0) = 0$ for $j = 0, 1, \ldots, n - 1$ and $E(x, y)$ is analytic in x and y and $E(0, 0) \neq 0$. Thus the problem of finding the curves described by equation (3.21) is reduced to the problem of solving the equation

$$y^n + a_{n-1}(x)y^{n-1} + \cdots + a_1(x)y + a_0(x) = 0$$

for y as a function of x. (To solve this problem if each $a_j(x)$ is a polynomial, we can use Puiseux series. See Siegel [1969, Chapter 2].) An exactly similar analysis can be made of equation (3.22).

Rather than assume that X and Y are analytic we will only assume for our purposes, following Lefschetz, that the following hypotheses are satisfied.
(a) X and Y have continuous first derivatives in a neighborhood N bounded by a circle C with center $(0, 0)$ and radius $r_0 > 0$. (Thus the Existence Theorem 1.1 can be applied to (3.20).)
(b) Equation (3.21)

$$r\frac{dr}{dt} = xX(x, y) + yY(x, y) = 0 \tag{3.21}$$

describes at most a finite set of curves (we will call them branch curves)

$$y = f_j(x) \qquad (j = 1, \ldots, k)$$

Each f_j has domain $[0, d_j]$ or $[-d_j, 0]$ where $d_j > 0$. Also each f_j is continuously differentiable on its domain and hence each branch curve has a continuously turning tangent at its points. (We can be assured that branch curves can be described this way by rotating the coordinate axes if necessary so that no branch curve has the y-axis as its tangent at $(0, 0)$.) Also the functions $f_j(x)$ are distinct in a neighborhood of $(0, 0)$. That is, there exists $d > 0$ such that if $0 < |x| < d$ and if f_p and f_q are defined at x and $p \neq q$, then

$$f_p(x) \neq f_q(x).$$

(c) The same kind of hypotheses placed on equation (3.21) are also imposed on equation (3.22)

$$r^2 \frac{d\theta}{dt} = xY(x, y) - yX(x, y) = 0 \tag{3.22}$$

Note that the branch curves of (3.21) and (3.22) are also the branch curves of

$$\frac{dr}{dt} = 0$$

and

$$\frac{d\theta}{dt} = 0$$

respectively.

Remark.   Consider a neighborhood $\mathcal{N}$ of $(0, 0)$ where $\mathcal{N}$ has as its boundary the circle C with center $(0, 0)$ and radius r_0.
1. If K is a branch curve described by (3.21) and L is a branch curve described by (3.22), then if r_0 is sufficiently small, the branch curves K and L do not intersect except at $(0, 0)$. For suppose there exists $(x_0, y_0) \neq (0, 0)$ such that

$$r \frac{dr}{dt} = x_0 X(x_0, y_0) + y_0 Y(x_0, y_0) = 0$$

and

$$r^2 \frac{d\theta}{dt} = -y_0 X(x_0, y_0) + x_0 Y(x_0, y_0) = 0$$

Since

$$\det \begin{bmatrix} x_0 & y_0 \\ -y_0 & x_0 \end{bmatrix} = x_0^2 + y_0^2$$

then

$$X(x_0, y_0) = Y(x_0, y_0) = 0$$

This contradicts the hypothesis that $(0, 0)$ is an isolated equilibrium point.

2. If r_0 is sufficiently small, each orbit which intersects the branch curve K crosses K in the same direction, that is, $\frac{d\theta}{dt}$ has the same sign at each point of K. For if $\frac{d\theta}{dt}$ changes sign on K, then by continuity there exists a point of K in $N - \{(0, 0)\}$ where $\frac{d\theta}{dt} = 0$. This contradicts Remark 1.

3. Given $\varepsilon > 0$, then if r_0 is sufficiently small and if an orbit intersects K in N at a point $P = (\bar{x}, \bar{y})$ and if m_1 and m are the slopes of the orbit and K, respective at the point P, then

$$m\, m_1 = -1 + \eta$$

where $|\eta| < \varepsilon$. That is, the orbit and K are almost orthogonal at P. To show this, we note first that with a proper choice of coordinates the slope at $(0, 0)$ of the curve K is finite and nonzero. Hence the branch curve K can be described by

$$y = f(x) = mx + h(x)$$

where $m \neq 0$ and $h(x)$ is differentiable and

$$\lim_{x \to 0} \frac{h(x)}{x} = 0$$

Also the slope of the tangent to the orbit at P is the same as the slope of the tangent at P of the circle with center $(0, 0)$ which passes through P because $\frac{dr}{dt} = 0$ at P. That is,

$$m_1 = -\frac{x}{y} = \frac{-x}{mx + h(x)}$$

$$= \frac{-1}{m + \frac{h(x)}{x}}$$

$$m_1\, m = \frac{-1}{1 + \frac{1}{m}\frac{h(x)}{x}}$$

4. If r_0 is sufficiently small, the branch curves of (3.21) do not intersect one another except at $(0, 0)$ and each branch curve is a single connected arc in N. First, if $|x|$ is small enough the arcs of the branch curves which are described by

$$y = f_j(x) \qquad (j = 1, \ldots, k)$$

do not intersect because the f_j's are distinct. The fact that each branch curve is a single connected arc in N follows from the fact (noted in the proof of Remark 3) that K can be described by

$$y = mx + h(x)$$

where $m \neq 0$ and $h(x)$ is higher order. $\quad\square$

To obtain the portrait, we start with the set of branch curves described in (b).

From Remark 1, if C is a circle with center $(0, 0)$ and sufficiently small radius r_0, the branch curves described by equation (3.21) divide C and its interior into triangular "sectors" as sketched in Figure 3.13. (In what follows, we sometimes denote the

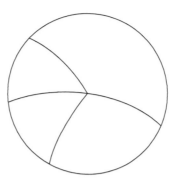

Figure 3.13

origin $(0, 0)$ by the letter O.) Each sector is bounded by an arc of circle C and two curves on which $\frac{dr}{dt} = 0$. Strictly speaking, these "sectors" are not the usual sectors of geometry because the curves described by $\frac{dr}{dt} = 0$ are not line segments. However if r_0 is sufficiently small, these curves are almost straight line segments. Also as orbits cross these curves, they are nearly perpendicular to the curves. (See Remark 3 above.)

It should be emphasized at the outset that the sectors just described are entirely different from the parabolic, hyperbolic, and elliptic sectors pictured earlier. Our procedure is to study these sectors in order to determine the existence and position of the parabolic, hyperbolic, and elliptic sectors. First we obtain three theorems which facilitate this study.

Theorem 3.7   *Let* OAB *denote a sector in which* $\frac{dr}{dt} < 0$. *(See Figure 3.14.) Let* $(\bar{x}(t), \bar{y}(t))$ *denote a solution of* (3.20) *such that for all* $t \geq t_0$, *a fixed value,* $(\bar{x}(t), \bar{y}(t))$ *is a point in* OAB. *Then if* $\bar{r}(t)$ *denotes*

$$\left\{ [\bar{x}(t)]^2 + [\bar{y}(t)]^2 \right\}^{\frac{1}{2}}$$

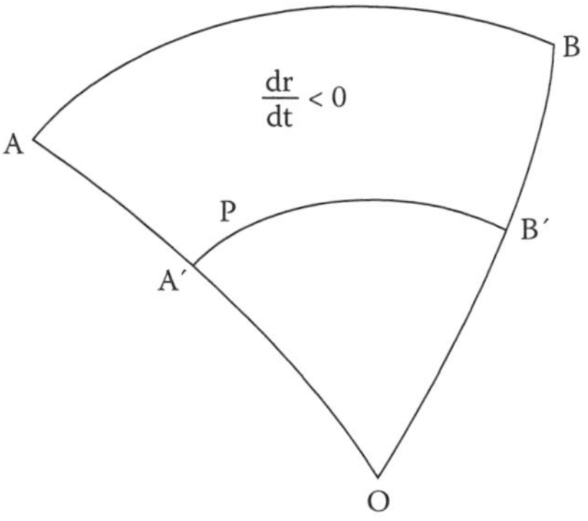

Figure 3.14

we have

$$\lim_{t \to \infty} \bar{r}(t) = 0$$

Proof   Since $\frac{dr}{dt} < 0$ in sector OAB, then

$$\lim_{t \to \infty} \bar{r}(t) = \operatorname*{glb}_{t \geq t_0} \bar{r}(t)$$

Suppose

$$\operatorname*{glb}_{t \geq t_0} \bar{r}(t) = r_0 > 0$$

Let $A'\ B'$ be the arc in OAB of the circle with center 0 and radius r_0. Since

$$\lim_{t \to \infty} \bar{r}(t) = r_0$$

then there exists a point $P \in A'B'$ such that P is an ω-limit point of solution $(\bar{x}(t), \bar{y}(t))$. We assume first that P is a point in the interior of OAB. (See Figure 3.14.) Denote the polar coordinates of P as $r(P) = r_0$ and $\theta(P)$. Since P is an ω-limit point of $(\bar{x}(t), \bar{y}(t))$ then there exists sequence $\{t_n\}$ such that $t_n \to \infty$ and

$$\bar{r}(t_n) \to r_0$$
$$\bar{\theta}(t_n) \to \theta(P)$$

where

$$\bar{\theta}(t_n) = \arctan \left[\frac{\bar{y}(t_n)}{\bar{x}(t_n)} \right]$$

We may assume that the sequence $\bar{\theta}(t_n)$ approaches $\theta(P)$ monotonically (i.e., the sequence $\{|\bar{\theta}(t_n) - \theta(P)|\}$ approaches 0 monotonically). Since P is a point in the interior of OAB, then at P

$$\frac{dr}{dt} = -m < 0$$

Hence if

$$\mathcal{M} = \{(r, \theta)/|r - r_0| < \delta, \ |\theta - \theta(P)| < \delta\}$$

where δ is sufficiently small, then in $\mathcal{M}$, we have

$$\frac{dr}{dt} < -\frac{m}{2}$$

It follows that if n and q are large enough,

$$|r(t_{n+q}) - r(t_n)| = \left| \left[\frac{dr}{dt} \right] [t_{n+q} - t_n] \right| \geq \frac{m}{2} |t_{n+q} - t_n|$$

Since $t_n \to \infty$, then $\{r(t_n)\}$ is not a Cauchy sequence and therefore cannot converge to r_0 which is a contradiction.

If $P = A'$, then at P, we have $\frac{dr}{dt} = 0$ and hence $\frac{d\theta}{dt} \neq 0$. (Otherwise P would be an equilibrium point.) Suppose that at P,

$$\frac{d\theta}{dt} = m > 0$$

Then in a sufficiently small neighborhood $\mathcal{M}$ of P

$$\frac{d\theta}{dt} \geq \frac{m}{2}$$

We may assume that $\theta(t_n) \to \theta(P)$ monotonically. Then if n and q are large enough

$$|\theta(t_{n+q}) - \theta(t_n)| = \left| \left[\frac{d\theta}{dt} \right] [t_{n+q} - t_n] \right|$$

$$\geq \frac{m}{2} |t_{n+q} - t_n|$$

Since $t_n \to \infty$, this contradicts the convergence of the sequence $\{\theta(t_n)\}$. A similar argument holds if $P = B'$. $\qquad\square$

It will be often convenient, if $(x(t), y(t))$ is a solution of (3.20), to describe the solution in terms of polar coordinates, that is, to use

$$r(t) = \left\{ [x(t)]^2 + [y(t)]^2 \right\}^{1/2}$$

and

$$\theta(t) = \arctan \frac{y(t)}{x(t)}$$

where

$$r \frac{dr}{dt} = x X(x, y) + y Y(x, y) \tag{3.23}$$

and

$$r^2 \frac{d\theta}{dt} = x Y(x, y) - y X(x, y) \tag{3.24}$$

We shall refer to $(r(t), \theta(t))$ as a solution of (3.20).

Theorem 3.8 *Let γ be the orbit of a solution $(r(t), \theta(t))$ of (3.20) such that*

$$\lim_{t \to \infty} r(t) = 0 \qquad \left[\lim_{t \to -\infty} r(t) = 0 \right]$$

Then there exists

$$\lim_{t\to\infty} \theta(t) \qquad \left[\lim_{t\to-\infty} \theta(t)\right]$$

where the limit is a finite value or $|\theta(t)|$ *increases without bound as* $t \to \infty[t \to -\infty]$.

Proof   It is sufficient to prove that if $\theta(t)$ remains bounded, then $\theta(t)$ cannot have two limiting values. Suppose $\theta(t)$ has two such values θ_1 and θ_2 with $\theta_1 < \theta_2$. That is, we suppose that there exist two sequences $\{t_{1q}\}$ and $\{t_{2q}\}$ such that

$$\lim_{q\to\infty} t_{1q} = \infty$$

$$\lim_{q\to\infty} t_{2q} = \infty$$

and

$$\lim_{q\to\infty} \theta(t_{1q}) = \theta_1$$

$$\lim_{q\to\infty} \theta(t_{2q}) = \theta_2$$

We may assume that for all q,

$$t_{1q} < t_{2q} < t_{1q+1}$$

If $r^2 \frac{d\theta}{dt} \geq 0$ in the sector S bounded by $\theta = \theta_1$, $r = r_0$, $\theta = \theta_2$, then we have an immediate contradiction because if the solution gets close to θ_2, it cannot get back near θ_1 since $\frac{d\theta}{dt} \geq 0$. A similar contradiction occurs if $r^2 \frac{d\theta}{dt} \leq 0$ in S.

If there exists a branch curve of

$$r^2 \frac{d\theta}{dt} = 0$$

in S, and if $r^2 \frac{d\theta}{dt}$ changes sign at that branch curve, we obtain a similar contradiction.

$\Box$

It is convenient to introduce the following definition.

Definition   If $(r(t), \theta(t))$ is a solution such that

$$\lim_{t\to\infty} r(t) = 0 \qquad \left[\lim_{t\to-\infty} r(t) = 0\right]$$

and

$$\lim_{t\to\infty} \theta(t) = \theta_0 \qquad \left[\lim_{t\to-\infty} \theta(t) = \theta_0\right]$$

where θ_0 is a finite value, then the solution $(r(t), \theta(t))$ approaches $(0, 0)$ *in the direction* θ_0 *as* $t \to \infty[t \to -\infty]$.

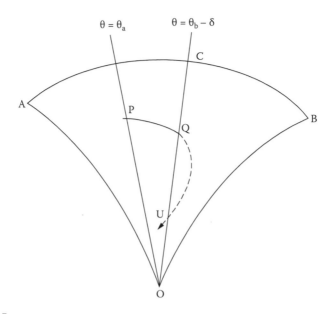

Figure 3.15

Theorem 3.9   *Suppose that solution* $(r(t), \theta(t))$ *approaches* $(0, 0)$ *in the direction* θ_0 *as* $t \rightarrow \infty[t \rightarrow -\infty]$. *Then* $\theta = \theta_0$ *is the tangent at* $(0, 0)$ *of a branch curve of*

$$\frac{d\theta}{dt} = 0$$

Proof   Suppose $\theta = \theta_0$ is not tangent to a branch curve of

$$\frac{d\theta}{dt} = 0$$

Then if r is small enough, $\frac{d\theta}{dt}$ is nonzero and has the same sign at each point of the line segment $\theta = \theta_0$. Suppose $\frac{d\theta}{dt} < 0$ at such a point P. (See Figure 3.15.) Let $(r(t), \theta(t))$ denote the solution whose orbit passes through the point P.

Let OA and OB be adjacent branch curves of $\frac{d\theta}{dt} = 0$ such that OP is contained in the sector OAB. (See Figure 3.15.) Thus $\frac{d\theta}{dt} < 0$ in OAB. Then there exists $\delta > 0$ such that the segment OC of the line segment $\theta = \theta_0 - \delta$ is contained in OAB. Since $\frac{d\theta}{dt} < 0$ in OAB then the orbit of $(r(t), \theta(t))$ intersects OC, say at a point Q. Thus the orbit exits OPQ. But since $(r(t), \theta(t))$ approaches $(0, 0)$ in the direction θ_0 then the orbit must reenter the sector OPQ as shown in Figure 3.15. But then $\frac{d\theta}{dt} > 0$ at some point U in the sector OAB. Thus we have a contradicton.   ☐

   Now we are ready to look at how orbits behave in a sector such as shown in Figure 3.16. We consider first a sector in which $\frac{dr}{dt} < 0$. By Remark 2, the orbits which intersect a side of the sector all cross the side in the same direction. We look first at the case shown in Figure 3.16 where any orbit which intersects OA or OB

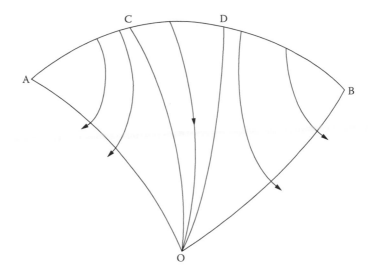

Figure 3.16

leaves the sector. Following in the negative direction the orbits which intersect the side OA (i.e., following in the direction as t decreases), as indicated by the dashed curves, we conclude that the dashed curves intersect the arc AB as shown. It is readily proved (see Exercise 5) that we define in this way a $1-1$ continuous mapping M of the side $[A, 0)$ onto $[A, C)$, the arc shown in Figure 3.16. Similarly there is a point D on arc AB such that $[B, 0)$ is mapped onto $[B, D)$. The orbits which intersect the arc CD must, by Theorem 3.7, approach O. If $C \neq D$, this set of orbits which intersect CD is called a *fan* or *parabolic sector*. If $C = D$, then the orbit which passes through C is called a *separatrix*. (The separatrix is characterized by the fact that neighboring orbits have very different behavior from that of the separatrix itself.)

The structure of the fan OCD in Figure 3.16 is described more accurately in light of Theorem 3.9. First by Theorem 3.8 the solutions in Figure 3.16 whose orbits contain C and D respectively approach O in directions, say, θ_1 and θ_2. If $\theta_1 = \theta_2 = \theta_0$, then all the solutions with orbits in the fan OCD must approach O in the direction θ_0. If $\theta_1 \neq \theta_2$, say $\theta_1 > \theta_2$, and if there is no line $\theta = \bar{\theta}$ with $\theta_1 > \bar{\theta} > \theta_2$ such that $\theta = \bar{\theta}$ is tangent at the origin to a branch curve of $\frac{d\theta}{dt} = 0$, then any solution with orbit in the fan OCD must, by Theorem 3.9, approach O in the direction θ_1 or θ_2. In Figure 3.17, we show the case in which $\frac{d\theta}{dt} < 0$ in OCD. If there is a line $\theta = \bar{\theta}$ with $\theta_1 > \bar{\theta} > \theta_2$ such that $\bar{\theta}$ is tangent at the origin to a branch curve of $\frac{d\theta}{dt}$, then one possible portrait of the fan would be as shown in Figure 3.18. (In this case, $\frac{d\theta}{dt}$ would change sign at the branch curve.) Clearly, if there is more than one such tangent line $\theta = \bar{\theta}$, the portrait of the fan will be more complicated. The point to be emphasized is that from fundamental assumption (c) and Theorem 3.9, there is only a finite set of directions of approach to O of the solutions.

Next consider the case in which every orbit which intersects OA or OB enters the sector (Figure 3.19). Since $\frac{dr}{dt} < 0$ on the arc AB, then no orbit can leave the sector;

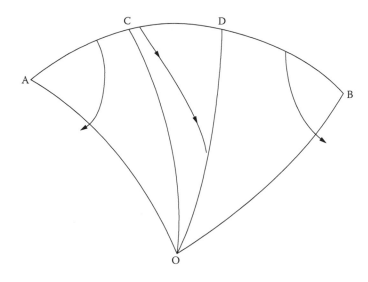

Figure 3.17

hence by Theorem 3.7, every orbit in the sector approaches O in one of a finite set of directions of approach. By Theorem 3.9, each such direction of approach, say θ_0, is tangent to a branch curve of $\frac{d\theta}{dt} = 0$. Since $\frac{d\theta}{dt} < 0$ on OA and $\frac{d\theta}{dt} > 0$ on OB, there is at least one such θ_0. In Figure 3.19, the case in which there is exactly one such θ_0 is illustrated.

Now we consider the case in which orbits cross OA and OB in the same direction, say from left to right, as shown in Figure 3.20. We consider the orbit in Figure 3.20

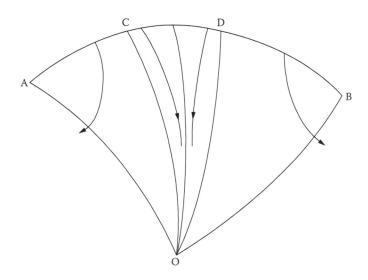

Figure 3.18

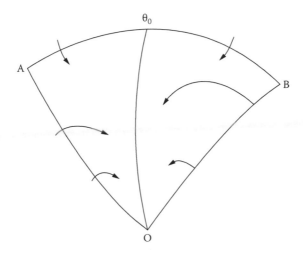

Figure 3.19

which passes through the point *A*. If, as shown in Figure 3.20, this orbit approaches *O*, then every orbit which crosses (*O*, *A*] must, by Theorem 3.7, approach *O*. Next we follow backward the orbits which intersect (*O*, *B*] (sketched with dashed lines) and conclude as in the case of Figure 3.16 that there is a point *D* on arc *AB* such that (*OB*] is mapped onto (*D*, *B*]. Then the orbits which pass through the arc [*AD*] constitute a fan. Now suppose the orbit which passes through the point *A* intersects *OB* and suppose this is true no matter how small the radius of the original neighborhood $\mathcal{N}$. Then we have the situation shown in Figure 3.21. (Note that this can occur only if $\frac{d\theta}{dt} \neq 0$ in the sector.)

The only remaining cases of $\frac{dr}{dt} < 0$ are the mirror images of Figures 3.20 and 3.21 in which the crossings of *OA* and *OB* are from right to left.

To obtain pictures of the orbits for a sector in which $\frac{dr}{dt} > 0$, it is only necessary to perform simple transformations of the results already obtained. (See Exercise 6.)

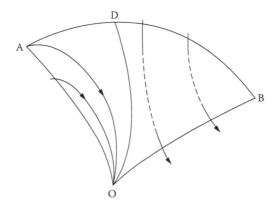

Figure 3.20

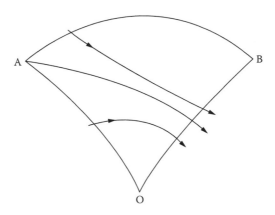

Figure 3.21

We need still to consider the case in which there is just one branch curve *OA* for

$$\frac{dr}{dt} = 0$$

Then $\frac{dr}{dt}$ has the same sign at all points of the complement of *OA* because the complement is a connected set. If there is just one branch curve *OB* of

$$\frac{d\theta}{dt} = 0,$$

then $\frac{d\theta}{dt}$ has the same sign at all points of the complement of *OB*. For definiteness, assume $\frac{dr}{dt} < 0$ on the complement of *OA* and $\frac{d\theta}{dt} < 0$ on the complement of *OB*. Then the orbits move toward 0 as shown in Figure 3.22. If there is more than one branch curve of

$$\frac{d\theta}{dt} = 0$$

the situation tends to be more complicated as in, for example, Figure 3.23.

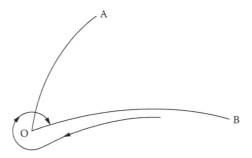

Figure 3.22

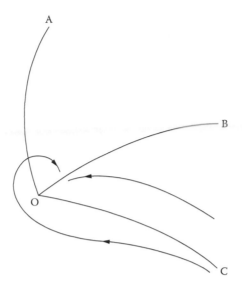

Figure 3.23

If there are no branch curves for

$$\frac{dr}{dt} = 0$$

that is, $\frac{dr}{dt} < 0$ for all $r \neq 0$, then each orbit approaches O. Just as in the preceding case, how the various orbits approach O depends on the existence of branch curves of $\frac{d\theta}{dt}$.

If we put together the sectors we have analyzed (they are sectors in which $\frac{dr}{dt} > 0$ or $\frac{dr}{dt} < 0$ and they are bounded by branch curves of $\frac{dr}{dt} = 0$) we obtain the elliptic, hyperbolic, and parabolic sectors sketched in Figure 3.12. (These sectors are bounded by orbits.) For example, if we match the sector in Figure 3.16 to the left of the same sector only with $\frac{dr}{dt} > 0$, we obtain two parabolic sectors and a hyperbolic sector. (See Figure 3.24.)

This completes our discussion of the general case. It should be noted again that the only novel behavior that occurs in the nonlinear system is the elliptic sector. (In the section that follows, we will describe examples of elliptic sectors.)

B. The case in which $X(x, y)$ and $Y(x, y)$ have lowest order terms of the same degree

The portrait we have given of the neighborhood of a critical point is informative and elegant, but it has the drawback of not being very constructive. In order to apply this theory, it is necessary to determine the branch curves of

$$\frac{dr}{dt} = 0$$

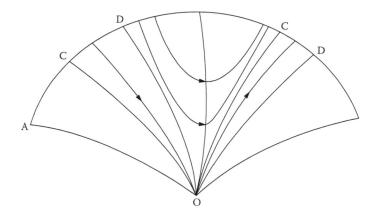

Figure 3.24

and

$$\frac{d\theta}{dt} = 0$$

This is, in general, a serious problem, and we turn now to a constructive approach for solving it.

We impose the same hypotheses (a), (b) and (c) used in Section A, but in addition we assume that the lowest order terms of $X(x, y)$ and $Y(x, y)$ are nonlinear and are of the same degree.

First we introduce a notion (discussed by Nemytskii and Stepanov [1960, pp. 89 ff]) which plays an important role in this approach. Let $(\bar{x}, \bar{y}) = (\bar{r}, \bar{\theta})$ be a point different from $(0, 0)$ and let $F(\bar{r}, \bar{\theta})$ be the vector field defined by the vector $(X(\bar{x}, \bar{y}), Y(\bar{x}, \bar{y}))$ with initial point at $(\bar{r}, \bar{\theta})$. See Figure 3.25. Let the ray

$$\mathcal{R} = \{(r, \theta)|r > 0,\ \theta = \bar{\theta}\}$$

have positive direction away from the origin and let $n(\bar{r}, \bar{\theta})$ denote the normal to $\mathcal{R}$ with positive direction in the direction of increasing θ. Let R denote the unit vector on $\mathcal{R}$ in the positive direction and starting at $\bar{r}$, and let N be the unit vector on $n(\bar{r}, \bar{\theta})$ in the positive direction and starting at $(\bar{r}, \bar{\theta})$. Then

$$F(\bar{r}, \bar{\theta}) = a\,R + b\,N$$

where

$$a = \frac{dr}{dt}(\bar{r}, \bar{\theta})$$

and

$$b = \bar{r}\,\frac{d\theta}{dt}(\bar{r}, \bar{\theta})$$

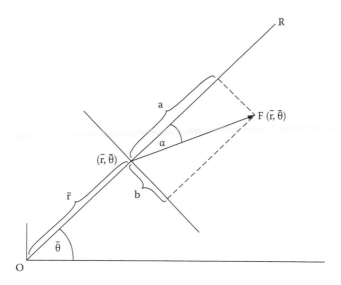

Figure 3.25

(See Exercise 7.) Finally, let $A(\bar{r}, \bar{\theta})$ denote the tangent of the angle α from R to $F(\bar{r}, \bar{\theta})$ as shown in Figure 3.25. Then

$$A(\bar{r}, \bar{\theta}) = \frac{b}{a} = \frac{\bar{r}\frac{d\theta}{dt}}{\frac{dr}{dt}}$$

Definition The direction $\bar{\theta}$ of the ray

$$\{(r, \theta)/r > 0, \ \theta = \bar{\theta}\}$$

is a *critical direction* of the vector field $F(\bar{r}, \bar{\theta})$ if there exists a sequence of points $\{(r_m, \theta_m)\}$ such that

$$\lim_{m \to \infty} r_m = 0 \tag{3.25}$$

$$\lim_{m \to \infty} \theta_m = \bar{\theta} \tag{3.26}$$

$$\lim_{m \to \infty} A(r_m, \theta_m) = 0 \tag{3.27}$$

Note that in its definition a critical direction is a property only of the vector field defined by (E_2). For example, the points (r_m, θ_m) may be in different orbits of (E_2). (This ocurs in showing that in Figure 3.7, $\theta = 0$ is a critical direction.) However we have the following useful relation between solutions and critical direction.

Theorem 3.10 *Suppose that $(r(t), \theta(t))$ is a solution of (E_2) which approaches $(0, 0)$ in the direction θ_0 as $t \to \infty$. Then θ_0 is a critical direction.*

Proof   By Theorem 3.9, $\theta = \theta_0$ is the tangent at O of a branch curve of

$$\frac{d\theta}{dt} = 0$$

Let $\{(r_m, \theta_m)\}$ be a sequence of points on the branch curve such that

$$\lim_{m \to \infty} r_m = 0 \tag{3.25}$$

$$\lim_{m \to \infty} \theta_m = \theta_0 \tag{3.28}$$

Since each point (r_m, θ_m) is on the branch curve, then for each m,

$$r_m^2 \frac{d\theta}{dt}(r_m, \theta_m) = 0$$

Hence

$$A(r_m, \theta_m) = 0 \tag{3.29}$$

Thus θ_0 is a critical direction.   ⬚

We turn now to some simple computational methods for determining the existence and location of critical directions. For this we will require that (E_2) satisfy the basic hypotheses (a), (b), (c), and we will also assume that in equation (E_2)

$$X(x, y) = X_m(x, y) + R_{m+1}(x, y) \tag{3.30}$$
$$Y(x, y) = Y_m(x, y) + S_{m+1}(x, y) \tag{3.31}$$

where

$$X_m(x, y) = a_0 x^m + a_1 x^{m-1} y + \cdots + a_m y^m$$
$$Y_m(x, y) = b_0 x^m + b_1 x^{m-1} y + \cdots + b_m y^m$$

That is, X_m and Y_m are homogeneous polynomials of degree m in x and y, where $m \geq 1$. Our major interest will be the case $m > 1$. However we will use the case $m = 1$ later in studying the Bendixson theory. Also we require that each of the polynomials

$$x X_m + y Y_m$$

and

$$x Y_m - y X_m$$

be not identically zero. The functions R_{m+1} and S_{m+1} are remainder terms such that

$$\lim_{r \to 0} \frac{R_{m+1}(x, y)}{r^m} = 0 \tag{3.32}$$

and

$$\lim_{r \to 0} \frac{S_{m+1}(x, y)}{r^m} = 0 \tag{3.33}$$

Since

$$x = r \cos \theta$$
$$y = r \sin \theta$$

then

$$X_m = r^m[a_0 \cos^m \theta + a_1(\cos^{m-1} \theta)(\sin \theta) + \cdots + a_m \sin^m \theta]$$
$$Y_m = r^m[b_0 \cos^m \theta + b_1(\cos^{m-1} \theta)(\sin \theta) + \cdots + b_m \sin^m \theta]$$

Hence

$$\begin{aligned}
\frac{dr}{dt} &= \frac{1}{r}(X_m + R_{m+1})x + \frac{1}{r}(Y_m + S_{m+1})y \\
&= r^m[a_0 \cos^m \theta + \cdots + a_m \sin^m \theta] \cos \theta \\
&\quad + r^m[b_0 \cos^m \theta + \cdots + b_m \sin^m \theta] \sin \theta \\
&\quad + r^m \left\{ (\cos \theta)\frac{(R_{m+1})}{r^m} + (\sin \theta)\frac{(S_{m+1})}{r^m} \right\}
\end{aligned} \tag{3.34}$$

We will use the following notation:

$$\begin{aligned}
N(\theta) &= [a_0 \cos^m \theta + \cdots + a_m \sin^m \theta] \cos \theta \\
&\quad + [b_0 \cos^m \theta + \cdots + b_m \sin^m \theta] \sin \theta
\end{aligned} \tag{3.35}$$

Similar to (3.34), we have

$$\begin{aligned}
r\frac{d\theta}{dt} &= \frac{1}{r}[xY_m + xS_{m+1} - yX_m - yR_{m+1}] \\
&= r^m[(\cos \theta)(b_0 \cos^m \theta + b_1 \cos^{m-1} \theta \sin \theta + \cdots b_m \sin^m \theta) \\
&\quad - (\sin \theta)(a_0 \cos^m \theta + a_1 \cos^{m-1} \theta \sin \theta + \cdots + a_m \sin^m \theta)] \\
&\quad + r^m \left[(\cos \theta)\frac{(S_{m+1})}{r^m} - (\sin \theta)\frac{(R_{m+1})}{r^m} \right]
\end{aligned} \tag{3.36}$$

We use the following notation:

$$\begin{aligned}
M(\theta) &= \cos \theta(b_0 \cos^m \theta + \cdots + b_m \sin^m \theta) \\
&\quad - (\sin \theta)(a_0 \cos^m \theta + \cdots + a_m \sin^m \theta)
\end{aligned} \tag{3.37}$$

Later we will be concerned with the zeros of $N(\theta)$ and $M(\theta)$, and it is convenient to point out here how these can be determined. By (3.34) and (3.35)

$$N(\theta) = \frac{1}{r^{m+1}}[xX_m(x, y) + yY_m(x, y)]$$

and since X_m, Y_m are homogeneous of degree m in x and y, then $xX_m + yY_m$ is homogeneous of degree $m + 1$ in x and y. Hence

$$xX_m + yY_m = (\alpha_1 x + \beta_1 y)(\alpha_2 x + \beta_2 y) \cdots (\alpha_q x + \beta_q y)\mathcal{H}(x, y)$$

where $(\alpha_1 x + \beta_1 y), \ldots, (\alpha_q x + \beta_q y)$ are the real linear factors of $xX_m + yY_m$ and where $\mathcal{H}(x, y)$ is nonzero and homogeneous of degree $m + 1 - q$ and is either positive definite or negative definite. Setting $\alpha_1 x + \beta_1 y$ equal to 0, we have

$$0 = \alpha_1 x + \beta_1 y = \alpha_1 r \cos \theta + \beta_1 r \sin \theta$$

Then if $\beta_1 = 0$, we obtain

$$\cos \theta = 0$$

and $\theta = \frac{\pi}{2}, -\frac{\pi}{2}$ are both zeros of $N(\theta)$. If $\beta_1 \neq 0$, then

$$\tan \theta = -\frac{\alpha_1}{\beta_1}$$

Let

$$\theta_0 = \arctan\left[-\frac{\alpha_1}{\beta_1}\right]$$

where $\theta_0 \in (-\frac{\pi}{2}, \frac{\pi}{2})$. Then θ_0 and $\theta_0 + \pi$ are zeros of $N(\theta)$.

Thus to find the zeros of $N(\theta)$, it is sufficient to find the real linear factors of $xX_m + yY_m$. Similarly, to find the zeros of $M(\theta)$, it is sufficient to find the real linear factors of $xY_m - yX_m$.

Note also that if $(\bar{x}, \bar{y}) \neq (0, 0)$ and is a zero of $xX_m + yY_m$ and is also a zero of $XY_m - yX_m$, that is, if

$$\bar{x}X_m(\bar{x}, \bar{y}) + \bar{y}Y_m(\bar{x}, \bar{y}) = 0$$

and

$$\bar{x}Y_m(\bar{x}, \bar{y}) - \bar{y}X_m(\bar{x}, \bar{y}) = 0$$

then

$$X_m(\bar{x}, \bar{y}) = 0$$

and

$$Y_m(\bar{x}, \bar{y}) = 0$$

because

$$\det\begin{bmatrix} \bar{x} & \bar{y} \\ \bar{y} & -\bar{x} \end{bmatrix} = -\bar{x}^2 - \bar{y}^2 \neq 0$$

The converse also holds. Thus $N(\theta)$ and $M(\theta)$ have a common real zero if and only if $X_m(x, y)$ and $Y_m(x, y)$ have a common zero.

From the definition of $A(r, \theta)$ and (3.34) to (3.37) it follows that

$$A(r, \theta) = \frac{r \frac{d\theta}{dt}}{\frac{dr}{dt}}$$

$$= \frac{M(\theta) + \frac{1}{r^m}[(\cos \theta)(S_{m+1}) - (\sin \theta)(R_{m+1})]}{N(\theta) + \frac{1}{r^m}[(\cos \theta)(R_{m+1}) + (\sin \theta)(S_{m+1})]} \tag{3.38}$$

Now suppose that θ_0 is a critical direction. Then there exists a sequence of points (r_j, θ_j) such that

$$\lim_{j \to \infty} r_j = 0$$

$$\lim_{j \to \infty} \theta_j = \theta_0$$

$$\lim_{j \to \infty} A(r_j, \theta_j) = \lim_{j \to \infty} \frac{r_j \frac{d\theta}{dt}(r_j, \theta_j)}{\frac{dr}{dt}(r_j, \theta_j)} = 0 \tag{3.39}$$

Assume that

$$N(\theta_0) \neq 0$$

Then by (3.32), (3.33), and (3.38) we have

$$\lim_{j \to \infty} A(r_j, \theta_j) = \frac{M(\theta_0)}{N(\theta_0)} \tag{3.40}$$

From (3.39) and (3.40), we have

$$M(\theta_0) = 0$$

Thus, if $N(\theta_0) \neq 0$ and θ_0 is a critical direction, then $M(\theta_0) = 0$.

Conversely, suppose $N(\theta_0) \neq 0$ and $M(\theta_0) = 0$. Then if $\{r_j\}$ and $\{\theta_j\}$ are sequences such that

$$\lim_{j \to \infty} r_j = 0$$

$$\lim_{j \to \infty} \theta_j = \theta_0$$

we have by (3.38),

$$\lim_{j \to \infty} A(r_j, \theta_j) = \frac{M(\theta_0)}{N(\theta_0)} = 0$$

and hence θ_0 is a critical direction.

Thus we have proved:

Theorem 3.11 *If $N(\theta_0) \neq 0$, then a necessary and sufficient condition that θ_0 be a critical direction is that $M(\theta_0) = 0$.*

Next let $B(\bar{r}, \bar{\theta})$ denote cot α where α is the angle defined earlier (Figure 3.25). That is,

$$B(\bar{r}, \bar{\theta}) = \frac{\frac{dr}{dt}(\bar{r}, \bar{\theta})}{\bar{r}\frac{d\theta}{dt}(\bar{r}, \bar{\theta})} = [A(\bar{r}, \bar{\theta})]^{-1}$$

Note that if $(\bar{r}, \bar{\theta})$ is a point on a branch curve of

$$\frac{dr}{dt} = 0$$

then

$$B(\bar{r}, \bar{\theta}) = 0 \tag{3.41}$$

and thus $\alpha = \frac{\pi}{2}$ or $-\frac{\pi}{2}$. From the definition of $B(\bar{r}, \bar{\theta})$ and equation (3.38), we have

$$B(r, \theta) = \frac{N(\theta) + \frac{1}{r^m}[(\cos\theta)(R_{m+1}) + (\sin\theta)(S_{m+1})]}{M(\theta) + \frac{1}{r^m}[(\cos\theta)(S_{m+1}) - (\sin\theta)(R_{m+1})]}$$

If $N(\theta_0) \neq 0$ and $M(\theta_0) \neq 0$, then

$$\lim_{\substack{r \to 0 \\ \theta \to \theta_0}} B(r, \theta) = \frac{N(\theta_0)}{M(\theta_0)} \neq 0 \tag{3.42}$$

Hence in a neighborhood of $r = 0$, $\theta = \theta_0$, $B(r, \theta)$ is bounded and also bounded away from zero.

Now if there were a branch curve of $\frac{dr}{dt} = 0$ with tangent at the origin described by $\theta = \theta_0$ then (3.41) would hold and we would obtain a contradiction to (3.42). Thus we have obtained

Theorem 3.12   *If $M(\theta_0) \neq 0$ then a necessary condition that there exist a branch curve of $\frac{dr}{dt} = 0$ with tangent at the origin described by $\theta = \theta_0$ is that $N(\theta_0) = 0$.*

Theorem 3.13   *If $M(\theta_0) \neq 0$ and if $N(\theta_0) = 0$ and $N'(\theta_0) \neq 0$, then there exists a unique branch curve of*

$$\frac{dr}{dt} = 0$$

such that the tangent to the branch curve approaches $\theta = \theta_0$ as $r \to 0$. Also $\frac{dr}{dt}$ changes sign at the branch curve.

Proof   From (3.32) to (3.35) we may write

$$\frac{dr}{dt} = r^m[N(\theta) + r\,H(r, \theta)]$$

where H has continuous first derivatives in r and θ. Hence since $M(\theta_0) \neq 0$, then to find the branch curves, it is sufficient to study the equation

$$N(\theta) + r\, H(r, \theta) = 0 \tag{3.43}$$

Since $r = 0$, $\theta = \theta_0$ is an initial solution of this equation and since

$$\frac{\partial}{\partial\theta}[N(\theta) + r\, H(r, \theta)]_{\theta=\theta_0, r=0} = N'(\theta_0) \neq 0$$

then we can apply the implicit function theorem and conclude that equation (3.43) has a unique solution in a neighborhood of $r = 0$, $\theta = \theta_0$,

$$\theta = T(r)$$

such that

$$\theta_0 = T(0)$$

The curve

$$\theta = T(r) \qquad r \geq 0$$

is a branch curve of $\frac{dr}{dt} = 0$ and $\theta \to \theta_0$ as $r \to 0$.
  Since

$$\frac{\partial}{\partial\theta}[N(\theta) + r\, H(r, \theta)] = N'(\theta) + r\,\frac{\partial H}{\partial\theta}(r, \theta)$$

then if r is sufficiently small and θ is sufficiently close to θ_0, we have, since $N'(\theta_0) \neq 0$,

$$\frac{\partial}{\partial\theta}[N(\theta) + r\, H(r, \theta)] \neq 0$$

and hence $\frac{dr}{dt}$ changes sign as it crosses the branch curve

$$\theta = T(r).$$

This completes the proof of Theorem 3.13. $\qquad\qquad\qquad\qquad\qquad$ ∏

  Parallel arguments yield the following theorem.

Theorem 3.14   *If $N(\theta_0) \neq 0$ and if $M(\theta_0) = 0$ and $M'(\theta_0) \neq 0$, then there exists a unique branch curve of*

$$\frac{d\theta}{dt} = 0$$

such that the tangent to the branch curve approaches $\theta = \theta_0$ as $r \to 0$. Also $\frac{d\theta}{dt}$ changes sign at the branch curve.

  As we shall see, Theorems 3.13 and 3.14 are the basis for finding an explicit portrait of the orbits in a neighborhood of $(0, 0)$ provided we impose the appropriate hypotheses on $M(\theta)$ and $N(\theta)$.

In order to apply the preceding theory to a system (3.20) in which $X(x, y)$ and $Y(x, y)$ are given by (3.23) and (3.24), we require that $M(\theta)$ and $N(\theta)$ have no common zeros and that each zero of $M(\theta)$ and each zero of $N(\theta)$ have multiplicity one. As already shown, the functions $M(\theta)$ and $N(\theta)$ have no common real zeros if and only if $X_m(x, y)$ and $Y_m(x, y)$ have no common real linear factors.

We proceed as in the general case (Section A) to find the sectors bounded by branch curves of $\frac{dr}{dt} = 0$. Since the case in which $\frac{dr}{dt} < 0$ except at the origin was dealt with in our discussion of the general case, we assume at the outset that there exist branch curves of $\frac{dr}{dt} = 0$. Let $\theta_1, \ldots, \theta_n$ be the zeros of $N(\theta)$ and suppose $N'(\theta_j) \neq 0$ for $j = 1, \ldots, n$. Then by Theorem 3.13 there is a unique branch curve $\mathcal{C}_j$ of $\frac{dr}{dt} = 0$ with tangent at the origin described by $\theta = \theta_j$ $(j = 1, \ldots, n)$ at which $\frac{dr}{dt}$ changes sign. Also by Theorem 3.12 there are no branch curves other than $\mathcal{C}_1, \ldots, \mathcal{C}_n$. Thus we obtain the sectors in which $\frac{dr}{dt} > 0$ and those in which $\frac{dr}{dt} < 0$.

Next we assume that $\tilde{\theta}_1, \ldots, \tilde{\theta}_m$ are the zeros of $M(\theta)$ and that $M'(\tilde{\theta}_k) \neq 0$ for $k = 1, \ldots, m$.

By Theorem 3.14 there is a unique branch curve of

$$\frac{d\theta}{dt} = 0$$

with tangent at the origin described by $\theta = \tilde{\theta}_k (k = 1, \ldots, m)$ at which $\frac{d\theta}{dt}$ changes sign. It follows by Theorem 3.11 that there are no other curves of $\frac{d\theta}{dt} = 0$.

If there is no $\tilde{\theta}_k$ in a sector Σ in which $\frac{dr}{dt} < 0$, then solutions just cross Σ, that is, each solution which enters Σ just crosses and then exits Σ as in, for example, Figure 3.21. For if a solution remains in Σ then by Theorems 3.7 and 3.8, the solution approaches O in a direction, say θ_0, and by Theorem 3.10, that θ_0 is a critical direction. Then by Theorem 3.11, it follows that $M(\theta_0) = 0$. But this contradicts our assumption that there is no $\tilde{\theta}_k$ in sector Σ. If there is at least one critical direction in the sector, we obtain the kinds of results illustrated in Figures 3.16, 3.17, 3.18, 3.19, and 3.20. If $\frac{dr}{dt} > 0$ in the sector, we make the change of variables $\tau = -t$ as done earlier. The complete portrait is obtained by putting together the sectors as described in Section A.

The use of functions $M(\theta)$ and $N(\theta)$ makes it simple to construct examples with particular properties. Now as promised earlier, we give an example of a system in which elliptic sectors occur. Our aim is to find a system which has a portrait near O of the kind sketched in Figure 3.26 where there are two hyperbolic sectors and two elliptic sectors. There are four critical directions: $\theta = 0, \frac{\pi}{2}, \pi$, and $\frac{3}{2}\pi$. Thus we choose $M(\theta)$ equal to zero for $\theta = 0, \frac{\pi}{2}, \pi, \frac{3}{2}\pi$ and nonzero elsewhere. Also since $\frac{d\theta}{dt} < 0$ in quadrants I and III and $\frac{d\theta}{dt} > 0$ in quadrants II and IV, we choose $M < 0$ in quadrants I and III and $M > 0$ in quadrants II and IV. Thus a simple form for $M(\theta)$ is

$$M(\theta) = -2\cos\theta\sin\theta = -\sin 2\theta$$

Note that $M'(\theta) \neq 0$ for $\theta = 0, \frac{\pi}{2}, \pi, \frac{3}{2}\pi$. Since $\frac{dr}{dt} = 0$ on the lines $x = y$ and $x = -y$ then we want to choose $N(\theta)$ to be zero for $\theta = \frac{\pi}{4}, \frac{3}{4}\pi, \frac{5}{4}\pi, \frac{7}{4}\pi$ and nonzero

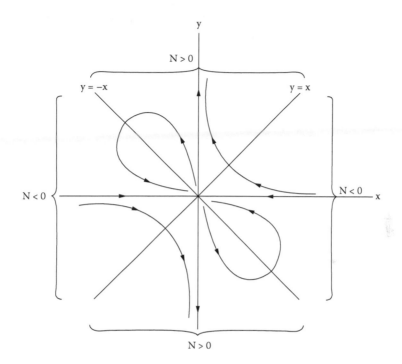

Figure 3.26

elsewhere. A simple form for $N(\theta)$ is

$$N(\theta) = (\sin\theta - \cos\theta)(\sin\theta + \cos\theta)$$
$$= \sin^2\theta - \cos^2\theta = -\cos 2\theta$$

(Note that $N'(\theta) \neq 0$ for $\theta = \frac{\pi}{4}, \frac{3}{4}\pi, \frac{5}{4}\pi, \frac{7}{4}\pi$). From the definitions, we have in general

$$M(\theta) = (-\sin\theta)X_m(\cos\theta, \sin\theta) + (\cos\theta)Y_m(\cos\theta, \sin\theta)$$
$$N(\theta) = (\cos\theta)X_m(\cos\theta, \sin\theta) + (\sin\theta)Y_m(\cos\theta, \sin\theta)$$

or

$$\begin{bmatrix} M \\ N \end{bmatrix} = \begin{bmatrix} -\sin\theta & +\cos\theta \\ \cos\theta & \sin\theta \end{bmatrix} \begin{bmatrix} X_m \\ Y_m \end{bmatrix}$$

Since

$$\begin{bmatrix} -\sin\theta & +\cos\theta \\ \cos\theta & \sin\theta \end{bmatrix}^{-1} = \begin{bmatrix} -\sin\theta & \cos\theta \\ \cos\theta & \sin\theta \end{bmatrix}$$

then

$$\begin{bmatrix} -\sin\theta & \cos\theta \\ \cos\theta & \sin\theta \end{bmatrix} \begin{bmatrix} M \\ N \end{bmatrix} = \begin{bmatrix} X_m \\ Y_m \end{bmatrix}$$

Hence we have

$$X_3(\cos\theta, \sin\theta) = (-\sin\theta)(M(\theta)) + (\cos\theta)(N(\theta))$$
$$= (-\sin\theta)(-2\cos\theta\sin\theta) + (\cos\theta)(\sin^2\theta - \cos^2\theta)$$
$$= 3\sin^2\theta\cos\theta - \cos^3\theta$$
$$Y_3(\cos\theta, \sin\theta) = (\cos\theta)(M(\theta)) + (\sin\theta)(N(\theta))$$
$$= \cos\theta(-2\cos\theta\sin\theta) + (\sin\theta)(\sin^2\theta - \cos^2\theta)$$
$$= -3\cos^2\theta\sin\theta + \sin^3\theta$$

Thus the system

$$\frac{dx}{dt} = -x^3 + 3xy^2$$
$$\frac{dy}{dt} = -3x^2y + y^3$$

has Figure 3.26 as its portrait near the origin.

The same portrait holds, if r is sufficiently small, for the system

$$\frac{dx}{dt} = (-x^3 + 3xy^2)F_K(x, y) + R_{K+3+1}(x, y)$$
$$\frac{dy}{dt} = (-3x^2y + y^3)F_K(x, y) + S_{K+3+1}(x, y)$$

where $F_K(x, y)$ is a nonzero positive definite homogeneous form of degree K and $R_{K+3+1}(x, y)$ and $S_{K+3+1}(x, y)$ are remainder terms of order higher than $K + 3$. That is,

$$\lim_{r \to 0} \frac{R_{K+3+1}(x, y)}{r^{K+3}} = 0$$

and

$$\lim_{r \to 0} \frac{S_{K+3+1}(x, y)}{r^{K+3}} = 0$$

This completes our general discussion of equilibrium points. The discussion is clearly not complete. (For example, we have obtained a constructive analysis only for the case in which the lowest order terms of $X(x, y)$ and $Y(x, y)$ have the same order.) For more extensive analysis and references to further work, see Nemytskii and Stepanov [1960, Chapter 2].

C. The index of an equilibrium point

A beautiful and useful concept that is introduced in the study of equilibrium points is the index of the equilibrium point. We will give here an intuitive version of the definition of index for equilibrium points in the two-dimensional case and leave to the Appendix a rigorous definition for the n-dimensional case in terms of the notion of topological degree.

Definition   Let $X(x, y)$, $Y(x, y)$ be real-valued continuous functions on an open set D in the xy-plane. Let $V(x, y)$ be the vector field

$$V(x, y) = (X(x, y), Y(x, y))$$

that is, V is a function with domain D and range contained in R^2. If $(x, y) \in D$, then $V(x, y)$ is the point or vector $(X(x, y), Y(x, y))$. An *equilibrium point* of $V(x, y)$ is a point $(x_0, y_0) \in D$ such that

$$X(x_0, y_0) = Y(x_0, y_0) = 0$$

Let C be a simple closed curve in D such that no point of C is an equilibrium point of V. We study how the vector associated with a point $(x_1, y_1) \in C$ changes as (x_1, y_1) is moved counterclockwise around C. The vector is rotated through an angle $j(2\pi)$ where j is an integer, positive or negative or zero. The number j is the *index of C with respect to* $V(x, y)$.

It is easy to see that this "definition" is not precisely formulated because several of the terms used have only an intuitive meaning. For example, what does it mean to "move counterclockwise" on C? As long as C is an easily visualized curve like a circle or ellipse, it is easy to describe precisely what we mean by "moving counterclockwise." But if C is a sufficiently "messy" curve, then it is not at all clear which is the counterclockwise direction on the curve. Secondly, although for easily visualized curves C, it is intuitively clear how to measure the angle through which the vector is rotated, we have not given a general and precise method for describing how to measure the angle.

Thus our "definition" is far from rigorous. An efficient way to give a rigorous definition is to formulate the definition in terms of topological degree. This is done in the Appendix of this book. The advantage of giving the definition in this way is that one obtains also a definition of index for the n-dimensional case where n is arbitrary. However, it should be pointed out that it is not difficult to give a rigorous definition of index for the two-dimensional case which is much shorter than the definition for the n-dimensional case. Such a definition can be obtained directly from the definition of winding number as given by Ahlfors [1966].

Having obtained the definition of the index of C with respect to $V(x, y)$, one can then prove that if C is continuously moved or deformed in such a way that it does not cross any equilibrium point of V during the continuous deformation, then the index j remains constant during the deformation. (In order to prove this result, one must first, of course, give a precise definition of "continuous deformation." This is done in the Appendix.)

Now suppose that (x_0, y_0) is an isolated equilibrium point of V, that is, suppose that there exists a circular neighborhood N of (x_0, y_0) such that there are no equilibrium points of V in N except (x_0, y_0). Let C_1, C_2 be simple closed curves in N such that (x_0, y_0) is in the interior of C_1 and is in the interior of C_2. It is not difficult to show that C_1 can be continuously deformed into C_2 without crossing any equilibrium point of V. Hence the index of C_1 equals the index of C_2, and indeed the index is independent of the simple closed curve C provided C is contained in N and (x_0, y_0) is in the interior of C. Hence we may formulate the following definition.

Definition   The *index of the isolated equilibrium point* (x_0, y_0) is the index of a simple closed curve C such that C is contained in N and (x_0, y_0) is contained in the interior of C.

Notice that in obtaining this definition we have used the concept of the interior of a simple closed curve. This is a serious notion which requires careful definition. We discuss it in the Appendix.

In the Appendix, we shall see how to compute, on a rigorous basis, the indices of the equilibrium points at the origin of various two-dimensional systems. However, using our intuitive definition of index of C with respect to V, it is easy to see, unrigorously, that the index of a node, stable or unstable, is $+1$ and the index of a saddle point is -1.

Now we list, without proof, some important properties of the index.

1. The index of a simple closed curve which contains no equilibrium points in its interior is zero.

2. The index of a simple closed curve which has a finite set of equilibrium points in its interior is the sum of the indices of the equilibrium points.

3. Let C be a simple closed curve such that if $(x_1, y_1) \in C$, then the vector $V(x_1, y_1)$ is parallel to the tangent to C at (x_1, y_1). Then the index of C is ± 1. (In order to be able to speak of the tangent to C, we must, of course, assume that C is differentiable.)

4. Let C be a simple closed curve such that if $(x_1, y_1) \in C$ the vectors $V(x_1, y_1)$ point into the exterior of C. Then the index of C is $+1$. Also if the vectors point into the interior of C, the index of C is $+1$.

Bendixson Formula. The index of the equilibrium point $(0, 0)$ of

$$\frac{dx}{dt} = X(x, y)$$
$$\frac{dy}{dt} = Y(x, y) \qquad\qquad (3.20)$$

is

$$1 + \frac{N_E - N_H}{2}$$

where N_E is the number of elliptic sectors and N_H is the number of hyperbolic sectors.

Informal proof: First suppose that $\frac{dr}{dt} \geq 0$ for all (x, y) in $N - \{(0, 0)\}$ where

$$N = \{(x, y)/x^2 + y^2 \leq r_0^2\}$$

and r_0 is sufficiently small. Then it follows that if r_0 is fixed, there exists $m > 0$ such that on the circle

$$C = \{(r, \theta)/r = r_0, \theta \in [(0, 2\pi]\}$$

we have

$$\frac{dr}{dt} \geq m > 0$$

Hence by property 4 of the index, the index of C is $+1$. (Part of the informality of this discussion is the fact that property 4 has not been proved.)

Now consider the case in which $\frac{dr}{dt} = 0$ has at least 2 branch curves at which $\frac{dr}{dt}$ changes sign. Then if r_0 is sufficiently small, the set N is the union of a finite set of elliptic, hyperbolic, and parabolic sectors. Each sector is bounded by an arc of C and two branch curves of

$$\frac{d\theta}{dt} = 0$$

If r_0 is sufficiently small, each of these branch curves can be made as close as desired to a straight line, that is, the tangent at $(0, 0)$ of the branch curve. (Since we are giving an "informal proof," that is, a geometric indication of why the result holds, we will identify each branch curve with its tangent at $(0, 0)$.)

Let $E_1, \ldots, E_k$ denote the elliptic sectors; $H_1, \ldots, H_m$ the hyperbolic sectors; and $S_1, \ldots, S_n$ the parabolic sectors. Let $\alpha_1, \ldots, \alpha_k$ denote the angles determined by the tangents (at $(0, 0)$) to the branch curves of the elliptic sectors. (See Figure 3.27.) Let $\beta_1, \ldots, \beta_m$ denote the angles determined by the hyperbolic sectors (see Figure 3.28), and let $\gamma_1, \ldots, \gamma_m$ denote the angles determined by the parabolic sectors (Figure 3.29). Now we apply the informal definition of index and consider how the field vector

$$V(x, y) = (X(x, y), Y(x, y))$$

is changed as (x, y) is moved counterclockwise on $C \cap E_j$, where $j = 1, \ldots, k$. There are two possible cases Figures 3.27a and 3.27b. In Figure 3.27a, the field vector rotates in the positive direction (counterclockwise) through the angle $\pi + \alpha$. In Figure 3.27b, we have the same conclusion. Similarly if we consider a hyperbolic sector (Figures 3.28a and 3.28b) the field vector rotates in the negative directions (clockwise) through the angle

$$-(\pi - \beta) = \beta - \pi$$

Finally it is easily seen that the field vector rotates through a nonnegative angle γ in the parabolic sector (Figures 3.29a and 3.29b).

Thus as (x, y) is moved counterclockwise around circle C through the angle

$$\sum_{j=1}^{k} \alpha_j + \sum_{j=1}^{m} \beta_j + \sum_{j=1}^{n} \gamma_j = 2\pi$$

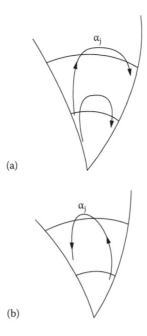

(a)

(b)

Figure 3.27

the field vector is rotated through the angle

$$\sum_{j=1}^{k} (\pi + \alpha_j) + \sum_{j=1}^{m} (\beta_j - \pi) + \sum_{j=1}^{n} \gamma_j$$

$$= k\pi + \sum_{j=1}^{k} \alpha_j + \sum_{j=1}^{m} \beta_j - m\pi + \sum_{j=1}^{n} \gamma_j$$

$$= (k - m)\pi + \sum_{j=1}^{k} \alpha_j + \sum_{j=1}^{m} \beta_j + \sum_{j=1}^{n} \gamma_j$$

$$= (k - m)\pi + 2\pi$$

$$= \left[\frac{(k - m)}{2} + 1 \right] 2\pi$$

That is, the index is $\frac{k-m}{2} + 1$.

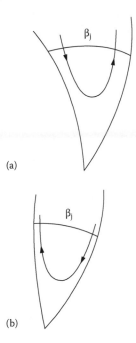

(a)

(b)

Figure 3.28

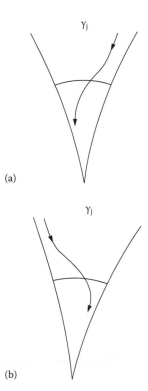

(a)

(b)

Figure 3.29

D. The Bendixson Theory

Using the index makes possible a treatment of the case in which the matrix

$$
\begin{bmatrix}
\frac{\partial X}{\partial x}(0,0) & \frac{\partial X}{\partial y}(0,0) \\[2mm]
\frac{2Y}{2x}(0,0) & \frac{\partial Y}{\partial y}(0,0)
\end{bmatrix}
$$

has one eigenvalue equal to 0 and one nonzero eigenvalue. Besides giving a desirable completeness to the earlier discussion in this chapter, this result is useful in applications. We follow the geometric treatment given by Lefschetz [1962, Chapter 10] with some changes. We choose this approach because it is explicit and can be easily applied. A conceptually clear and considerably briefer treatment along the same lines can be given, but it is less explicit and hence harder to apply. (See Exercise 9b.)

The original study (see Bendixson [1901]) uses a different approach, a construction called polar blow-up. For a description of this and further references, see Arnold [1988, pp. 80–82]. See especially Dumortier [1977] where smooth vector fields are treated.

We write equation (3.20) in the form

$$
\frac{dx}{dt} = ax + by + \sum_{m \geq 2} X_m(x, y)
$$

$$
\frac{dy}{dt} = cx + dy + \sum_{m \geq 2} Y_m(x, y) \tag{3.44}
$$

where X_m, Y_m denote the homogeneous forms of degree m in x and y, and the matrix

$$
\begin{bmatrix}
a & b \\
c & d
\end{bmatrix}
$$

has eigenvalues λ (nonzero) and 0. With standard changes of variables (see Chapter 2) equation (3.44) becomes

$$
\frac{dx}{dt} = \sum_{j \geq 2} P_j(x, y)
$$

$$
\frac{dy}{dt} = y + \sum_{j \geq 2} Q_j(x, y) \tag{3.45}
$$

where, for convenience, we keep the same notation for the two dependent variables x and y and the independent variable t, and P_j, Q_j denote homogeneous forms of degree j in x and y. With a further change of variable which can be represented by the matrix

$$
\begin{bmatrix}
1 & 1 \\
0 & 1
\end{bmatrix}
$$

the system (3.45) becomes

$$\frac{dx}{dt} = y + \sum_{j \geq 2} P_j(x, y)$$

$$\frac{dy}{dt} = y + \sum_{j \geq 2} Q_j(x, y) \tag{3.46}$$

where again we retain the same notation for the dependent variables and the independent variable, and P_j and Q_j denote homogeneous forms of degree j in x and y.

Lemma 3.5 *There exist series*

$$A(x) = \sum_{m=2}^{\infty} a_m x^m$$

and

$$E_1(x, y) = 1 + \sum_{p+q=1}^{\infty} \alpha_{pq} x^p y^q$$

such that

$$y + \sum_{j=2}^{\infty} P_j(x, y) = [y - A(x)] E_1(x, y)$$

Similarly there exist series

$$B(x) = \sum_{m=2}^{\infty} b_m x^m$$

and

$$E_2(x, y) = 1 + \sum_{p+q=1}^{\infty} \beta_{pq} x^p y^q$$

such that

$$y + \sum_{j \geq 2} Q_j(x, y) = \{y - B(x)\} E_2(y)$$

All these series converge if $|x|$ and $|y|$ are sufficiently small.

Proof The cognoscenti among us will dismiss the proof of this lemma as a trivial application of the Weierstrass preparation theorem, that is, a proof which warrants no discussion. However in order to apply results in the earlier part of this chapter, we need to compare the magnitudes of $A(x)$ and $B(x)$. So we indicate a simpler proof (equating coefficients of like powers) which yields explicit values of the coefficients in $A(x)$ and $B(x)$.

First the equation

$$y + \sum_{j \geq 2} P_j(x, y) = 0$$

has the initial solution $x = 0$, $y = 0$, and at that initial solution

$$\frac{d}{dy}\left[y + \sum_{j \geq 2} P_j(x, y) \right] = 1$$

Hence by the implicit function theorem (for analytic functions), it follows that there exists a unique real analytic function

$$y(x) = \sum_{m=1}^{\infty} a_m x^m$$

such that $y(0) = 0$ and if $|x|$ is sufficiently small, then

$$y(x) + \sum_{j \geq 2} P_j(x, y(x)) = 0$$

or

$$\sum_{m \geq 1} a_m x^m + \sum_{j \geq 2} P_j \left(x, \sum a_m x^m \right) = 0 \qquad (3.47)$$

Since coefficients in P_j ($j \geq 2$) are all given, then equating like powers of x, we can solve for the coefficients a_m. Note that since $j \geq 2$, then $a_1 = 0$. That is,

$$y(x) = \sum_{m \geq 2} a_m x^m$$

For brevity, let

$$F(x, y) = \sum_{m \geq 2} P_j(x, y)$$

Then using (3.47), we may write

$$y + F(x, y) = y + F(x, y) - \sum_{m \geq 2} a_m x^m - F \left(x, \sum_{m \geq 2} a_m x^m \right)$$

$$= y - \sum_m a_m x^m + F(x, y) - F \left(x, \sum a_m x \right)$$

$$= y - \sum_m a_m x^m + \left[F_y \left(x, \sum a_m x^m \right) \right] \left(y - \sum a_m x^m \right)$$

$$+ \left[\frac{1}{2} F_{yy} \left(x, \sum a_m x^m \right) \right] \left(y - \sum a_m x^m \right)^2 + \cdots$$

$$= \left(y - \sum a_m x^m \right) [1 + h(x, y)]$$

where $h(x, y)$ is an analytic function such that $h(0, 0) = 0$. Thus the desired series is

$$A(x) = \sum_{m=2}^{\infty} a_m x^m$$

Similarly we calculate $B(x)$. This completes the proof of Lemma 3.5.   ☐

Remark. In proving Lemma 3.5, we have invoked the implicit function theorem. Another approach is to solve formally for the coefficients and then prove that the resulting series $\sum a_m x^m$ converge. See, for example, Goursat [1904].

Applying Lemma 3.5, we write system (3.46) as:

$$\frac{dx}{dt} = [y - A(x)]E_1(x, y)$$

$$\frac{dy}{dt} = [y - B(x)]E_2(x, y) \tag{3.48}$$

and

$$E_1(0, 0) = E_2(0, 0) = 1$$

We study (3.48) in the set

$$S = \{(x, y)/x^2 + y^2 \le r_0^2\}$$

where we require that r_0 be small enough so that the functions $A(x)$ and $B(x)$ do not have zeros in the set $(-r_0, 0) \cup (0, r_0)$ and the curves $y = A(x)$ and $y = B(x)$ have just one point of intersection in S, that is, the point $(0, 0)$. Also r_0 should be small enough so that $\frac{dx}{dt}$ and $\frac{dy}{dt}$ are approximated by $y - A(x)$ and $y - B(x)$, respectively that is, $E_1(x, y)$ and $E_2(x, y)$ are both close to 1.

Our first step is to determine the branch curves of $\frac{dr}{dt} = 0$ and $\frac{d\theta}{dt} = 0$. Using the notation in the Section B on critical directions, we have in equation (3.48)

$$X_m = X_m = y$$
$$-yX_m + xY_m = -y^2 + xy = y(x - y)$$
$$xX_m + yY_m = xy + y^2 = y(x + y)$$

Hence

$$M(\theta) = (\sin\theta)(\cos\theta - \sin\theta)$$
$$N(\theta) = (\sin\theta)(\sin\theta + \cos\theta)$$

and

$$M'(\theta) = (\sin\theta)(-\sin\theta - \cos\theta) + (\cos\theta)(\cos\theta - \sin\theta)$$
$$= -\sin 2\theta + \cos 2\theta$$
$$N'(\theta) = (\sin\theta)(\cos\theta - \sin\theta) + (\cos\theta)(\sin\theta + \cos\theta)$$
$$= \sin 2\theta + \cos 2\theta$$

The zeros of $M(\theta)$ are $\theta = 0, \frac{\pi}{4}, \pi, \frac{5}{4}\pi$ and the zeros of $N(\theta)$ are $\theta = 0, \frac{3}{4}\pi, \pi,$ $\frac{7}{4}\pi$. Then by Theorem 3.13 since $M\left(\frac{3}{4}\pi\right) \neq 0$ and $M\left(\frac{7}{4}\pi\right) \neq 0$ and $N'\left(\frac{3}{4}\pi\right) \neq 0$ and $N'\left(\frac{7}{4}\pi\right) \neq 0$, it follows that there exist two branch curves of

$$\frac{dr}{dt} = 0$$

such that the tangents to these branch curves approach $\theta = \frac{3}{4}\pi$ and $\theta = \frac{7}{4}\pi$ respectively as $r \to 0$. Also $\frac{dr}{dt}$ changes sign at each of the branch curves. Similarly by Theorem 3.14 since $N\left(\frac{\pi}{4}\right) \neq 0$ and $N\left(\frac{5}{4}\pi\right) \neq 0$ and $M'\left(\frac{\pi}{4}\right) \neq 0$ and $M'\left(\frac{5}{4}\pi\right) \neq 0$, it follows that there exist two branch curves of

$$\frac{d\theta}{dt} = 0$$

at which $\frac{d\theta}{dt}$ changes sign and the tangents to the branch curves approach $\theta = \frac{\pi}{4}$ and $\theta = \frac{5}{4}\pi$ respectively as $r \to 0$. If r is sufficiently small, the branch curves of

$$\frac{d\theta}{dt} = 0$$

are very close to the curves $\theta = \frac{\pi}{4}$ and $\theta = \frac{5}{4}\pi$, respectively. The signs of $\frac{d\theta}{dt}$ near the curve

$$\frac{d\theta}{dt} = 0$$

are as indicated in Figure 3.30 and the signs of $\frac{dr}{dt}$ are similarly indicated.

Now suppose θ_0 is such that $M(\theta_0) \neq 0$ and $N(\theta_0) \neq 0$. Then by Theorem 3.12, there is no branch curve of $\frac{dr}{dt} = 0$ with tangent at the origin described by $\theta = \theta_0$. By Theorem 3.11, the angle θ_0 is not a critical direction. However, as shown in the proof of Theorem 3.10, if an angle $\bar{\theta}$ is the direction of the tangent at O of a branch curve of $\frac{d\theta}{dt} = 0$, then $\bar{\theta}$ is a critical direction. Thus it follows that no branch curve of

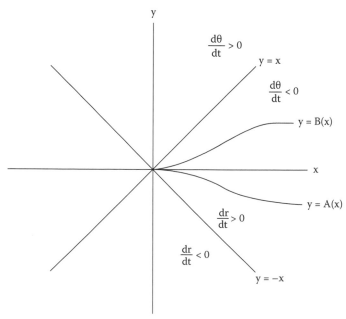

Figure 3.30

$\frac{d\theta}{dt} = 0$ has θ_0 as the direction for the tangent at O of a branch curve of $\frac{d\theta}{dt} = 0$. If r is sufficiently small, the branch curves of

$$\frac{d\theta}{dt} = 0$$

are very close to the curves $\theta = \frac{\pi}{4}$ and $\theta = \frac{5}{4}\pi$, respectively. The signs of $\frac{d\theta}{dt}$ near the curve

$$\frac{d\theta}{dt} = 0$$

are as indicated in Figure 3.30, and the signs of $\frac{dr}{dt}$ are similarly indicated.

It remains to study the status of $\theta = 0$ and $\theta = \pi$. (At each of these values, $M(\theta)$ and $N(\theta)$ are both zero.) For the investigation, we use a geometric view in which the information already obtained is utilized. We sketch the tangents at 0 of the branch curves already determined and the curves $y = A(x)$ and $y = B(x)$. From these two curves we can find the slopes of the tangents to the orbits because, by (3.48), if r_0 is sufficiently small, the slopes of the orbits are given approximately by

$$\frac{y - B(x)}{y - A(x)}$$

As we shall see, these slopes are determined essentially by the relative positions of the curves $y = A(x)$ and $y = B(x)$. Consequently our arguments are independent of the particular curves sketched.

We investigate first the orbits in the set

$$S_R = S \cap \{(x, y)/y < x\}$$

Case I: $B(x) > A(x)$ for $x > 0$.

In Figures 3.31 and 3.31a, we have sketched two examples of curves $y = A(x)$ and $y = B(x)$ satisfying this condition. Since $A(x)$ and $B(x)$ both start with higher order terms, then the curves $y = A(x)$ and $y = B(x)$ have slope zero at $x = 0$. Since $|x|$ is small, we represent the branch curves of $\frac{dr}{dt} = 0$ and $\frac{d\theta}{dt} = 0$ by their tangents at the origin. In Figure 3.31 if we consider a point P_1 between $y = B(x)$ and $y = x$ then the tangent at P_1 of an orbit which passes through P_1 is as sketched because $y - B(x)$ and $y - A(x)$ are both positive at P_1. The tangents of orbits at P_2 and P_3 are similarly sketched. The tangents at points on $y = B(x)$ and $y = A(x)$ are horizontal and vertical, respectively. Note that if $A(x) > 0$ (unlike Figure 3.31), then if (x, y) is such that

$$y < A(x) < B(x)$$

then $[y - B(x)]$ and $[y - A(x)]$ are both negative and

$$|y - B(x)| > |y - A(x)|$$

Hence any orbit through (x, y) proceeds downward and if y decreases toward 0, then the orbit proceeds to cross the x-axis downward.

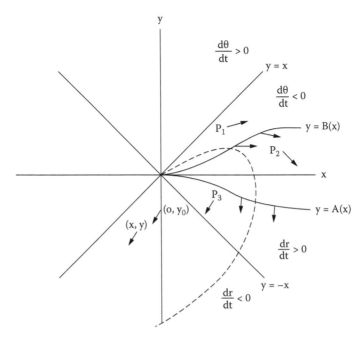

Figure 3.31

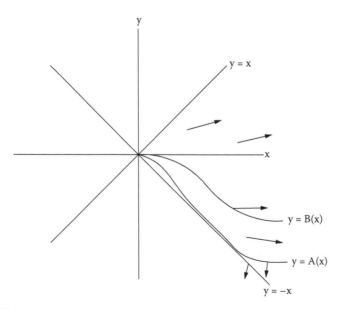

Figure 3.31 a

Thus we have a picture of the orbit in the region

$$\{(x, y)/x > 0, y < x\}$$

On the set

$$\{(x, y)/x = 0, y < 0\}$$

the orbit which passes through a point $(0, y_0)$, where $y_0 < 0$, has the tangent sketched in Figure 3.31 since

$$\frac{dy}{dx} = \frac{y_0}{y_0}$$

If we consider a point (x, y) such that

$$(x, y) \in \{(x, y)/y < x, x < 0\}$$

then since $A(x)$ and $B(x)$ each consist of higher order terms, we have

$$y < x < A(x)$$

and

$$y < x < B(x)$$

Hence

$$y - A(x) < 0$$
$$y - B(x) < 0$$

and the tangent at (x, y) of the orbit which passes through (x, y) is as sketched in Figure 3.31.

Thus the portrait in S_R is a fan bounded by the two branch curves of $\frac{d\theta}{dt} = 0$ which have as tangents at O the lines $\theta = \frac{\pi}{4}$ and $\theta = \frac{5}{4}\pi$.

In Figure 3.31a, we consider another example of the case $B(x) > A(x)$ which, although $A(x)$ and $B(x)$ are somewhat diferent, shows that the same condition is reached.

Case II: $A(x) > B(x)$ for $x > 0$

In this case, we obtain two hyperbolic sectors which are separated by a separatrix. (See Figure 3.32.) To show this, let P_0 be a point on the positive x-axis and let P_1 be the point where the vertical line through P_0 intersects the curve $y = B(x)$. Let OP_0 be the segment of the x-axis between O and P_0, let OP_1 denote the part of the curve $y = B(x)$ betwen O and P_1, and let $P_0 P_1$ be the vertical line segment joining P_0 and P_1. As follows from the directions of the tangents to orbits passing through OP_0, OP_1, and $P_0 P_1$, there exists a fan in the sector bounded by OP_0, OP_1, and $P_0 P_1$ (cf. Figure 3.32). Moreover we will prove later (see Lemma 3.6) that this fan is a separatrix (denoted in Figure 3.32 by S).

Thus S_R is the union of two hyperbolic sectors, the lower one bounded by $\theta = \frac{5}{4}\pi$ (more precisely, the branch curve of $\frac{d\theta}{dt} = 0$ which has tangent $\theta = \frac{5}{4}\pi$ at the origin) and S and the upper hyperbolic sector bounded by S and $\theta = \frac{\pi}{4}$.

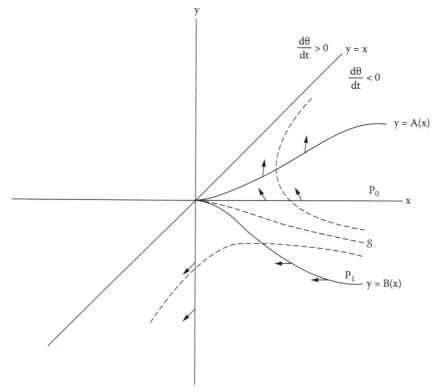

Figure 3.32

Now we study the orbits in the set

$$S_L = S \cap \{(x, y)/y > x\}$$

by utilizing the results already obtained. We reduce the problem to the previous result for S_R by taking mirror images in the line $y = x$. More explicitly, let (x, y) be a point in S_L and let

$$x = -\xi$$
$$y = -\eta$$

Then $\xi > \eta$ and substituting in (3.48), we obtain

$$-\frac{d\xi}{dt} = [-\eta - A(-\xi)]E_1(-\xi, -\eta)$$
$$-\frac{d\eta}{dt} = [-\eta - B(-\xi)]E_2(-\xi, -\eta)$$

or

$$\frac{d\xi}{dt} = [\eta + A(-\xi)]E_1$$
$$\frac{d\eta}{dt} = [\eta + B(-\xi)]E_2 \qquad (3.49)$$

Let

$$-A(-\xi) = \mathcal{A}(\xi)$$
$$-B(-\xi) = \mathcal{B}(\xi)$$

Then (3.49) becomes

$$\frac{d\xi}{dt} = [\eta - \mathcal{A}(\xi)]E_1$$
$$\frac{d\eta}{dt} = [\eta - \mathcal{B}(\xi)]E_2 \qquad (3.50)$$

where $\xi > \eta$.

Now suppose that for $x < 0$,

$$A(x) < B(x)$$

Then

$$\mathcal{A}(\xi) = -A(-\xi) = -A(x) > -B(x) = \mathcal{B}(\xi)$$

Similarly if for $x < 0$

$$A(x) > B(x)$$

then

$$\mathcal{A}(\xi) < \mathcal{B}(\xi).$$

Thus the study of orbits in S_L is reduced to the study of (3.50) where $\xi > \eta$. This is the problem we have already dealt with.

We summarize our results for equation (3.48) in four theorems. In each theorem the description of the portrait of the orbits in a small neighborhood follows directly from the preceding discussion. The index of O in each case is easily obtained by following how the value of $\frac{y - B(x)}{y - A(x)}$ changes as one proceeds counterclockwise around a circle with center O. We leave the details of this to Exercise 9.

Theorem B1. *Suppose that if $x > 0$*

$$B(x) > A(x)$$

and if $x < 0$

$$B(x) < A(x)$$

Then the portrait of the orbits of (3.48) in a sufficiently small neighborhood of O consists of two fans (Figure 3.33). The index of O is $+1$.

Theorem B2. *Suppose that if $x > 0$*

$$A(x) > B(x)$$

and if $x < 0$

$$A(x) < B(x)$$

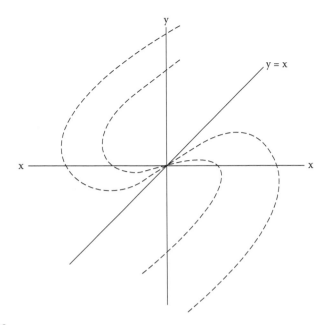

Figure 3.33

Then the portrait consists of four hyperbolic sectors (Figure 3.34). The index of O is −1.

Theorem B3. *Suppose that if* $x > 0$

$$A(x) > B(x)$$

and if $x < 0$

$$A(x) > B(x)$$

Then the portrait consists of two hyperbolic sectors in S_R *and a fan in* S_L *(Figure 3.35). The index of O is zero.*

Theorem B4. *Suppose that if* $x > 0$

$$B(x) > A(x)$$

and if $x < 0$

$$B(x) > A(x)$$

then the portrait consists of a fan in S_R *and two hyperbolic sectors in* S_L *(Figure 3.36). The index of O is zero.*

Definition   If the portrait consists of two hyperbolic sectors and a fan, the equilibrium point O is a *saddle node*.

To complete the discussion we give the promised proof that S is a separatrix.

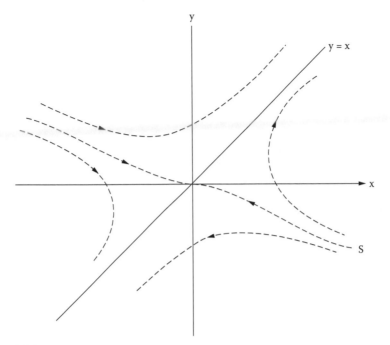

Figure 3.34

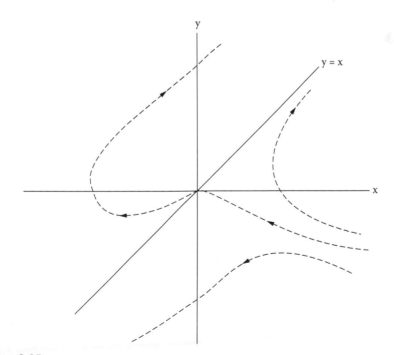

Figure 3.35

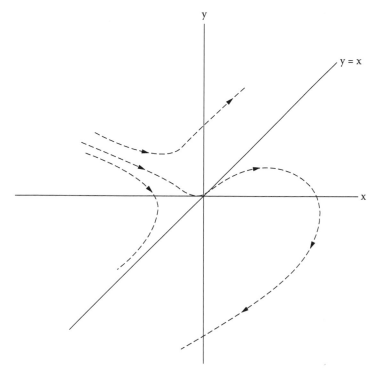

Figure 3.36

Lemma 3.6 *Orbit S is a separatrix, that is, there is no fan which contains S.*

Proof Suppose there exists such a fan. Let two of the orbits in the fan be denoted by $y(x)$ and $y(x) + u(x)$ where $u(x) > 0$ if $x > 0$. See Figure 3.37.

Then we have by equation (3.48),

$$\frac{d[y(x) + u(x)]}{dx} = \frac{[y(x) + u(x) - B]E(x, y(x) + u(x))}{y(x) + u(x) - A}$$

where $E(x, y) = \frac{E_2(x,y)}{E_1(x,y)}$ and hence $E(0, 0) = 1$. Also

$$\frac{dy(x)}{dx} = \frac{(y(x) - B)E}{y(x) - A}$$

and

$$\frac{du}{dx} = \frac{d(y + u)}{dx} - \frac{dy}{dx} \tag{3.51}$$

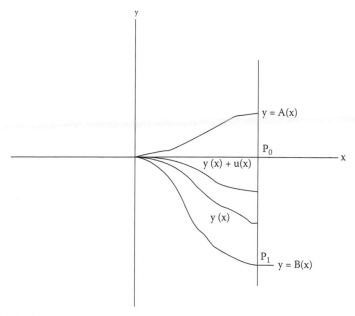

Figure 3.37

Let

$$Y(y) = \frac{(y - B)E}{y - A}$$

then (3.51) becomes

$$\frac{du}{dx} = Y(y + u) - Y(y) = \left[\frac{dY}{dy}(y)\right] u + H(u) \tag{3.52}$$

where $H(u)$ denotes higher order terms in u, that is,

$$\lim_{u \to 0} \frac{|H(u)|}{|u|} = 0$$

But

$$\frac{dY(y)}{dy} = \left[\frac{(y - A) - (y - B)}{(y - A)^2}\right] E + \left[\frac{y - B}{y - A}\right] \frac{\partial E}{\partial y} \tag{3.53}$$

Let $C = A - B$. Then (3.53) becomes

$$\frac{dY}{dy} = \frac{-CE + (y - A)(y - B)\frac{\partial E}{\partial y}}{(y - A)^2}$$

Thus if $u > 0$ and u is sufficiently small, the sign of $\frac{du}{dx}$ is determined by the sign of

$$-CE + (y - A)(y - B)\frac{\partial E}{\partial y}$$

Since

$$B(x) < y(x) < A(x)$$

then

$$|y - A| \le |A - B| \le |A| + |B| \tag{3.54}$$

and

$$|y - B| \le |A - B| \le |A| + |B| \tag{3.55}$$

We may write:

$$A(x) = ax^p \mathcal{E}_1(x)$$
$$B(x) = bx^q \mathcal{E}_2(x)$$

where $p \ge 2$, $q \ge 2$, $a \ne 0$, $b \ne 0$ and $\mathcal{E}_1(x)$, $\mathcal{E}_2(x)$ are power series in x and $\mathcal{E}_j(0) = 1$ for $j = 1, 2$.

Suppose $p < q$. Then if $a > 0$ and x is positive and sufficiently small

$$A(x) = ax^p \mathcal{E}_1(x) > bx^q \mathcal{E}_2(x) = B(x)$$

Also by (3.54) and (3.55),

$$|(y - A)(y - B)| \le (|A| + |B|)^2 = |A|^2 + 2|A||B| + |B|^2$$

Thus $|(y - A)(y - B)|$ is less than or equal to a power series in x, the lowest order term of which has the form

$$Kx^{2p}$$

where K is a positive constant.

Since $\frac{\partial E}{\partial y}$ is a convergent power series in x and y, then if $|x|$ and $|y|$ are sufficiently small, the sign of

$$-cE + (y - A)(y - B)\frac{\partial E}{\partial y}$$

is determined by the sign of

$$-c = B - A < 0$$

and thus $\frac{du}{dt} < 0$. But if $x > 0$ and sufficiently small, then from the definition of fan, it follows that given $\delta > 0$, there exists $x_\delta \in (0, \delta)$ such that $\frac{du}{dt}(x_\delta) > 0$. (Otherwise the fan would just be a separatrix.) Thus if $p < q$ and $a > 0$ the assumption that there exists a fan which contains S leads to a contradiction. The remainder of the proof of Lemma 3.6 consists in showing that each of the conditions (i) $p < q$ and $a < 0$, (ii) $p > q$, and (iii) $p = q$ leads to a contradiction with the hypothesis that $A(x) > B(x)$.

Now suppose $p < q$ and $a < 0$. Then

$$ax^p - bx^q = x^p(a - bx^{q-p})$$

Thus if x is sufficiently small, the expression

$$ax^p - bx^q$$

has the same sign as a which is negative. Thus

$$ax^p < bx^q$$

But this contradicts the hypothesis that $A(x) > B(x)$.

Next suppose that $p > q$. By hypothesis we have

$$ax^p \mathcal{E}_1(x) = A(x) > B(x) = bx^q \mathcal{E}_2(x) \tag{3.56}$$

Then dividing (3.56) by x^q (which is positive) we obtain

$$ax^{p-q} \mathcal{E}_1(x) > b\mathcal{E}_2(x) \tag{3.57}$$

But (3.57) cannot hold if x is sufficiently small because the limit of the left-hand side as $x \to 0$ is 0 while the limit of the right-hand side as $x \to 0$ is $b \neq 0$. Thus if $A(x) > B(x)$, $p > q$ leads to a contradiction.

Finally suppose $p = q$. Then we have

$$ax^p \mathcal{E}_1(x) = A(x) > B(x) = bx^p \mathcal{E}_2(x)$$

Dividing by $x^p > 0$, we have

$$a\mathcal{E}_1(x) > b\mathcal{E}_2(x)$$

The limit of the left-hand side as $x \to 0$ is a, and the limit of the right-hand side as $x \to 0$ is b. Therefore $a = b$. Since $A(x) > B(x)$, then

$$A(x) - x^p > B(x) - ax^p \tag{3.58}$$

Let

$$\mathcal{E}_1(x) = 1 + \alpha_1 x + \alpha_2 x^2 + \cdots$$
$$\mathcal{E}_2(x) = 1 + \beta_1 x + \beta_2 x^2 + \cdots$$

Then

$$A(x) = ax^p (1 + \alpha_1 x + \alpha_2 x^2 + \cdots)$$
$$B(x) = ax^p (1 + \beta_1 x + \beta_2 x^2 + \cdots)$$

and

$$A(x) - ax^p = ax^p \alpha_1 x + ax^p \alpha_2 x^2 + \cdots = a[\alpha_1 x^{p+1} + \alpha_2 x^{p+2} + \cdots]$$
$$B(x) - ax^p = ax^p \beta_1 x + ax^p \beta_2 x^2 + \cdots = a[\beta_1 x^{p+1} + \beta_2 x^{p+2} + \cdots]$$

Substituting in (3.58), we have

$$a[\alpha_1 x^{p+1} + \alpha_2 x^{p+2} + \cdots] > a[\beta_1 x^{p+1} + \beta_2 x^{p+2} + \cdots]$$

Dividing by x^{p+1} we have

$$a[\alpha_1 + \alpha_2 x + \cdots] > a[\beta_1 + \beta_2 x + \cdots] \qquad (3.59)$$

Taking the limit as $x \to 0$ in equation (3.59), we obtain

$$a\alpha_1 = a\beta_1$$

and therefore, dividing by a, we obtain

$$\alpha_1 = \beta_1$$

Continuing in this way we can prove by induction that

$$\alpha_j = \beta_j, \quad j = 2, \ldots$$

and hence that

$$\mathcal{E}_1(x) = \mathcal{E}_2(x)$$

But then since $a = b$ (as already shown)

$$A(x) = B(x)$$

Thus if $A(x) > B(x)$, then $p = q$ leads to a contradiction.

This completes the proof of Lemma 3.6. □

Note.

In order to apply the Bendixson theory, it is only necessary to compare the values of $A(x)$ and $B(x)$ for small values of $|x|$, and this is done by comparing the lowest order terms in $A(x)$ and $B(x)$. Once this comparison is made, Theorems B1, B2, B3, and B4 are immediately applicable.

The Poincaré-Bendixson Theorem

We have been studying solutions which, for the most part, approach equilibrium points or become unbounded as $t \to \infty$ or $t \to -\infty$. So a natural question that arises is: What happens if the solution remains bounded but does not approach an equilibrium point? Of course we have seen examples of such solutions in both the linear case (see Figure 3.8) and the nonlinear case (see Figure 3.11).

But the general answer to this "natural question" is given by an old and widely applied result: the Poincaré-Bendixson theorem. Very roughly, this theorem says that if a two-dimensional autonomous system has a solution which stays in a bounded region and does not approach an equilibrium point, then the solution is itself periodic or it spirals toward a solution which is periodic. From an intuitive viewpoint this is a rather reasonable result. Since the solution stays in a bounded region and does not approach an equilibrium point, then it has to "pile up" some place; so it piles up on a periodic solution. However, a rigorous proof of the theorem is fairly lengthy and requires the full force of the Jordan curve theorem. As we will see, part of the difficulty in proving the theorem lies in the fact that all the considerations take place in the Euclidean plane and distinguishing between intuitive and rigorous arguments in the plane is sometimes difficult.

We start with a precise statement of the Poincaré-Bendixson theorem. For this, we need the notion of the ω-limit set of a solution S, denoted by $\Omega(S)$. We remind the reader that if the solution S is bounded, then $\Omega(S)$ is nonempty, bounded, connected, closed, and invariant (Theorem 3.4). We also use the notation $O(S)$ to denote the orbit of solution S.

Poincaré-Bendixson Theorem. *Given the autonomous system*

$$x' = P(x, y)$$
$$y' = Q(x, y) \tag{3.60}$$

where P, Q are continuous and satisfy a local Lipschitz condition at each point of an open set in R^2, suppose that the solution $S = (x(t), y(t))$ of (3.60) is defined for all $t \geq t_0$, where t_0 is a fixed value, and is such that there exists a number M such that for all $t \geq t_0$

$$|x(t)| + |y(t)| < M$$

Suppose also that $\Omega(S)$ contains no equilibrium points of (3.60).
    Then one of the following alternatives holds. Either

(1) *$(x(t), y(t))$ is a periodic solution (in which case $O(S) = \Omega(S))$;*
    or

(2) *$\Omega(S)$ is the orbit of a periodic solution and solution S approaches $\Omega(S)$ "spirally from the inside" or "spirally from the outside." (The sense in which the words in quotes are used will be described in the proof of the theorem.)*

Definition   The orbit $\Omega(S)$ in alternative (2) is called a *limit cycle*. (For an example which sheds light on the hypothesis that $\Omega(S)$ contains no equilibrium points of (3.60), see Exercise 13.)

In order to prove the Poincaré-Bendixson theorem, we need several preliminary results.

Jordan Curve Theorem. *Let C be a simple closed curve in R^2. Then*

$$R^2 - C = O_1 \cup O_2$$

where O_1, O_2 are disjoint nonempty connected open sets such that:

(1) *For $i = 1, 2$, the boundary of O_i is C.*

(2) *One of the open sets, say O_1, is bounded (it is called the* interior *of C) and the other, O_2, is unbounded (it is called the* exterior *of C).*

(3) *If $p \in O_2$, then*

$$i(C, p) = 0$$

For all $\in O_1$, the index $i(C, p)$ has the same value, either ± 1 or -1. (The sign depends on the orientation assigned to C.)

A rigorous definition of index and a reference for a proof of the Jordan curve theorem are given in the Appendix.

Definition   A point $(x_0, y_0) \in D$ which is not an equilibrium point of (3.60) is a *regular point.*

Definition   A *finite closed segment of a straight line* is a set of points of one of the following forms:

$$L = \{(x, y)/y = mx + b \quad \text{and} \quad c \leq x \leq d\}$$

or

$$L = \{(e, y)/f \leq y \leq g\}$$

where m, b, c, d, e, f, g are constants.

Definition   Let $V = V(x, y)$ denote the vector field $(P(x, y), Q(x, y))$ with domain D. A *transversal* (or *segment without contact*) of V is a finite closed segment of a straight line, say L, such that

(1) $L \subset D$;

(2) If $(x, y) \in L$, then (x, y) is a regular point of V;

(3) If $(x_1, y_1) \in L$, then the slope of $V(x_1, y_1)$ is not equal to the slope of L, that is, L is not tangent to an orbit of (3.60).

We list some properties of transversals which will be needed.

(1) If (x_0, y_0) is a regular point of $V(x, y)$ and if λ is a line which contains (x_0, y_0) and which is not parallel to $V(x_0, y_0)$, there exists a transversal $L \subset \lambda$ such that (x_0, y_0) is in the interior of L.

Proof   Since P, Q are continuous, there is a circular neighborhood $N(x_0, y_0)$ such that if $(x, y) \in \overline{N(x_0, y_0)}$ then the vector $(P(x, y), Q(x, y))$ is not parallel to λ. Let

$$L = \left\{ \overline{N(x_0, y_0)} \right\} \cap \lambda \qquad \qquad \square$$

(2) All orbits of (3.60) which intersect transversal L cross L in the same direction as t increases.

Proof If orbits C_1 and C_2 cross L in different directions at points p_1 and p_2, then there exists $p_3 \in L$ such that p_3 is between p_1 and p_2 and such that the orbit through p_3 is tangent to L. (This follows from the continuity of P and Q and the intermediate value theorem.) □

(3) If
$$F = \{(x(t), y(t))/t \in [a, b]\}$$

is a finite arc of the orbit C of a solution of equation (3.60), and if L is a transversal, then F can cross L only a finite number of times.

Proof Suppose this is not true. Then there exists a monotonic sequence $\{t_n\} \subset [a, b]$ such that
$$\lim_{n \to \infty} t_n = t_0 \in [a, b]$$
and $(x(t_n), y(t_n)) \in L$ for all n, and such that for all n,
$$(x(t_n), y(t_n)) \neq (x(t_0), y(t_0))$$

Let
$$A_0 = (x(t_0), y(t_0))$$

and
$$A_n = (x(t_n), y(t_n))$$

Since
$$\lim_{n \to \infty} A_n = A_0$$

then the limiting direction for the secant $\overline{A_0 A_n}$ of orbit C is the direction of the tangent to C at $(x(t_0), y(t_0))$. But for all n, $\overline{A_0 A_n}$ is contained in L. Hence L is tangent to C at $(x(t_0), y(t_0))$. This contradicts the condition that L is a transversal. □

(4) Let A be an interior point of transversal L. Then, given $\varepsilon > 0$, there exists $r > 0$ such that if Δ is a disc with center A and radius less than or equal to r, then if orbit C (described by solution $(x(t), y(t))$) is in disc Δ at $t = 0$ (i.e., $(x(0), y(0)) \in \Delta$) there exists t_1 such that $|t_1| < \varepsilon$ and $(x(t_1), y(t_1)) \in L$. (See Figure 3.38.)

Proof We assume that the coordinate axes have been rotated and translated so that A is the origin of the coordinate system and L is contained in the x-axis. From the basic Existence Theorem 1.1, there is a unique solution $(x(t, 0, 0), y(t, 0, 0))$ of (3.60) such that
$$x(0, 0, 0) = 0, \quad y(0, 0, 0) = 0$$

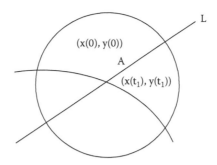

Figure 3.38

and there exists a neighborhood N of $(0, 0)$ such that if $(x_0, y_0) \in N$, then there is a solution $(x(t, x_0, y_0), y(t, x_0, y_0))$ such that

$$x(0, x_0, y_0) = x_0$$
$$y(0, x_0, y_0) = y_0$$

If

$$\frac{\partial}{\partial t} y(0, 0, 0) = 0$$

then the solution $(x(t, 0, 0), y(t, 0, 0))$ is tangent to the x-axis (transversal L) at the origin A. This contradicts the fact that L is a transversal. Hence

$$\frac{\partial}{\partial t} y(0, 0, 0) \neq 0$$

and we can apply the implicit function theorem to solve the equation

$$y(t, x_0, y_0) = 0$$

uniquely for t as a function of (x_0, y_0) in a neighborhood of $t = x_0 = y_0 = 0$.   ▯

Now we obtain some lemmas which are needed to prove the Poincaré-Bendixson theorem.

Lemma 1P. *Suppose $\Omega(S)$ contains a regular point A, and let L be a transversal such that A is an interior point of L. Then there exists a monotonic sequence $\{t_m\}$ such that $t_m \to \infty$ and such that if*

$$A_m = (x(t_m), y(t_m))$$

then if L is contained in a sufficiently small neighborhood of A, it follows that

$$A \cup (\cup_m A_m) = [O(S)] \cap L$$

If $A_1 = A_2$, then $A = A_m$ for all m and $O(S)$ is a simple closed curve. If $A_1 \neq A_2$, then all the A_m's are distinct (i.e., if $i \neq j$, then $A_i \neq A_j$) and for all m, A_{m+1} is between A_m and A_{m+2} on L.

Proof Note first that a transversal L exists by the first of the properties of transversals that were listed earlier. By property (4) of transversals and from the fact that $A \in \Omega(S)$, it follows that there exists a monotonic sequence $\{t_m\}$ such that $t_m \to \infty$ and for each m,

$$(x(t_m), y(t_m)) \in O(S) \cap L$$

By property 3 of transversals, the sequence $\{t_m\}$ can be chosen so that if

$$\bar{t} \in (t_m, t_{m+1})$$

then

$$(x(\bar{t}), y(\bar{t})) \notin L.$$

Let $A_m = (x(t_m), y(t_m))$. Then by (2), either $\{A_m\}$ is a finite set or A is a limit point of the set $\{A_m\}$. If $A_1 = A_2$, then by Lemma 3.3, $O(S)$ is a simple closed curve and $A_2 = A_3 = A_m = A$ for all m because if $s > t_2$,

$$(x(s), y(s)) \in \{(x(t), y(t))/t \in [t_1, t_2]\}$$

Now suppose $A_1 \neq A_2$. If $t \in (t_1, t_2)$, then $(x(t), y(t)) \notin L$. Hence the line segment $\overline{A_1 A_2}$ and the curve

$$\{(x(t), y(t))/t \in [t_1, t_2]\}$$

form a simple closed curve C.

The conventional argument for the proof of Lemma 1P then proceeds as follows. We consider two cases.

Case I. There exists $\varepsilon > 0$ such that if $t \in (t_2, t_2 + \varepsilon)$ then $(x(t), y(t))$ is an element of the interior of C. (See Figure 3.39.) Then for all $t > t_2$, $(x(t), y(t))$ is an element of the interior of C. In order to show this, assume it is not true and let

$$t' = \text{glb}\{t > t_2/(x(t), y(t)) \text{ is not in the interior of } C\}$$

Then $(x(t'), y(t')) \in C$. Since $O(S)$ cannot cross itself (by Theorem 3.2, then $(x(t'), y(t'))$ is a point in the interior of the line segment $\overline{A_1 A_2}$. But then $(x(t), y(t))$ crosses transversal L at $(x(t'), y(t'))$ in the direction opposite to the direction of the crossing at $(x(t_2), y(t_2))$. This contradicts property 2 of transversals and we conclude that for all $t > t_2$, $(x(t), y(t))$ is an element of the interior of C.

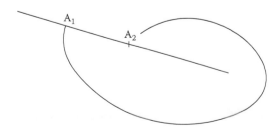

Figure 3.39

Figure 3.40

Next, $A_3 \neq A_2$ because otherwise $O(S)$ would intersect itself but would not be a closed curve. By the argument in the preceding paragraph, $A_3 \neq \overline{A_1 A_2}$. All points on L "to the left of" A_1 are in the exterior or C. Hence, A_3 is "to the right of" A_2. (We will indicate below how the phrases enclosed in quotation marks can be replaced by rigorous language.) The remainder of the proof follows by induction since Case I also holds at A_3, that is, there exists $\varepsilon > 0$ such that if $t \in (t_2, t_2 + \varepsilon)$ then $(x(t), y(t))$ is an element in the interior of the simple closed curve formed by $\overline{A_2 A_3}$ and the curve

$$\{(x(t), y(t))/t \in [t_2, t_3]\}$$

Case II. There exists $\varepsilon > 0$ such that if $t \in (t_2, t_2 + \varepsilon)$, then $(x(t), y(t))$ is an element of the exterior of C. (See Figure 3.40.) By the same kind of argument as for Case I, the set

$$\{(x(t), y(t))/t > t_2\}$$

is in the exterior of C. The remaining steps are also parallel to those in Case I except that at some stage, we may be reduced from Case II to Case I (as, for example, in Figure 3.41). This completes the proof of Lemma 1P. □

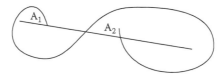

Figure 3.41

Notice that in this proof, we use only Parts (1) and (2) in the statement of the Jordan curve theorem.

Now we indicate how to complete the proof of Lemma 1P starting from the beginning of Case I without splitting the proof into cases and without making any appeal to a geometric pictures, that is, Figures 3.39, 3.40, and 3.41. For this proof, we require Part (3) of the statement of the Jordan curve theorem. Let C be the simple closed curve formed by the segment $\overline{A_1 A_2}$ and the curve

$$\{(x(t), y(t))/t \in [t_1, t_2]\}$$

If for some $t > t_2$, the solution crosses $\overline{A_1 A_2}$, then by property (2) of transversals, it must cross as indicated by the dashed arrow in Figure 3.5, that is, in the same direction as the crossing at A_2. Let

$$\tilde{t} = \text{lub}\{t > t_2\}/(x(s), y(s)) \notin \overline{A_1 A_2} \quad \text{for} \quad t_2 \leq s \leq t$$

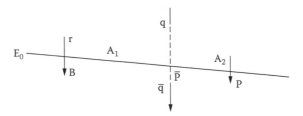

Figure 3.42

and let

$$p = (x(t_0), y(t_0))$$

where

$$t_2 < t_0 < \tilde{t}$$

and let

$$\tilde{p} = (x(\tilde{t}), y(\tilde{t}))$$

Then it is easy to show that there exist points q and $\bar{q}$, as shown in Figure 3.42, such that

$$i(C, q) \neq i(C, \bar{q}) \quad \text{(by the Jordan curve theorem)} \tag{3.61}$$

and

$$i(C, p) = i(C, \bar{q}) \tag{3.62}$$

But

$$i(C, p) = i(C, q) \tag{3.63}$$

because the subset of the solution curve between p and q does not cross $\overline{A_1 A_2}$. Equations (3.62) and (3.63) contradict (3.61). Hence if $t > t_2$, the solution does not cross $\overline{A_1 A_2}$.

Let E_0 be the endpoint of L, that is on the other side of A_1 from A_2. (It is straightforward to translate this statement into rigorous language.) It remains to show that for $t > t_2$, the solution curve does not cross the segment $\overline{E_0 A_1}$. Suppose the curve does cross $\overline{E_0 A_1}$. Let

$$t_3 = \min\{t / t > t_2 \text{ and } (x(t), y(t)) \in \overline{E_0 A_1}\}$$

and let

$$B = (x(t_3), y(t_3))$$

Since the crossing must take place in the direction indicated in Figure 3.42, then if the point r is as indicated in Figure 3.42,

$$i(C, p) = i(C, r) \tag{3.64}$$

(since the part of the orbit joining the points p and r does not intersect the curve C) and

$$i(C, r) = i(C, q) \tag{3.65}$$

if the segment rq is close enough to the segment $\overline{A_1 A_2}$. But by (3.61) and (3.62),

$$i(C, q) \neq i(C, p) \tag{3.66}$$

But (3.64) to (3.66) yield a contradiction.

Lemma 2P. *If L is a transversal of sufficiently short length, then $L \cap \Omega(S)$ contains at most one point.*

Proof   Since L is a transversal, L contains no equilibrium points. By Lemma 1P, $L \cap \Omega(S)$ contains at most one point because if $L \cap \Omega(S)$ contains two points B_1 and B_2, then by Lemma 1P there exist sequences $\{A_m^{(1)}\}$ and $\{A_m^{(2)}\}$ such that $\lim A_m^{(1)} = B_1$ and $\lim A_m^{(2)} = B_2$, and

$$O(S) \cap L = \cup_m A_m^{(1)} \cup B_1 \tag{3.67}$$

and

$$O(S) \cap L = \cup_m A_m^{(2)} \cup B_2 \tag{3.68}$$

Since $B_1 \neq B_2$, then

$$\cup_m A_m^{(1)} \neq \cup_m A_m^{(2)}$$

Hence (3.67) and (3.68) provide a contradiction.   ☐

Lemma 3P. *If $O(S)$ is a closed curve, then $O(S) = \Omega(S)$.*

Proof   First we show that $O(S) \supset \Omega(S)$. Since $O(S)$ is a closed curve, there exist numbers t_1, t_2 such that $t_1 < t_2$ and

$$O(S) = \{(x(t), y(t))/t \in [t_1, t_2]\}$$

Thus $O(S)$ is compact (because it is the continuous image of a compact set) and hence contains its limit points. But an ω-limit point of S is a limit point of $O(S)$.

Now we show that $O(S) \subset \Omega(S)$. Let T be the period of $(x(t), y(t))$, and let $(x(t_0), y(t_0)) \in O(S)$. Then

$$\lim_{m \to \infty} (x(t_0 + mT), y(t_0 + mT)) = \lim_{m \to \infty} (x(t_0), y(t_0)) = (x(t_0), y(t_0))$$

□

Lemma 4P. *If $\Omega(S) \cap O(S) \neq \phi$, then $O(S)$ is a closed curve.*

Proof If $A \in \Omega(S) \cap O(S)$ then A is a regular point because each point of $O(S)$ is a regular point. If $O(S)$ is not a closed curve, then by Lemma 1P there is a sequence of distinct points $\{A_n\}$ such that for all n, $A_n \in O(S) \cap L$ where L is a transversal which has A as an interior point. But $O(S) \subset \Omega(S)$ because $\Omega(S)$ is invariant (by Theorem 3.4). Hence

$$\{A_n\} \subset O(S) \cap L \subset \Omega(S) \cap L$$

This contradicts Lemma 2P.

□

Lemma 5P. *If $\Omega(S)$ contains no equilibrium points and if $\Omega(S) \supset O(S_1)$ where S_1 is a periodic solution, then $\Omega(S) \subset O(S_1)$.*

Proof Suppose the set

$$Q = [\Omega(S)] \cap [O(S_1)]^c$$

is nonempty. Since the set $O(S_1)$ is a closed set, the set Q is not closed because otherwise

$$\Omega(S) = O(S_1) \cup Q$$

is the union of two disjoint closed bounded sets and this contradicts the connectedness of $\Omega(S)$ (given by Theorem 3.4).

Next we show that there exists a limit point p of Q such that $p \in O(S_1)$ by the following argument: first, since $Q \neq \phi$ and is not closed, then Q is infinite. Also, Q is bounded because $\Omega(S)$ is bounded since solution S is bounded for all $t > t_0$. Hence Q has a limit point, and since Q is not closed there is a limit point p such that $p \notin Q$. Hence

$$p \in Q^c = [\Omega(S)]^c \cup O(S_1)$$

Since p is a limit point of Q, then p is a limit point of $\Omega(S)$. But $\Omega(S)$ is closed (by Theorem 3.4). Hence $p \in \Omega(S)$. Since

$$p \in [\Omega(S)]^c \cup O(S_1)$$

it follows that $p \in O(S_1)$.

Let L be a transversal such that p is an interior point of L. If $N_\varepsilon(p)$ is a circular neighborhood of p of radius $\varepsilon > 0$, then the set

$$N_\varepsilon(p) \cap Q = [N_\varepsilon(p)] \cap \{\Omega(S) \cap [O(S_1)]^c\}$$

is nonempty (because p is a limit point of Q) and consists of regular points because p is regular. Let $q \in N_\varepsilon(p) \cap Q$. By property (4) of transversals, L is intersected at point $\bar{p}$ by an orbit $O(S_2)$ through q. $O(S_2)$ is contained in $\Omega(S)$ because $q \in \Omega(S)$ and $\Omega(S)$ is invariant (by Theorem 3.4). Since $q \in [O(S_1)]^c$, then by Theorem 3.1,

$$O(S_2) \cap O(S_1) = \phi$$

Hence since

$$p \in O(S_1) \cap L \subset \Omega(S) \cap L$$

and

$$\bar{p} \in O(S_2) \cap L \subset \Omega(S) \cap L$$

the points p and $\bar{p}$ are distinct points in $\Omega(S) \cap L$. This contradicts Lemma 2P and hence completes the proof of Lemma 5P. $\square$

Proof of Poincaré-Bendixson Theorem. Let $p \in \Omega(S)$. Since $\Omega(S)$ contains no equilibrium points, there is a solution $\bar{S}$ with orbit $O(\bar{S})$ such that $p \in O(\bar{S})$. Since $\Omega(S)$ is invariant, $O(\bar{S}) \subset \Omega(S)$. If $\bar{p} \in \Omega(\bar{S})$, then $\bar{p} \in O(\bar{S})$ or $\bar{p}$ is a limit point of $O(\bar{S})$. Hence, since $\Omega(S)$ is closed, then

$$\Omega(\bar{S}) \subset \Omega(S)$$

Also, since $\Omega(S)$ contains no equilibrium points, then $\bar{p}$ is regular. Now by property (1) of transversals, there is a transversal L such that $\bar{p}$ is an interior point of L, and by Lemma 2P

$$L \cap \Omega(S) = \bar{p}$$

Since $O(\bar{S}) \subset \Omega(S)$, then $L \cap [O(\bar{S})]$ contains at most one point. Hence, by Lemma 1P, $\bar{S}$ is periodic. Since $O(\bar{S}) \subset \Omega(S)$, then by Lemma 5P,

$$\Omega(S) = O(\bar{S})$$

If S is periodic, then by Lemma 3P

$$O(S) = \Omega(S) = O(\bar{S})$$

If S is not periodic, then by Lemma 4P

$$\Omega(S) \cap O(S) = \phi$$

and since $O(\bar{S}) = \Omega(S)$, then

$$O(\bar{S}) \cap O(S) = \phi$$

$O(\bar{S})$ is a simple closed curve, and $O(S)$ is connected. Hence, $O(S)$ is in the interior of $O(\bar{S})$ or in the exterior of $O(\bar{S})$. Let $q \in O(\bar{S})$. Since q is regular, there is a transversal L such that q is an interior point of L. Since S is not periodic, then by Lemma 1P

$$[O(S)] \cap L = \{A_m\} \tag{3.69}$$

where $\{A_m\}$ is a sequence of distinct points linearly ordered on L by subscript and $\lim_{m \to \infty} A_m = q$. Also, since $O(\bar{S}) = \Omega(S)$, then if $\mathcal{U}$ is an open set such that

$$O(\bar{S}) \subset \mathcal{U}$$

then there is a number τ_0, such that if $t \geq \tau_0$, then

$$(x(t), y(t)) \in \mathcal{U} \tag{3.70}$$

Solution S spirals toward $\Omega(S) = O(\bar{S})$ in the sense described by (3.69) and (3.70). This completes the proof of the Poincaré-Bendixson theorem. $\square$

Extensions of Poincaré theory

Poincaré initiated much further work on limit cycles. He studied conditions under which the number of limit cycles is finite if the vector field $(X(x, y), Y(x, y))$ is such that X and Y are polynomials. Dulac [1923] extended greatly Poincaré's work but in 1981, an error was discovered in his work. A complete study (including history) is given by Il'yashenko [1991]. See also Anasov and Arnold [1988, Volume 1].

Application of the Poincaré-Bendixson Theorem

Next we discuss a little of the significance, limitations, and applications of the Poincaré-Bendixson Theorem.

As pointed out earlier, the proof of the Poincaré-Bendixson theorem depends heavily on use of the Jordan curve theorem, which is a theorem in the xy-plane. Also, the intuitive idea (described earlier), that a bounded solution would tend to "pile up" on a periodic solution, no longer has much validity if the solution has an n-dimensional space, where $n > 2$, in which to "move about." Consequently, it is natural to expect that there is no n-dimensional generalization of the Poincaré-Bendixson theorem.

(This suggests that if $n > 2$, then the behavior of orbits may be much more complicated than in the case $n = 2$. This is the first hint we have of how complicated and disorderly the behavior of orbits can be, especially if $n > 2$. In other words, we have our first harbinger of chaos, a subject to which we will return in Chapter 6.)

The Poincaré-Bendixson theorem has been used to find periodic solutions in physics and engineering for many years. (See Andronov and Chaikin [1949] and Farkas [1994].) It is worth observing that although Poincaré developed his theory in papers published in the 1880s (see Poincaré [1881] and later), the theory was not applied to oscillation theory until 1929 (by Andronov [1929]). This was pointed out by Malkin [1956 (translated in 1959)]. We shall discuss some of this use in Chapter 6 after we have obtained a formal description of stability. For now, we describe a couple of conditions under which hypotheses of the Poincaré-Bendixson theorem are satisfied.

Often instead of showing that a particular solution is bounded, it is shown that there exists a bounded open set such that no solution whose orbit contains a point in the bounded open set "escapes" the open set. For example, suppose $f(r)$ is continuously differentiable for $r \geq 0$ and there exists $r_0 > 0$ such that

$$f(r_0) < 0$$

then no solution of the system

$$\frac{dr}{dt} = f(r)$$
$$\frac{d\theta}{dt} = g(r, \theta)$$

where g is continuously differenitable in (r, θ), "escapes" the disc

$$\{(r, \theta)/r < r_0\}$$

because $\frac{dr}{dt} > 0$ at $r = r_0$.

The problem of showing that $\Omega(S)$ contains no equilibrium points is often resolved by studying the equilibrium points themselves. For example, if $(0, 0)$ is an equilibrium point and $\frac{dr}{dt} > 0$ (except at $(0, 0)$) in a neighborhood, however small, of $(0, 0)$ then no solution can approach $(0, 0)$.

Exercises

1. Consider the nonautonomous system

$$\frac{dx}{dt} = f(t, x, y)$$
$$\frac{dy}{dt} = g(t, x, y) \qquad (3.71)$$

where x, y are scalars. Then a solution $(x(t), y(t))$ describes a curve in the xy-plane. Show, with an example, that two such curves may intersect one another.

2. Give an example of (3.71) in Exercise 1 for which there is a solution which describes a curve which crosses itself.

3. Prove the following theorem.

   Theorem (Bendixson Criterion). *Given the system*

$$x' = P(x, y)$$
$$y' = Q(x, y) \qquad (3.60)$$

where P and Q have continuous first partial derivatives with respect to x and y at each point of the (x, y)-plane. Then if the function

$$\frac{\partial P}{\partial x} + \frac{\partial Q}{\partial y}$$

is nonzero at each point of the (x, y)-plane, system (3.60) has no nontrivial periodic solutions.

Hint: Use Green's theorem.

4. Find the general solution of the system

$$\frac{dx}{dt} = y + x(1 - x^2 - y^2)$$

$$\frac{dy}{dt} = -x + y(1 - x^2 - y^2)$$

Show that the circle $x^2 + y^2 = 1$ is the orbit of a solution of the system.

Hint: Transform the system into polar coordinates.

5. Prove that the mapping introduced in the case considered in Figure 3.16 is $1-1$ and continuous.

6. Find the portrait of the orbits in a sector OAB in which $\frac{dr}{dt} > 0$ and such that all orbits which intersect OA or OB exit the sector.

7. Derive the formula

$$F(\bar{r}, \bar{\theta}) = \frac{dr}{dt}(\bar{r}, \bar{\theta})R + \bar{r}\frac{d\theta}{dt}(\bar{r}, \bar{\theta})N$$

(used in the definition of critical direction).

8. Give an example of a system in which there are two adjacent elliptic sectors.

9a. Find the indices of the various equilibrium points obtained in the Bendixson theory.

9b. A conceptually clear and shorter account of the Bendixson theory can be obtained by applying the Weierstrass preparation theorem to $r\frac{dr}{dt} = 0$ and $r^2\frac{d\theta}{dt} = 0$. (It is, however, less explicit and therefore harder to apply.) The approach proceeds as follows:

$$r\frac{dr}{dt} = xy + y^2 + H_r(x, y)$$

where

$$H_r(x, y) = \sum_{p+q \geq 3} h_{pq}x^p y^q$$

then applying the Weierstrass preparation theorem to $r\frac{dr}{dt}$, we obtain

$$r\frac{dr}{dt} = \{y^2 + [b(x)]y + c(x)]\}E_r(x, y)$$

Thus to find the zeros of $r\frac{dr}{dt}$, we need investigate only an equation which is quadratic in y. Similarly for $r^2\frac{d\theta}{dt}$.

10. Give examples to show why the hypothesis that $\Omega(S)$ does not contain an equilibrium point is needed in the Poincaré-Bendixson theorem and why the hypothesis $n = 2$ is needed.

   One important step in the study of a nonlinear autonomous equation is to determine the equilibrium points and how solutions near them behave. We do this now and at the end of Chapter 4 for some of the examples described at the end of Chapter 1.

11. Show that the equilibrium points of the Volterra equations

$$x' = ax - Ax^2 - cxy$$
$$y' = -dy + exy \tag{3.72}$$

are

$$(0, 0)$$
$$\left(\frac{a}{A}, 0\right)$$
$$\left(\frac{d}{e}, \frac{a}{c} - \frac{A}{c}\frac{d}{e}\right)$$

12. Finding the equilibrium points of the Hodgkin-Huxley equation is a serious computation. Setting the right-hand sides of (H-H) equal to zero, we obtain at once

$$m = m_\infty(V)$$
$$h = h_\infty(V)$$
$$n = n_\infty(V)$$

But when these functions of V are substituted into the right-hand side of the first equation in (H-H), we obtain the following messy equation which must be solved for V.

$$\frac{I}{C} - \frac{1}{C}\{\bar{g}_{N_a}[m_\infty(V)]^3[h(V)][(V - V_{N_a})$$
$$+ g_K[n_\infty(V)]^3(V - V_n) + \bar{g}_L(V - V_L)\} = 0$$

In the study by FitzHugh [1969], it is assumed that h and n are constants h_0 and n_0, respectively, and that the phenomena are approximately described by the first two equations:

$$V' = \frac{I}{C} - \frac{1}{C}\left[\bar{g}_{N_a} m^3 h_0(V - V_{N_a}) + \bar{g}_K n_0^4(V - V_K) + \bar{g}_L(V - V_L)\right]$$

$$m' = \frac{m_\infty(V) - m}{\tau_m(V)} \tag{3.73}$$

This assumption is based on the idea that since τ_h and τ_n are much larger than τ_m the system can be approximated by taking τ_h and τ_n to be "infinite." But even finding the equilibrium points of (3.73) is nontrivial. See FitzHugh [1969].

13. Show that there are exactly two equilibrium points in the first octant of the Field-Noyes equations:

$$x' = k_1 A y - k_2 xy + k_3 A x - 2k_4 x^2$$
$$y' = -k_1 A y - k_2 xy + k_5 f z$$
$$z' = k_3 A x - k_5 z$$

and determine their coordinates in terms of A, f, k_1, k_2, k_3, k_4, k_5. (One of the two equilibrium points is the origin.)

14. Show that the Goodwin equations have an equilibrium point in the set

$$P = \{(x_1, \ldots, x_n)/x_i \geq 0, \ i = 1, \ldots, n\}$$

Determine the coordinates of the equilibrium point if $\rho = 1$.

Chapter 4

Stability

Introduction

The material in the previous chapters is basic to all further study of differential equations. The topic of this chapter, stability, is certainly fundamental, but our emphasis and comparatively lengthy treatment are partly motivated by applications to problems in the physical world, especially biological problems. To some extent, by placing a strong emphasis on stability, we are choosing now a particular path in our study of differential equations.

The subject of stability can be approached from two viewpoints. The less important and less interesting viewpoint is that of pure mathematics. That is, a reasonable mathematical problem is to generalize or extend some of the results we obtained concerning the orbits of two dimensional linear homogeneous systems to orbits of nonlinear systems of dimension $n > 2$. It is easy to see that if we attempt as fine an analysis in the more general situation, the results become extremely complicated. The example given by equation (3.19) in Chapter 3 shows how complicated the results can become in the nonlinear two dimensional case, and if we consider even the linear problem in the n-dimensional case, where $n > 2$, the results become quite complicated (see Exercise 1). We are forced to ask for a more modest result than a detailed description of the orbits. Indeed, one of the few questions we can ask which has a reasonable and uncomplicated answer is: Under what conditions do solutions approach the equilibrium point or stay close to the equilibrium point for all sufficiently large t? Actually, the question can be made somewhat more general. We can ask under what conditions solutions approach or stay close to a given solution? Thus our first step is to say precisely what we mean by "approach" or "stay close to." These are stability properties. Then we seek sufficient conditions that these stability properties hold.

Far more interesting and important is the approach to stability theory from the viewpoint of applications to problems in the physical world. If we assume that some physical system is described with a fair degree of accuracy by a system of ordinary differential equations, then the next question is: How are the solutions of the system of ordinary differential equations reflected in the actual behavior of the physical system? For example, suppose the system of equations has an equilibrium point, that is, a solution in which all the components are constants. This corresponds to a state of the physical system in which all the significant quantities are constant. If the system of differential equations has an equilibrium point, can we expect that the physical system will display corresponding behavior, that is, are all significant quantities constant? It is fairly clear that we cannot expect such behavior from the physical system unless the equilibrium point of the system of differential equations has some additional properties. If, for example, all solutions "approach" the equilibrium point, then it would be reasonable to expect corresponding equilibrium behavior of the physical system. On the other hand, if all solutions "go away from" the equilibrium point, then it seems highly unlikely that the existence of the equilibrium point would be reflected by corresponding behavior of the physical system. Thus, we are led again to the question: Under what conditions do solutions approach or stay close to a given solution?

Definition of Stability

Our first step is to formulate this question precisely. This formulation and the theory based on it are due to the great Russian mathematician A. M. Lyapunov.

Definition   Given the n-dimensional system

$$x' = f(t, x) \tag{4.1}$$

where f has domain D, an open set in (t, x)-space (i.e., Euclidean $(n + 1)$-space) which includes the positive t-axis and f is continuous on D; suppose that the solution $x(t)$ of (4.1) is defined for all $t > \tau$. Then solution $x(t)$ is *stable (on the right) in the sense of Lyapunov* if there exists $t_0 > \tau$ such that, if $x(t_0) = x^0$ and if $x(t)$ is denoted by $x(t, t_0, x^0)$, the following conditions are satisfied:

1. There exists a positive constant b such that if

$$|x^1 - x^0| < b$$

then the solution $x(t, t_0, x^1)$ of (4.1) is defined for all $t \geq t_0$;

2. Given $\varepsilon > 0$, then there exists $\delta > 0$, where $\delta = \delta(\varepsilon, f, t_0, x^0)$, that is, δ depends on ε, f, t_0, x^0, such that $\delta \leq b$ and such that if

$$|x^1 - x^0| < \delta$$

then for all $t \geq t_0$,
$$|x(t, t_0, x^1) - x(t, t_0, x^0)| < \varepsilon$$

   Solution $x(t, t_0, x^0)$ is *asymptotically stable (on the right) in the sense of Lyapunov* if conditions (1) and (2) hold and, in addition, we have:

3. There exists $\bar{\delta} > 0$, where $\bar{\delta} = \bar{\delta}(f, t_0, x^0)$, that is, $\bar{\delta}$ depends on f, t_0, x^0, such that $\bar{\delta} < b$ and such that if

$$|x^1 - x^0| < \bar{\delta}$$

then

$$\lim_{t \to \infty} |x(t, t_0, x^1) - x(t, t_0, x^0)| = 0$$

Definition   If conditions (1) and (2) [conditions (1), (2), and (3)] are satisfied by the solutions $x(t, t_0, x^1)$ in a given nonempty subset M of the solutions of (E), then $x(t, t_0, x^0)$ is *conditionally stable* [*asymptotically conditionally stable*] (*on the right*) *in the sense of Lyapunov.*

Definition   Solution $x(t)$ is *unstable* if it is not stable.

Remarks. 1. Condition (1) says roughly that if a solution gets close enough to $x(t, t_0, x^0)$, then the solution is defined for all sufficiently large t. Condition (2) says roughly that if a solution gets close enough to $x(t, t_0, x^0)$, then it stays close to $x(t, t_0, x^0)$ for all later t. It is easy to show (see Example 2 below) that conditions (1) and (2) do not imply (3). Also there are examples (see Cesari [1971]) for which conditions (1) and (3) are satisfied, but for which condition (2) is not satisfied. Thus if condition (1) holds, conditions (2) and (3) are independent.

2. In the definition of stability, the initial condition (t_0, x^0) seems to play a prominent role. However, under reasonable hypotheses, it can be shown (Exercise 2) that if there exists an initial condition (t_0, x^0) such that conditions (1) and (2) [conditions (1), (2), and (3)] in the definition of stability [asymptotic stability] are satisfied then if $\bar{t}_0 > t_0$ and $x(\bar{t}_0) = \bar{x}^0$, then $(\bar{t}_0, \bar{x}^0)$ is an initial condition for which conditions (1) and (2) [conditions (1), (2), and (3)] are satisfied. Also if $\bar{t}_0 < t_0$ and solution $x(t)$ is defined for $t \geq \bar{t}_0$, the same statement holds. Hence stability [asymptotic stability] does not depend on the point t_0.

3. Stability (on the left) and asymptotic stability (on the left) are defined in almost parallel ways. The only difference is that t is decreasing instead of increasing.

4. The word "stable" used in denoting the various kinds of equilibrium points of linear homogeneous two dimensional systems studied in Chapter 3 (e.g., stable node) is not used in the same sense as in the definition of stable given above. All the equilibrium points in Chapter 3 that were termed "stable" are asymptotically stable in the sense of the definition given above.

5. Since we will be using the definitions of stability and asymptotic stability given above most of the time, we will generally omit "(on the right) in the sense of Lyapunov."

6. There are many definitions of stability. (For more extensive accounts of the theory see Lefschetz [1962] and Hahn [1967].) In formulating a definition of stability, one seeks a concept which seems to agree to some extent with an intuitive picture of the physical situation and is at the same time a condition that can be verified in particular cases and can be used as the basis for a coherent mathematical theory. The Lyapunov concepts of stability have these properties and hence have been studied extensively. However, the physical interpretation of the theory is far from satisfactory. At the end of this chapter, we will point out some of the difficulties.

Examples

1. Among the equilibrium points of linear homogeneous two dimensional systems studied in Chapter 3, the stable [unstable] node and spiral are asymptotically stable on the right [left], the center is stable, but not asymptotically stable, and the saddle point is conditionally asymptotically stable. For other examples of conditional asymptotic stability, we can look at the notion of fan (or parabolic sector) introduced in Chapter 3. Suppose F is a parabolic sector in which all the solutions approach the equilibrium point, say $(0, 0)$. Then $(0, 0)$ is asymptotically conditionally stable where the set M is the fan F.

2. For the (scalar) equation

$$x' = 0$$

every solution has the form

$$x(t) = k$$

where k is a constant. Hence every solution is stable but not asymptotically stable.

3. The (scalar) equation

$$x' = x^2$$

is easily solved by separating variables as follows:

$$\frac{dx}{x^2} = dt$$

$$-\frac{1}{x} = t + C$$

If, for $t = 0$, we require that $x = x^0 > 0$, then

$$-\frac{1}{x^0} = C \qquad \text{and}$$

$$-\frac{1}{x} = t - \frac{1}{x^0} = \frac{x^0 t - 1}{x^0}$$

or

$$x(t) = \frac{x^0}{1 - x^0 t}$$

Thus the solution is not defined for $t = 1/x^0$. Thus $x(t) \equiv 0$ is a solution which is not stable because condition (1) is not satisfied.

Stability of Solutions of Linear Systems

We begin with a remark about linear systems (Theorem 4.1 below). Roughly, it states that if one solution of a linear system is stable [asymptotically stable], then all solutions are stable [asymptotically stable]. Conversely, if one solution is unstable, they all are.

Theorem 4.1 *Given the linear system*

$$x' = A(t)x + f(t) \tag{4.2}$$

where $A(t)$ and $f(t)$ are continuous for each $t > t_0 - \delta$, where $\delta > 0$; if there exists a solution $x(t, t_0, u^0)$ of (4.2) which is [asymptotically] stable then every solution of (4.2) is [asymptotically] stable.

Proof Since $x(t, t_0, u^0)$ is stable, then $\varepsilon > 0$ implies there exists $\delta > 0$ such that if $|\Delta u| < \delta$, then for $t \geq t_0$ the solution $x(t, t_0, u^0 + \Delta u)$ is defined as

$$|x(t, t_0, u^0 + \Delta u) - x(t, t_0, u^0)| < \varepsilon \tag{4.3}$$

But

$$x(t, t_0, u^0 + \Delta u) - x(t, t_0, u^0)$$

and

$$x(t, t_0, u + \Delta u) - x(t, t_0, u)$$

are solutions of

$$x' = A(t)x$$

which both have value Δu at $t = t_0$. Hence by the uniqueness of solution Theorem 2.1 and by (4.3), we have for $t \geq t_0$

$$[x(t, t_0, u + \Delta u) - x(t, t_0, u)] = |x(t, t_0, u^0 + \Delta u) - x(t, t_0, u^0)| < \varepsilon$$

Thus $x(t, t_0, u)$ is stable. A slight extension of the argument shows that if $x(t, t_0, u^0)$ is asymptotically stable, $x(t, t_0, u)$ is asymptotically stable. $\square$

Now we state our main results.

Stability Theorem for Homogeneous Linear Systems. *Given the linear homogeneous system*

$$x' = A(t)x \tag{4.4}$$

where $A(t)$ is continuous for all $t > \bar{t}$, a fixed value, let 0 denote the identically zero solution of (4.4). Then the following conclusions hold.

(1) *If A is a constant matrix, then 0 is stable iff*

    (i) $R(\lambda) \leq 0$ *for all eigenvalues λ of A, and*

    (ii) *if $R(\lambda) = 0$, then λ appears only in one-dimensional blocks in the Jordan canonical form of A.*

(2) *If A is constant, then 0 is asymptotically stable iff $R(\lambda) < 0$ for all eigenvalues λ of A.*

(3) *If A(t) has period T, then 0 is stable iff*

    (i) $R(\rho) \leq 0$ *for each characteristic exponent ρ of A(t), and*

    (ii) *if $R(\rho) = 0$, then ρ appears only in one-dimensional boxes in the Jordan canonical form of R (where R is the matrix introduced in Theorem 2.10).*

(4) *If A(t) has period T, then 0 is asymptotically stable iff $R(\rho) < 0$ for each characteristic exponent ρ of A(t).*

Proof   Follows from Theorems 2.8 and 2.13 in Chapter 2.   ⬜

Stability Theorem for Inhomogeneous Linear Systems. *Suppose $\bar{x}(t)$ is a solution of*

$$\frac{dx}{dt} = A(t)x + f(t) \tag{4.5}$$

where A(t) and f(t) are continuous for all $t > \bar{t}$, a fixed value. Then conclusions exactly parallel to those of the preceding theorem hold except that we have $\bar{x}(t)$ and equation (4.5) instead of solution 0 and equation (4.4).

Proof   We indicate the proof for one conclusion. The other conclusions follow similarly. Assume $A(t)$ is constant. Suppose $y(t)$ is a solution of equation (4.5) and suppose

$$y(t) = \bar{x}(t) + u(t)$$

Then

$$\frac{dy}{dt} = A(t)y + f(t)$$

and since $\bar{x}(t)$ is a solution of equation (4.5) we have

$$\frac{du}{dt} = Au$$

and we may apply the preceding theorem.   ⬜

     The stability theorem for homogeneous linear systems reduces the problem of determining whether the solution 0 of equation (4.4) is stable or asymptotically stable to the problem of studying the real parts of the roots of polynomial equations. This is by no means a simple problem, especially if the polynomial is of high degree, say, for example, degree 67. The problem has been discussed at length by Marden [1966]. Practical topological criteria have been obtained by Cesari [1971] and Cronin [1971]. Here we will just state one important and widely used test.

Routh-Hurwitz Criterion

If

$$P(z) = z^n + a_1 z^{n-1} + \cdots + a_{n-1} z + a_n$$

is a polynomial with real coefficients, let $D_1 \ldots, D_n$ denote the following determinants:

$$D_1 = a_1$$

$$\cdots$$

$$D_k = \begin{vmatrix} a_1 & a_3 & a_5 & \cdots & a_{2k-1} \\ 1 & a_2 & a_4 & \cdots & a_{2k-2} \\ 0 & a_1 & a_3 & \cdots & a_{2k-3} \\ 0 & 1 & a_2 & \cdots & a_{2k-4} \\ & \cdot & \cdot & \cdot & \\ 0 & 0 & 0 & \cdot & a_k \end{vmatrix} \quad (k = 2, 3, \ldots, n)$$

where $a_j = 0$ if $j > n$. If $D_k > 0, k = 1, \ldots, n$, then all the solutions of the equation

$$P(z) = 0$$

have negative real parts.

Proof See Marden [1966, Chapter 9]. ▯

Theorem 4.2 *A necessary condition that all the solutions of $P(z) = z^n + a_1 z^{n-1} + \cdots + a_{n-1} z + a_n = 0$, where $a_1, a_2, \ldots, a_n$ are real, have negative real parts is that $a_1, a_2, \ldots, a_n$ are positive.*

Proof If the solutions all have negative real parts, then

$$P(z) = \prod_{q=1}^{k} (z + \alpha_q) \prod_{j=1}^{m} (z + \beta_j - i\gamma_j)(z + \beta_j + i\gamma_j)$$

$$= \prod_{q=1}^{k} (z + \alpha_q) \prod_{j=1}^{m} (z^2 + 2\beta_j z + \beta_j^2 + \gamma_j^2)$$

where $\alpha_q > 0$ for $q = 1, \ldots, k$, and $\beta_j > 0$ for $j = 1, \ldots, m$. ▯

In order to obtain a few more results for linear systems and also to obtain in the next section fundamental results for nonlinear systems, we will use:

Gronwall's Lemma   *If u, v are real-valued nonnegative continuous functions with domain $\{t/t \geq t_0\}$ and if there exists a constant $M \geq 0$ such that for all $t \geq t_0$*

$$u(t) \leq M + \int_{t_0}^{t} u(s)v(s)\,ds \qquad (4.6)$$

then

$$u(t) \leq M \exp \int_{t_0}^{t} v(s)\,ds \qquad (4.7)$$

Proof   See Exercise 12 of Chapter 1.   ⬜

Theorem 4.3   *If the identically zero solution of the linear equation*

$$x' = Ax \qquad (4.8)$$

where A is a constant matrix, is asymptotically stable, then there exists $\varepsilon_0 > 0$ such that if the matrix $C(t)$ is continuous for all $t > -\delta$, where $\delta > 0$, and if $|C(t)| < \varepsilon_0$ for all $t > -\delta$, then the equilibrium point $x = 0$ of the equation

$$x' = [A + C(t)]x \qquad (4.9)$$

is asymptotically stable.

Proof   Let $x(t)$ be an arbitrary fixed solution of (4.9). By the variation of constants formula we may write:

$$x(t) = y(t) + X(t) \int_{0}^{t} [X(s)]^{-1} C(s)x(s)\,ds \qquad (4.10)$$

where $y(t)$ is the solution of $y' = Ay$ such that $y(0) = x(0)$ and $X(t)$ is the fundamental matrix of $y' = Ay$ such that $X(0) = I$, the identity matrix. Then if s is a positive constant, the matrix

$$X(t)[X(s)]^{-1}$$

is a fundamental matrix and at $t = s$,

$$X(t)[X(s)]^{-1} = I$$

The matrix $X(t - s)$ is also a fundamental matrix such that at $t = s$, the matrix is the identity matrix. Hence by uniqueness of solutions, we can conclude that for all real t,

$$X(t)[X(s)]^{-1} = X(t - s)$$

So (4.10) can be written:

$$x(t) = y(t) + \int_0^t X(t-s)C(s)x(s)\,ds \tag{4.11}$$

Since the solution of $x' = Ax$ is asymptotically stable then by (4.3) of the stability theorem for linear systems the eigenvalues of A all have negative real parts. Hence there exist positive constants a and k such that for all $t \geq 0$,

$$|X(t)| \leq ke^{-at}$$

It follows that if $|y(0)| = k_1$, then for all $t \geq 0$, $|y(t)| \leq k_1 ke^{-at}$. Then from (4.11), we have:

$$|x(t)| \leq |y(t)| + \int_0^t |X(t-s)|\,|C(s)|\,|x(s)|\,ds$$

$$\leq k_1 ke^{-at} + \int_0^t \varepsilon_0 ke^{-a(t-s)}|x(s)|\,ds$$

and

$$|x(t)|e^{at} \leq k_1 k + \int_0^t \varepsilon_0 ke^{as}|x(s)|\,ds$$

By Gronwall's lemma,

$$|x(t)|e^{at} \leq k_1 k \exp\int_0^t \varepsilon_0 k\,ds = k_1 k \,\exp(\varepsilon_0 kt)$$

or

$$|x(t)| \leq k_1 k \,\exp[(\varepsilon_0 k - a)t]$$

Then if

$$\varepsilon_0 < \frac{a}{k}$$

the desired result holds. Note that the bound on ε_0 is independent of the solution $x(t)$. This completes the proof of Theorem 4.3. □

Theorem 4.4 *If all the solutions of*

$$x' = Ax$$

where A is a constant matrix, are bounded on the right, if the matrix $C(t)$ is continuous for $t > -\delta$ and if $\int_0^\infty |C(t)|\, dt < \infty$, then all the solutions of

$$x' = [A + C(t)]x$$

are bounded on the right.

Proof   As in the proof of Theorem 4.3, we have:

$$x(t) = y(t) + \int_0^t X(t-s)C(s)x(s)\, ds$$

By hypothesis, there exist positive constants c_1 and c_2 such that for all $t \geq 0$,

$$|y(t)| < c_1$$

and

$$|X(t)| < c_2$$

Since

$$|x(t)| \leq |y(t)| + \int_0^t |X(t-s)|\, |C(s)|\, |x(s)|\, ds$$

or

$$|x(t)| \leq c_1 + \int_0^t c_2|C(s)|\, |x(s)|\, ds$$

then by Gronwall's lemma, for all $t \geq 0$,

$$|x(t)| \leq c_1 \exp \int_0^t c_2|C(s)|\, ds$$

$$\leq c_1 \exp \left\{ c_2 \int_0^\infty |C(s)|\, ds \right\}$$

$\square$

Stability of Solutions of Nonlinear Systems

The results we obtain in this chapter for nonlinear systems are essentially obtained by approximating the nonlinear system with a linear system. Thus the stability theorem for linear systems obtained in the preceding section plays an important role.

We consider the n-dimensional equation

$$x' = f(t, x) \tag{4.1}$$

where $f(t, x)$ has domain D, an open set in (t, x)-space, that is, R^{n+1}, such that

$$D \supset \{(t, 0) \in R^{n+1} / t \geq 0\}$$

and if f_i denotes the ith component of f, then $\partial f_i / \partial x_j$ $(i, j = 1, \ldots, n)$ exists and is continuous at each point of D. Suppose

$$x(t) = x(t, t_0, x^0)$$

is a solution of (4.1) which is defined for all $t \geq t_0 \geq 0$. If $\bar{x}(t)$ is another solution of (4.1) let the function $u(t)$ be defined by

$$\bar{x}(t) = x(t) + u(t)$$

Since $\bar{x}(t)$ is a solution of (4.1), then

$$x' + u' = f(t, x + u) \tag{4.12}$$

Since

$$x' = f(t, x) \tag{4.1}$$

then subtracting (4.1) from (4.12), we obtain

$$u' = f(t, x + u) - f(t, x) \tag{4.13}$$

If we denote the ith component of the right hand side of (4.13) by $F_i(t, u)$, then

$$F_i(t, u) = \sum_j a_{ij}(t) u_j + X_i(t, u)$$

where

$$a_{ij}(t) = \frac{\partial f_i}{\partial x_j}[t, x(t)]$$

u_j is the jth component of $u(t)$ and $X_i(t, u)$ is continuous in t and u and

$$\lim_{|u| \to 0} \frac{X_i(t, u)}{|u|} = 0$$

for each $t \geq t_0$. Thus (4.13) becomes:

$$u' = [a_{ij}(t)]u + X(t, u) \tag{4.14}$$

where $X(t, u)$ has components $X_1(t, u), \ldots, X_n(t, u)$.

Definition   Equation (4.14) is the *variational system of* (4.1) *relative to solution* $x(t)$. The linear equation

$$u' = [a_{ij}(t)]u \tag{4.15}$$

is the *linear variational system* of (4.1) relative to $x(t)$.

Notice that $x(t)$ is a stable [asymptotically stable] solution of (4.1) iff $u(t) \equiv 0$ is a stable [asymptotically stable] solution of (4.14).

Definition   If $u(t) \equiv 0$ is a stable solution of the linear variational system (4.15), then solution $x(t)$ of (4.1) is *infinitesimally stable*.

Infinitesimal stability refers to the stability of the linear approximation of (4.1) that holds in a neighborhood of the given solution $x(t)$. It is rather reasonable to conjecture that infinitesimal stability implies stability. But this reasonable conjecture is not always valid as the following examples show:

1. The linear variational system of the equation $x' = x^2$ relative to the solution $x(t) \equiv 0$ is $u' = 0$. The solution $u(t) \equiv 0$ of $u' = 0$ is a stable solution, but $x(t) \equiv 0$ is an unstable solution of $x' = x^2$. (See Example 3 earlier in this chapter.)

2. The linear variational system of the equation

$$\begin{aligned} x' &= -y + x^3 + xy^2 \\ y' &= x + y^3 + x^2 y \end{aligned} \tag{4.16}$$

relative to the solution $x(t) \equiv 0$, $y(t) \equiv 0$ is

$$\begin{aligned} x' &= -y \\ y' &= x \end{aligned}$$

Since the equilibrium point $(0, 0)$ of this system is a center, the solution $x(t) \equiv 0$, $y(t) \equiv 0$ of (4.16) is infinitesimally stable, but $x(t) \equiv 0$, $y(t) \equiv 0$ is an unstable solution of equation (4.16) because

$$rr' = x(-y + x^3 + xy^2) + y(x + y^3 + x^2 y) = (x^2 + y^2)^2 = r^4$$

(Since $r' > 0$ for all $(x, y) \neq (0, 0)$, all solutions "move away" from $(0, 0)$ with increasing time.)

These examples show that infinitesimal stability does not imply stability. But we will show that if $[a_{ij}(t)]$ is a constant matrix or a periodic matrix, then if $u(t) \equiv 0$ is asymptotically stable, the solution $x(t)$ is asymptotically stable; and if $u(t) \equiv 0$ is unstable, then $x(t)$ is unstable. In order to state these results, it is convenient to introduce the following "little o" definition.

Definition   Suppose the vector function $h(t, x)$ is defined on an open set D in (t, x)-space such that

$$D \supset \{(t, 0) \mid t \in E\}$$

where E is a subset of the t-axis. Then $h(t, x)$ *satisfies the condition*

$$|h(t, x)| = o(|x|)$$

as $|x| \to 0$ *uniformly with respect* to $t \in E$ means: there exists a real-valued function $m(r)$ defined for $r > 0$ such that:

(1) $m(r) > 0$ for all $r > 0$;

(2) $\lim_{r \to 0} m(r) = 0$;

and there exists $\bar{\varepsilon} > 0$ such that if $|x| < \bar{\varepsilon}$, then for all $t \in E$,

$$|h(t, x)| \leq [m(|x|)]|x|$$

Asymptotic Stability Theorem for Nonlinear Systems. *Suppose that the eigenvalues of matrix A are $\lambda_1, \ldots, \lambda_m$ and that $R(\lambda_i) < 0$ for $i = 1, \ldots, m$. Suppose also that $h(t, x)$ is continuous on the set*

$$[(t, x)/t > -\delta, \ x \in U]$$

where $\delta > 0$ and U is a neighborhood in R^n of $x = 0$ and $h(t, x)$ satisfies a Lipschitz condition in x, and

$$|h(t, x)| = o(|x|)$$

as $|x| \to 0$ *uniformly with respect to t where $t > -\delta$. Then the solution $x(t) \equiv 0$ of the equation*

$$x' = Ax + h(t, x) \tag{4.17}$$

is asymptotically stable.

Proof   Let $a < 0$ be a negative number such that

$$R(\lambda_i) < a \qquad (i = 1, \ldots, m)$$

and let $Y(t)$ be the fundamental matrix of $y' = Ay$ such that $Y(0) = I$, the identity matrix. Then there is a positive number C such that for all $t \geq 0$,

$$|Y(t)| \leq Ce^{at}$$

Suppose a solution $x(t)$ of (4.17) is defined in an open interval I containing $t = 0$. By the variation of constants formula, if $t \in I$,

$$x(t) = y(t) + \int_0^t Y(t - s)h[s, x(s)]\, ds$$

But

$$|y(t)| = |Y(t)y(0)| = |Y(t)x(0)| \leq Ce^{at}|x(0)|$$

and hence

$$|x(t)| \leq Ce^{at}|x(0)| + \int_0^t |Y(t-s)|\,|h[s, x(s)]|\,ds$$

Now given a constant $M > 0$, then there exists $d = d(M) > 0$ such that if $t \in I$ and $|x| \leq d$, then

$$|h(t, x)| \leq M|x|$$

If $|x(0)| < d$, then there exists $\delta = \delta(d)$ such that if $t \in (-\delta, \delta)$, then

$$|x(t)| < d$$

Hence if $t \in (-\delta, \delta)$, then

$$|x(t)| \leq C|x(0)|e^{at} + \int_0^t Ce^{a(t-s)}M|x(s)|\,ds$$

Since $a = -b$ where $b > 0$, we may write:

$$|x(t)|e^{bt} \leq C|x(0)| + \int_0^t MCe^{bs}|x(s)|\,ds$$

and hence by Gronwall's lemma, if $0 \leq t < \delta$,

$$|x(t)|e^{bt} \leq C|x(0)|\exp\int_0^t MC\,ds$$

or

$$|x(t)| \leq C|x(0)|\exp[(MC - b)t]$$

and if

$$M < \frac{b}{C}$$

then

$$|x(t)| \leq C|x(0)|$$

Thus we have proved the statement: if M is a positive constant such that $M < b/C$, then there exists $d = d(M) > 0$ and $\delta > 0$ such that if $|x(0)| < d$ and $t \in [0, \delta)$, then $|x(t)| < d$ and

$$|x(t)| < C|x(0)|$$

From this statement we obtain at once the following lemma. $\qquad$ $\square$

Lemma 4.1 *Let* $d = d(M)$*, and let*

$$r = \min\left(d, \frac{d}{2C}\right)$$

Let $|x(0)| < r$*, and let* $\delta_1 > 0$ *be such that if* $t \in [0, \delta_1)$ *then*

$$|x(t)| < d$$

Then if $t \in [0, \delta_1)$*,*

$$|x(t)| < C|x(0)| < Cr \leq \frac{d}{2}$$

Now we prove that $x(t) \equiv 0$ is asymptotically stable. To verify conditions (1) and (2), consider solution $x(t)$ such that $|x(0)| < r$, where $r = \min\left(d, \frac{d}{2C}\right)$ and $d \leq d(M)$ and $M < \frac{b}{C}$. Let

$$t_0 = \text{lub}\{t_1/t_1 \geq 0 \text{ and if } t \in [0, t_1) \text{ then solution } x(t) \text{ is defined and } |x(t)| < d\}$$

Then $t_0 > 0$ because $|x(0)| < d$, and $x(t)$ is continuous. Now suppose t_0 is finite. By Lemma 4.1, if $t \in (0, t_0)$, then

$$|x(t)| < \frac{d}{2}$$

It follows that the solution $x(t)$ can be extended so that its domain I contains t_0 and, by the continuity of the solution,

$$|x(t)| < d$$

for all $t \in I$. This contradicts the properties of t_0. Thus conditions (1) and (2) in the definition of stability are established. To verify condition (3), notice that we have just shown that if $|x(0)| < r$, then $|x(t)| < d$ for all $t \geq 0$. We proved earlier that if $|x(t)| < d$, then

$$|x(t)| < C|x(0)| \exp[(MC - b)t]$$

Hence if $|x(0)| < r$, then for all $t \geq 0$,

$$|x(t)| < C|x(0)| \exp[(MC - b)t]$$

Since $MC - b < 0$,

$$\lim_{t \to \infty} |x(t)| = 0$$

Thus condition (3) in the definition of asymptotic stability is satisfied. This completes the proof of the asymptotic stability theorem for nonlinear systems.

Instability Theorem for Nonlinear Systems. *If matrix A has at least one eigenvalue λ such that $\mathcal{R}(\lambda) > 0$, the solution $x(t) \equiv 0$ of*

$$x' = Ax + h(t, x)$$

is unstable.

Proof   We postpone the proof of the theorem until we can use a Lyapunov function. (See Chapter 5, p. 236.)   ⬚

A corollary of the stability and instability theorems in which the constant matrix A is replaced with a periodic matrix $A(t)$ is easily obtained by using the theory of characteristic multipliers and exponents developed in Chapter 2. We obtain:

Corollary to Stability and Instability Theorems for Nonlinear Systems. *Suppose the matrix $A(t)$ is continuous in t for all real t and has period T, and suppose that the hypotheses on $h(t, x)$ are the same as in the stability and instability theorems. Let $\lambda_1, \ldots, \lambda_m$ be the characteristic multipliers and $\rho_1, \ldots, \rho_m$ the corresponding characteristic exponents of $A(t)$. If $|\lambda_1| < 1$ $(\mathcal{R}(\rho_i) < 0)$ for $i = 1, \ldots, m$, the solution $x(t) \equiv 0$ of the equation*

$$x' = A(t)x + h(t, x)$$

is asymptotically stable. If there exists λ_j such that $|\lambda_j| > 1$ $(\mathcal{R}(\rho_j) > 0)$ then $x(t) \equiv 0$ is not stable.

Proof   Let

$$v = Zx$$

where

$$Z(t) = e^{tR}[X(t)]^{-1}$$

Matrix $Z(t)$ is defined in the proof of Theorem 2.13 and we obtain:

$$\frac{dv}{dt} = Rv + Zh[t, Z^{-1}v] \tag{4.18}$$

where the characteristic exponents are the eigenvalues of matrix R. From the properties of $Z(t)$, it follows easily that the stability and instability theorems are applicable to the solution $v(t) \equiv 0$ of (4.18). If all the eigenvalues of R (all the characteristic exponents) have negative real parts, then $v(t) \equiv 0$ is asymptotically stable and, again from the properties of $Z(t)$, it follows that

$$x(t) = Z^{-1}(t)v(t) \equiv 0$$

is asymptotically stable. If there is an eigenvalue of R with positive real part, a similar argument shows that $x(t) \equiv 0$ is not stable. This completes the proof of the corollary.

⬚

The preceding corollary makes possible the study of stability properties of periodic solutions of nonlinear systems. Suppose we consider the system

$$x' = f(t, x) \tag{4.1}$$

where f has continuous first derivatives in an open set D in (t, x)-space and f has period τ in T. Suppose that $\bar{x}(t)$ is a solution of (4.1) which has period T and is such that

$$D \supset \{(t, \bar{x}(0))/0 \le t \le T\}$$

The variational system of (4.1) relative to $\bar{x}(t)$ is:

$$u' = \left[\frac{\partial f_i}{\partial x_j} [t, \bar{x}(t)] \right] u + X(t, u) \tag{4.19}$$

where f_i is the ith component of f and $X(t, u)$ has the same meaning as in equation (4.14). Since $\bar{x}(t)$ has period T and $f(t, x)$ has period T in t, it follows that the matrix

$$M(t) = \left[\frac{\partial f_i}{\partial x_j} [t, \bar{x}(t)] \right]$$

has period T.

Stability Theorem for Periodic Solutions. *Let* $\lambda_1, \ldots, \lambda_m$ *be the characteristic multipliers and* $\rho_1, \ldots, \rho_m$ *the corresponding characteristic exponents of* $M(t)$. *If* $|\lambda_i| < 1$ $(R(\rho_i) < 0)$ *for* $i = 1, \ldots, m$, *then* $\bar{x}(t)$ *is asymptotically stable. If there exists* λ_j *such that* $|\lambda_j| > 1$ $(R(\rho_j) > 0)$ *then* $\bar{x}(t)$ *is not stable.*

Proof If $|\lambda_i| < 1$ for $i = 1, \ldots, m$, then by the corollary, $u(t) \equiv 0$ is an asymptotically stable solution of the variational system (4.19). Hence $\bar{x}(t)$ is asymptotically stable. A similar proof yields the second statement of the theorem. ☐

Some Stability Theory for Autonomous Nonlinear Systems

There are some unexpected complications when the stability theory developed thus far is applied to solutions of autonomous systems. First, it turns out that there is no possibility of applying the stability theorem for periodic solutions obtained above. For if the autonomous system

$$x' = f(x) \tag{4.20}$$

has a solution $\bar{x}(t)$ of period T, then substituting $\bar{x}(t)$ into (4.20) and differentiating, we obtain:

$$\frac{d}{dt} \left[\frac{d}{dt} \bar{x}_i \right] = \sum_{j=1}^{n} \frac{\partial f_i}{\partial x_j} [\bar{x}(t)] \frac{d\bar{x}_j}{dt} \quad (i = 1, \ldots, n) \tag{4.21}$$

where $\bar{x}_i$, f_i are the ith components of $\bar{x}$ and f, respectively. Equation (4.21) shows that $d\bar{x}/dt$ is a solution of the linear variational system of (4.20) relative to solution $\bar{x}(t)$. Since $\bar{x}(t)$ has period T, then $d\bar{x}/dt$ has period T. Hence by Theorem 2.14, at least one characteristic multiplier of the coefficient matrix

$$\left[\frac{\partial f_i}{\partial x_j} [\bar{x}(t)] \right]$$

of (4.21) has the value 1. Hence the hypothesis for asymptotic stability in the stability theorem for periodic solutions cannot be satisfied.

Actually we can go considerably further and show that there are no nontrivial asymptotically stable periodic solutions of an autonomous system. That is, we prove the following theorem.

Theorem 4.5   *If $x(t)$ is a nontrivial periodic solution of an autonomous system*

$$x' = f(x) \tag{4.20}$$

then $x(t)$ is not asymptotically stable.

Proof   Since $x(t)$ is a nontrivial periodic solution, then there exists t_0 such that $f[x(t_0)] \neq 0$. Then

$$\frac{dx}{dt}(t_0) \neq 0 \tag{4.22}$$

Suppose $x(t)$ is asymptotically stable. Then given $\varepsilon > 0$, let $\bar{\delta}(\varepsilon)$ denote the minimum of the δ in condition (2) of the definition of asymptotic stability and the $\bar{\delta}$ in condition (3) of the definition of asymptotic stability. If $|\bar{t}|$ is sufficiently small and fixed, then

$$|x(t_0) - x(t_0 + \bar{t})| < \bar{\delta}(\varepsilon) \tag{4.23}$$

Also by (4.22) there exists $r > 0$ such that

$$|x(t_0) - x(t_0 + \bar{t})| > r \tag{4.24}$$

By the asymptotic stability of $x(t)$ and the fact that $x(t + \bar{t})$ is a solution of (4.20) (by Lemma 3.1 of Chapter 3), it follows that

$$\lim_{t \to \infty} |x(t) - x(t + \bar{t})| = 0 \tag{4.25}$$

but since $x(t)$ has period $T > 0$, then by (4.23) for all integers n,

$$|x(t_0 + n\tau) - x(t_0 + n\tau + \bar{t})| > r$$

and hence (4.25) cannot be satisfied.   □

Theorem 4.5 shows that we must seek a weaker asymptotic stability condition for use in studying periodic solutions of autonomous systems. The following definitions are sometimes useful.

Definition   Let $x(t)$ be a solution of the system

$$x' = f(x) \tag{4.20}$$

where f is defined and satisfies a Lipschitz condition in an open set $D \subset R^n$, such that $x(t)$ has period T. Let C be the orbit of $x(t)$ and if $p \in R^n$, let

$$d(p, C) = \underset{q \in C}{g\ell b} \; d(p, q)$$

where $d(p, q)$ is the usual Euclidean distance between the points p and q in R^n. Then the orbit C is *orbitally stable* if: $\varepsilon > 0$ implies that there exists $\delta > 0$ such that for any solution $x^{(1)}(t)$ of (4.20) for which there exists t_0 such that

$$d(x^{(1)}(t_0), C) < \delta$$

it is true that $x^{(1)}(t)$ is defined for all $t \geq t_0$ and

$$d(x^{(1)}(t), C) < \varepsilon$$

for all t. Orbit C is *asymptotically orbitally stable* if C is orbitally stable and there exists $\varepsilon_0 > 0$ such that for any solution $x(t)$ for which there exists t_0 such that

$$d(x(t_0), C) < \varepsilon_0$$

it is true that

$$\lim_{t \to \infty} d(x(t), C) = 0$$

Orbital stability is a considerably weaker condition than stability. Roughly speaking, it says that solutions stay close together in a point set sense but the positions of the solutions for equal values of t may get very far apart. (For an example of a periodic solution which is orbitally stable but not stable, see Hahn [1967, p. 172]. In the example, the point set of any solution near a periodic solution spirals toward that periodic solution. So, the periodic solution is orbitally stable; and, in fact, it is asymptotically orbitally stable. However, the phase difference between the periodic solution and any nearby solution increases without bound as t increases. Consequently, the periodic solution is not stable.) The notion of orbital stability is useful only if the orbit C is a closed curve, that is, C is described by a periodic solution. Otherwise the orbit C may "cover so much territory" that orbital stability is meaningless.

A stronger stability condition which can be imposed on solutions of autonomous systems and which is somewhat closer to an intuitively desirable stability condition is the following:

Definition   Let $x(t)$ be a solution of the system

$$x' = f(x) \tag{4.20}$$

where f is defined and continuous on an open set $D \subset R^n$. Then $x(t)$ is *uniformly stable* if there exists a constant K such that $\varepsilon > 0$ implies there is a positive $\delta(\varepsilon)$ so that if $u(t)$ is a solution of (4.20) and if there exist numbers t_1, t_2 such that $t_2 \geq K$ and such that

$$|u(t_1) - x(t_2)| < \delta(\varepsilon)$$

then for all $t \geq 0$

$$|u(t + t_1) - x(t + t_2)| < \varepsilon$$

Solution $x(t)$ of (4.20) is *phase asymptotically stable* if there exists a constant K such that $\varepsilon > 0$ implies there is a positive $\delta(\varepsilon)$ so that if $u(t)$ is a solution of (4.20) and if there exist numbers t_1, t_2 such that $t_2 \geq K$ and such that

$$|u(t_1) - x(t_2)| < \delta(\varepsilon)$$

then for all $t \geq 0$,

$$|u(t + t_1) - x(t + t_2)| < \varepsilon$$

and there exists a number t_3 such that

$$\lim_{t \to \infty} |u(t) - x(t_3 + t)| = 0 \qquad (4.26)$$

Roughly speaking, solution $x(t)$ is phase asymptotically stable if $x(t)$ is uniformly stable and condition (4.26) holds.

Phase Asymptotic Stability Theorem for Periodic Solutions. *Suppose*

$$x' = g(x) \qquad (4.27)$$

is such that g has continuous third derivatives in an open set D in R^n and suppose that $\bar{x}(t)$ is a solution of (4.27) such that $\bar{x}(t)$ has period $T > 0$ and such that the linear variational system of (4.27) relative to $\bar{x}(t)$ has $(n-1)$ characteristic exponents $\rho_1, \ldots, \rho_{n-1}$ such that

$$R(\rho_j) < 0 \qquad (j = 1, \ldots, n-1)$$

Then $\bar{x}(t)$ is phase asymptotically stable.

Proof   We note first that, since (4.27) is autonomous, then as pointed out at the beginning of this section, at least one characteristic multiplier of the linear variational system equals one. Hence at least one characteristic exponent is zero and the hypothesis of the theorem could be stated as: Zero is a characteristic exponent of algebraic multiplicity one and all the other characteristic exponents have negative real parts.

By translation of coordinate axes, we may assume that $\bar{x}(0) = 0$ and by a rotation of coordinate axes, we may assume that $\bar{x}_1'(0) \neq 0$, $\bar{x}_j'(0) = 0$ for $j = 2, \ldots, n$. By a further change of coordinates using Floquet theory we may write the variational equation of (4.27) relative to $\bar{x}(t)$ as:

$$u' = Ru + f(t, u) \qquad (4.28)$$

where

$$R = \begin{bmatrix} 0 & 0 \\ 0 & B \end{bmatrix}$$

where the eigenvalues of B are the characteristic exponents which have negative real parts and

$$\lim_{|u-u^{(1)}| \to 0} \frac{|f(t, u) - f(t, u^{(1)})|}{|u - u^{(1)}|} = 0$$

uniformly in t.

In order to prove the theorem, we obtain first a preliminary result.

Suppose that vector a has the form $(0, a_2, \ldots, a_n)$ and consider the integral equation

$$
w(t, a) = e^{tR}a + \int_0^t \begin{bmatrix} 0 & 0 \\ 0 & e^{(t-s)B} \end{bmatrix} f[s, w(s, a)] ds
$$

$$
- \int_t^\infty \begin{bmatrix} 1 & 0 \\ 0 & 0 \end{bmatrix} f[s, w(s, a)] ds \tag{4.29}
$$

We show that if $|a|$ is sufficiently small and $t \geq 0$, then (4.29) can be solved for $w(t, a)$ and

$$\lim_{t \to \infty} w(t, a) = 0$$

uniformly in a. The proof is a standard use of successive approximations. We define the sequence

$$w_0(t, a) = 0$$

$$
w_{k+1}(t, a) = e^{tR}a + \int_0^t \begin{bmatrix} 0 & 0 \\ 0 & e^{(t-s)B} \end{bmatrix} f[s, w_k(s, a)] ds
$$

$$
- \int_t^\infty \begin{bmatrix} 1 & 0 \\ 0 & 0 \end{bmatrix} f[s, w_k(s, a)] ds
$$

Since $a = (0, a_2, \ldots, a_n)$, there exists $M > 1$ such that

$$|e^{tR}a| \leq M|a|e^{-\sigma t}$$

where $\sigma > 0$ and $R(\lambda) < -\sigma$ for all eigenvalues λ of B. Then it follows that

$$|w_1(t, a) - w_0(t, a)| \leq M|a|e^{-\sigma t}$$

Since f is a "higher order" term, uniformly in t, there exists $\delta > 0$ such that if $|u| < \delta$, $|u^{(1)}| < \delta$, then

$$|f(t, u) - f(t, u^{(1)})| < \frac{\sigma}{8M}|u - u^{(1)}|$$

It is a straightforward computation to prove by induction (see Exercise 8) that if $t \geq 0$ and $|a|$ is sufficiently small, $|a| < 1$, then for $k = 1, 2, \ldots$,

$$|w_{k+1}(t, a) - w_k(t, a)| \leq \frac{M|a|e^{-\frac{\sigma}{2}t}}{2^k}$$

It follows that the sequence $\{w_k(t, a)\}$ converges uniformly in t to a function $w(t, a)$ and

$$|w(t, a)| \leq 2M|a|e^{-\frac{\sigma}{2}t}$$

Thus $w(t, a)$ is a solution of (4.29) and

$$\lim_{t \to \infty} w(t, a) = 0$$

uniformly in a for $|a| < 1$. Straightforward computation shows that $w(t, a)$ is a solution of (4.28). From (4.29), we have:

$$w(0, a) = a - \int_0^\infty \begin{bmatrix} 1 & 0 \\ 0 & 0 \end{bmatrix} f[s, w(s, a)] \, ds \qquad (4.30)$$

Denoting $w(0, a)$ by $(v_1, \ldots, v_n)$, we may rewrite (4.30) as

$$v_1 = -\int_0^\infty f_1[s, w(s, a_2, \ldots, a_n)] \, ds$$

$$v_2 = a_2$$

$$v_n = a_n$$

Then we may restate the results concerning $w(t, a)$ as $\qquad$ ☐

Lemma 4.2 *Let $w(t, a)$ where $a = (0, a_2, \ldots, a_n)$ be the solution (obtained above) of (4.29) such that*

$$w(0, a) = (a_1, a_2, \ldots, a_n)$$

where

$$a_1 = -\int_0^\infty f_1[s, w(s, a_2, \ldots, a_n)] \, ds \qquad (4.31)$$

Then $w(t, a)$ is such that

$$\lim_{t \to \infty} w(t, a) = 0$$

uniformly for $|a| < 1$. Also $w(t, a)$ is a solution of (4.28).

Now we are ready to prove the theorem. Let Σ denote the surface in $(a_1, \ldots, a_n)$-space described by (4.31). Suppose $x(t)$ is a solution such that for some t_1,

$$|x(t_1) - \bar{x}(0)| = |x(t_1)| < \varepsilon$$

By reparameterizing $x(t)$, we may choose $t_1 = 0$. We prove first that if ε is sufficiently small, then $x(t)$ intersects the surface Σ. That is, we show that if ε is sufficiently small, then the equation

$$x_1(t) + \int_0^\infty f_1[s, w(s, a_2, \ldots, a_n)] \, ds = 0 \tag{4.32}$$

can be solved for t as a function of $a_2, \ldots, a_n$. Equation (4.32) has the initial solution

$$t = 0, \; a_2 = \cdots = a_n = 0$$

because $w(s, 0) = 0$. If ε is sufficiently small, then

$$x_1'(0) \neq 0.$$

Hence by the implicit function theorem, equation (4.32) can be solved locally for t as a function of $(a_2, \ldots, a_n)$. Thus we have

$$x_1[t(a_2, \ldots, a_n)] = \left(-\int_0^\infty f_1[s, w(s, a_2, \ldots, a_n)] \, ds, a_2, \ldots, a_n \right)$$

Now take a fixed $x(t)$. Reparameterize so that

$$x(0) = \left(-\int_0^\infty f_1[s, w(s, a_2 \ldots, a_n)] \, ds, a_2, \ldots, a_n \right)$$

Let

$$u(t) = x(t) - \bar{x}(t)$$

Then

$$u(0) = \left(-\int_0^\infty f_1[s, w(s, a_2, \ldots, a_n)] \, ds, a_2, \ldots, a_n \right) \tag{4.33}$$

Since $u(t)$ is a solution of (4.28), then (4.33) shows that

$$u(t) = w(t, a)$$

or

$$x(t) - \bar{x}(t) = w(t, a)$$

The proof of the theorem follows from the properties of $w(t, a)$.

It should be pointed out that the notion of phase asymptotic stability cannot be applied to Hamiltonian systems. More precisely, we have the following:

Definition   A *time-independent Hamiltonian system* is a $(2n)$-dimensional system

$$x' = H_y(x, y)$$
$$y' = -H_x(x, y) \tag{4.34}$$

where x, y are n-vectors, $H(x, y)$ is a real-valued function of x, y which has continuous first derivatives with respect to each of the components of x and y, $H_x(x, y)$ denotes the n-vector $(\partial H/\partial x_1, \ldots, \partial H/\partial x_n)$ where $x_1, \ldots, x_n$ are the components of x, and $H_y(x, y)$ has a similar meaning. Thus as a system of scalar equations, the Hamiltonian system is:

$$x_1' = \frac{\partial H}{\partial y_1} \qquad\qquad y_1' = -\frac{\partial H}{\partial x_1}$$

$$\cdots \qquad\qquad\qquad \cdots$$

$$x_n' = \frac{\partial H}{\partial h_n} \qquad\qquad y_n' = -\frac{\partial H}{\partial x_n}$$

Theorem 4.6   *If $(x(t), y(t))$ is a nontrivial periodic solution of a time-independent Hamiltonian system, then $(x(t), y(t))$ is not phase asymptotically stable.*

Proof   If $(x(t), y(t))$ is a nontrivial solution of (4.34), then

$$\begin{aligned}
\frac{d}{dt} H[(x(t), y(t))] &= \frac{\partial H}{\partial x}\frac{dx}{dt} + \frac{\partial H}{\partial y}\frac{dy}{dt} \\
&= H_x \frac{\partial H}{\partial y} + H_y \left[-\frac{\partial H}{\partial x} \right] \\
&= 0
\end{aligned}$$

Thus $H[(x(t), y(t))]$ is constant on the orbit of $(x(t), y(t))$. Suppose $(x(t), y(t))$ is phase asymptotically stable. Then if $(\bar{x}(t), \bar{y}(t))$ is a solution such that there exist t_1, t_2 with

$$|(\bar{x}(t_1), \bar{y}(t_1)) - (x(t_2,), yt_2))| < \delta$$

it follows that there exists t_3 such that

$$\lim_{t \to \infty} |(\bar{x}(t), \bar{y}(t)) - (x(t_3 + t), y(t_3 + t)| = 0$$

Hence

$$H[(\bar{x}(t), \bar{y}(t))] = H[(x(t), y(t)]$$

and it follows that there exists an open set

$$U = \{(u, v) \in R^{2n} / |u - x(t_2)| + |v - y(t_2)| < \delta\}$$

such that $H(x, y)$ is constant in U. Then every point of U is an equilibrium point of the time-independent Hamiltonian system. Since $(x(t_2), y(t_2) \in U$, this yields a contradiction. $\qquad\square$

Some Further Remarks Concerning Stability

Asymptotic stability of the kind discussed in this chapter seems an intuitively reasonable condition to impose. However, it should be pointed out that while the stability concepts introduced in this chapter have the advantage that they can be used as the basis for a coherent and attractive mathematical theory, they are not entirely satisfactory from the physical viewpoint.

An an example of this, consider a four-leg stool in which three of the legs have the same length and the fourth leg is slightly shorter. Then the stool is unstable: It teeters between two positions, neither of which is completely stable. Nevertheless if the fourth leg is only slightly shorter than the other three, the chair would be quite safe. On the other hand, if the legs of a three-leg stool all have about the same length, the stool is stable. But if the legs are each, say, four feet long but are very close to one another, say two inches, few people would risk sitting on it.

Lyapunov stability is neither necessary nor sufficient for stability in a physical sense. For example, suppose that a physical system is described by an equation

$$x' = f(t, x) \tag{4.1}$$

and $x(t)$ is a stable solution of the equation. Suppose that, in the definition of stability, for some reasonable value of ε, say $\varepsilon = 1/2$, it turns out that the corresponding $\delta(\varepsilon)$ is 10^{-25}. From the mathematical viewpoint, this is not important, but from the physical viewpoint, the solution $x(t)$ can hardly be regarded as stable. On the other hand, the solution $x(t)$ may possess useful stability properties even if it is not stable in the rigorous mathematical sense. For example, suppose equation (4.1) is a two dimensional autonomous system and $x(t)$ is an equilibrium point p, and the orbits near p are as sketched in Figure 4.1. that is, the orbits leave p and approach the

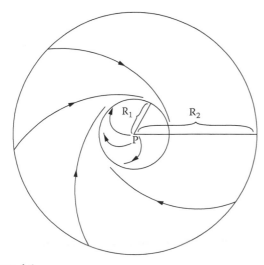

Figure 4.1

circle of radius R_1 which is the orbit of a periodic solution. Orbits which intersect the circle of radius R_2 all approach the circle of radius R_1. Then the equilibrium point p is unstable. But if R_1 is very small and R_2 is large, the point p has useful stability properties. The equilibrium point p is said to be practically stable. For a discussion of practical stability, see LaSalle and Lefschetz [1961], Hahn [1967], and, for an extensive treatment, Lakshmikantham, Leela, and Martynyuk [1990].

Throughout this chapter, we have taken the view that the solutions of the differential equation which have physical significance are those which have stability properties. Another viewpoint is to look at the differential equation itself. Instead of imagining a disturbance such that the physical system being described is moved from one solution onto another solution, we consider that the differential equation itself is perturbed, that is, the original differential equation

$$x' = f(x)$$

is replaced by the equation

$$x' = f(x) + p(x)$$

where $p(x)$ represents a perturbation, and we ask if there exists a homeomorphism from the orbits of the equation

$$x' = f(x)$$

to the orbits of the equation

$$x' = f(x) + p(x)$$

If for "small" perturbations $p(x)$, there exists such a homeomorphism, the equation

$$x' = f(x)$$

is said to be *structurally stable*. (For an enlightening detailed discussion of structural stability with references, see Farkas [1994], especially Chapter 7.)

However, it should be emphasized that neither the viewpoint of studying stable solutions nor the viewpoint of studying structurally stable differential equations is always satisfactory. Certainly for some purposes, the global property of structural stability is better, both for applications and for aesthetic reasons. On the other hand, there are problems important in applications in which the differential equation that must be studied is not structurally stable. This occurs, for example, in bifurcation problems which will be considered in Chapter 7.

Exercises

1. Sketch some of the orbits near the origin of solutions of the following systems:

   (a) $x' = 2x$
   $y' = -3y$
   $z' = -2z$

(b) $x' = 2x + y$
    $y' = -x + 2y$
    $z' = 3z$
(c) $x' = 2x + y$
    $y' = 2y + z$
    $z' = 2z$

2. Prove the statements in Remark 2 following the definitions of stability and asymptotic stability. (Hint: impose enough hypotheses so that the solution $x(t, t_0, x^0)$ is unique and is continuous in x^0. Then prove that if A is a real number and $B_r(x^0)$ is closed ball in R^n with center x^0 and radius r such that the mapping

$$M : \bar{x} \rightarrow x(t_0 + A, t_0, \bar{x})$$

is defined on B_r, then M is a homeomorphism on B_r. The continuity of M follows from the continuity of $x(t, t_0, x^0)$ in x^0. The fact that M is $1 - 1$ follows from the uniqueness of solution $x(t, t_0, x^0)$.

3. Show that Theorem 4.3 is not true if the term "asymptotically stable" is replaced with "stable" (in the hypothesis and the conclusion) by showing that two linearly independent solutions of the equation

$$x'' - \frac{2}{t + M} x' + x = 0 \qquad (t \geq 0)$$

where M is a positive constant, are

$$\sin t - (t + M) \cos t$$

and

$$\cos t + (t + M) \sin t$$

4. Given the linear homogeneous system

$$x' = A(t)x \qquad (4.4)$$

where $A(t)$ is continuous for all real t, prove that if there exists a stable solution $x(t)$ of (4.4), then all solutions of (4.4) are bounded for $t \geq t_0$, a fixed value. Prove that if (4.4) has a fundamental matrix which is bounded for $t \geq t_0$, then each solution of (4.4) is stable. Give an example of an equation (not necessarily linear homogeneous) for which each solution is stable, but no solution is bounded for $t \geq t_0$.

5. Investigate the stability of the equilibrium points of the Volterra equations. (See Exercise 11 at the end of Chapter 3.) That is, determine conditions on the coefficients a, A, c, d, e so that the equilibrium points are asymptotically stable or unstable.

6. Investigate the stability of the equilibrium point $(0, 0, 0)$ of the Field-Noyes equations.

7. Investigate the stability of the equilibrium point $(0, 0, 0)$ of the Goodwin equations if $n = 3$, $\rho = 1$, $\alpha = 1$.

8. Prove the inequality

$$|w_{k+1}(t, a) - w_k(t, a)| \le \frac{M|a|e^{-\frac{\sigma}{2}t}}{2^k}$$

which appears in the proof of the phase asymptotic stability theorem for periodic solutions.

Chapter 5

The Lyapunov Second Method

Definition of Lyapunov Function

We turn now to an entirely different method for studying the stability properties of a given solution. This method, the Lyapunov second method or direct method, uses an approach different from that used in the preceding chapter. It requires no study of the linear parts of equations (eigenvalues or characteristic exponents) and no knowledge is needed of solutions near the solution under study. Of course another price has to be paid: application of Lyapunov's second method demands the construction of the Lyapunov functions to be described later. Such constructions are generally difficult and have been made only for certain classes of equations.

Lyapunov's studies stemmed from consideration of astronomical problems; but the second method has proved very valuable in applications to other problems, for example, control theory. The basic idea of the second method is to generalize the statement that if the potential energy of a physical system is a minimum [maximum] at an equilibrium point, then the equilibrium point is stable [unstable]. A precise version of this statement is called Lagrange's theorem. For a discussion of the relation between the concept of stability and Lagrange's theorem and for a proof of Lagrange's theorem, see Lefschetz [1962]. In the Lyapunov second method, the potential energy function is replaced by a more general kind of function, the Lyapunov function. Our object here is simply to present detailed proofs of the basic theorems of the second method and indicate some applications. For a beautiful and suggestive account of the theory and some applications, especially to control theory, see LaSalle and Lefschetz [1961]. For an extensive treatment of applications, see Aggarwal and Vidyasagar [1977].

We consider an n-dimensional system

$$x' = f(t, x) \tag{5.1}$$

where f is defined and satisfies a local Lipschitz condition with respect to x at each point of the set

$$S = \{(t, x)/t_1 < t, \ |x| < a\}$$

where a is a positive constant and t_1 is a constant. We assume that $x(t) \equiv 0$ is a solution of (5.1), that is, we assume that for all $t > t_1$,

$$f(t, 0) = 0$$

We will describe how the Lyapunov second method can be used to study the stability properties of the solution $x(t) \equiv 0$ of (5.1). (To study the stability properties of a nonzero solution $\bar{x}(t)$, we investigate, as in Chapter 4, the variational system relative to solution $\bar{x}(t)$.) We need first some definitions. Let the real-valued function

$$V(t, x_1, \ldots, x_n) = V(t, x)$$

have the domain

$$S_1 = \{(t, x)/t_2 < t, \ |x| < A\}$$

where $t_2 \geq t_1$ and $0 < A < a$ and assume that:

(1) $V(t, x)$ has continuous first partial derivatives with respect to $t, x_1, \ldots, x_n$ at every point of S_1;

(2) $V(t, 0, \ldots, 0) = 0$ for $t > t_2$.

Definition   The function V is *positive [negative] semidefinite* on S_1 if for all $(t, x) \in S_1$,

$$V(t, x) \geq 0 \ [\leq 0]$$

The function $W(x) = W(x_1, \ldots, x_n)$, satisfying the same hypotheses as $V(t, x)$ but assumed to be independent of t, is *positive [negative] definite on S_1* if $W(x) > 0$ $[< 0]$ for all x such that

$$0 < |x| < A$$

and if $W(0) = 0$. The function $V(t, x)$ is *positive [negative] definite on S_1* if there exists a positive definite function $W(x)$ such that for all $(t, x) \in S_1$,

$$V(t, x) \geq W(x) \quad [-V(t, x) \geq W(x)]$$

The function $V(t, x)$ is *bounded* if there exists a constant $M > 0$ such that for all $(t, x) \in S_1$,

$$|V(t, x)| < M$$

The function $V(t, x)$ has an *infinitesimal upper bound* if $V(t, x)$ is bounded and $\varepsilon > 0$ implies there exists $\delta > 0$ such that if $|x| < \delta$ and $t > t_2$, then $|V(t, x)| < \varepsilon$.

Remark The term "infinitesimal upper bound" is misleading and confusing. The property thus termed is simply the property that $V(t, x)$ is bounded and

$$\lim_{|x| \to 0} V(t, x) = 0$$

uniformly in t for $t > t_2$. However, since the phrase "infinitesimal upper bound" sometimes occurs in the literature, it is necessary to be acquainted with it.

Definition   If $V(t, x)$ has domain S, the function

$$\dot{V} = \dot{V}(t, x_1, \ldots, x_n) = \dot{V}(t, x)$$

is defined to be:

$$\dot{V}(t, x) = \sum_{i=1}^{n} \frac{\partial V}{\partial x_i} f_i(t, x_1, \ldots, x_n) + \frac{\partial V}{\partial t} = \nabla V \cdot f + V_t$$

where $f_i(t, x_1, \ldots, x_n)$ is the i-th component of the function $f(t, x)$ in equation (5.1).

Remark   If $x(t)$ is a solution of (5.1), then

$$\dot{V}(t, x(t)) = \sum_{i=1}^{n} \frac{\partial V}{\partial x_i} x_1' + \frac{\partial V}{\partial t}$$

that is, $\dot{V}(t, x(t))$ is the derivative of $V(t, x)$ along the solution $x(t)$.

Theorems of the Lyapunov Second Method

If the function $V(t, x)$ satisfies a definiteness condition and the function $\dot{V}(t, x)$ satisfies a definiteness condition of the opposite sign, the function $V(t, x)$ is called a *Lyapunov function*. Our next step is to show how the existence of Lyapunov functions implies that stability conditions hold.

Lyapunov Stability Theorem.   *If there exists a function $V(t, x)$ which is positive definite on S and if $\dot{V}$ is negative semidefinite on S, then $x(t) \equiv 0$ is a stable solution of* (5.1)

Remark   The geometric meaning of this theorem is clearly seen if we think of a simple case in which $V(t, x)$ is independent of t and is a positive definite quadratic form $Q(x)$ in the components of X. Then if K is a positive constant the equation

$$Q(x) = K$$

is the equation of an ellipsoid. If K is increased the axes of the ellipsoid are increased. But $\dot{V}(t, x)$ is the rate of change of $Q(x)$ along a solution of (5.1). Hence, since $\dot{V}(t, x)$ is negative semidefinite, then if a solution enters an ellipsoid

$$Q(x) = K$$

it can never "escape" that ellipsoid. Hence, $x(t) \equiv 0$ is stable. The proof of the theorem essentially consists in describing these geometric ideas analytically and rigorously.

   Proof of Stability Theorem. By hypothesis there exists a function $W(x)$ such that if $t > t_1$ and $0 < |x| < a$, then

$$V(t, x) \geq W(x) > 0$$

and

$$\dot{V}(t, x) \le 0$$

Let $\bar{t} > t_1$ and let $x(t, \bar{t}, b)$ be the solution of (5.1) such that $x(\bar{t}, \bar{t}, b) = b$ where $|b| < a$. Then if solution $x(t, \bar{t}, b)$ is maximal, its domain contains $[\bar{t}, t_2)$ where $t_2 = \infty$ or if $t_2 < \infty$ and t_2 is the least upper bound of the domain of $x(t, \bar{t}, b)$, then by Extension Theorem 1.1 in Chapter 1,

$$\lim_{t \to t_2} |x(t, \bar{t}, b)| = a \tag{5.2}$$

If $t \in [\bar{t}, t_2]$

$$V[t, x(t, \bar{t}, b)] - V[\bar{t}, b] = \int_{\bar{t}}^{t} \dot{V}[s, x(s, \bar{t}, b)] \, ds \le 0$$

and hence

$$V[t, x(t, \bar{t}, b)] \le V[\bar{t}, b]$$

Let $\varepsilon > 0$ be such that $\varepsilon < A$ where $A \le a$ and let

$$I = \{x / \varepsilon \le |x| \le A\}$$

Then I is compact, and $0 \notin I$. Hence

$$\min_{x \in I} W(x) = \mu > 0$$

Since $V(t, 0, \ldots, 0) = 0$ for $t > t_1$, there is a number $\lambda > 0$ such that:

(i) $\lambda < \mu$

(ii) $|x| < \lambda$ implies $V(\bar{t}, x) < \mu$

Let b be such that $|b| < a$ and $|b| < \lambda$. Then if $t \in [\bar{t}, t_2)$

$$\mu > V(\bar{t}, b) \ge V(t, x(t, \bar{t}, b)) \ge W[x(t, \bar{t}, b)]$$

Then from the definition of μ, it follows that if $t \in [\bar{t}, t_2)$, then

$$|x(t, \bar{t}, b)| < \varepsilon < a \tag{5.3}$$

If $t_2 < \infty$, then (5.2) holds and (5.3) contradicts (5.2). Hence, $t_2 = \infty$. Thus if $|b| < \min(a, \lambda)$, then (5.3) holds for all $t > \bar{t}$. This completes the proof of the Lyapunov stability theorem.

Lyapunov Asymptotic Stability Theorem. *If there exists a function $V(t, x)$ which is positive definite on S, if $\dot{V}(t, x)$ has an infinitesimal upper bound and if $\dot{V}(t, x)$ is negative definite on S, then $x(t) \equiv 0$ is an asymptotically stable solution of (5.1).*

Remark The geometric meaning of this theorem is very similar to that of the stability theorem. The additional hypotheses that $\dot{V}(t, x)$ has an infinitesimal upper bound and that $\dot{V}(t, x)$ is negative definite rather than just negative semidefinite force a solution not only to stay inside an ellipsoid but to enter smaller ellipsoids.

Proof   From the stability theorem, solution $x(t) \equiv 0$ is stable. Hence, there is a positive constant L such that if $|b| < L$, then $x(t, \bar{t}, b)$ is defined for all $t > \bar{t}$ and $|x(t, \bar{t}, b)| < a$ for all $t > \bar{t}$. We complete the proof of the theorem by showing that if $|b| < L$,

$$\lim_{t \to \infty} |x(t, \bar{t}, b)| = 0$$

First suppose there exists a solution $\bar{x}(t, \bar{t}, b)$ such that $|b| < L$ and such that there exists a constant $m > 0$ and a number $t_0 > \bar{t}$ so that for all $t \geq t_0$

$$V[t, \bar{x}(t, \bar{t}, b)] \geq m > 0$$

Since V has an infinitesimal upper bound there is a number $h \in (0, a)$ such that if $|x| < h$, then $V(t, x) < m$ and hence for all $t > t_0$,

$$|\bar{x}(t, \bar{t}, b)| \geq h$$

By hypothesis, there exist positive definite functions $W(x)$, $W_1(x)$ such that if $t > t_1$ and $|x| < a$, then

$$V(t, x) \geq W(x)$$

and

$$-\dot{V}(t, x) \geq W_1(x)$$

Let

$$\mu_1 = \min_{|x| \in [h, A]} W_1(x)$$

Then for all $t \geq t_0$

$$W_1[\bar{x}(t, \bar{t}, b)] \geq \mu_1 > 0$$

and hence

$$-\dot{V}[t, \bar{x}(t, \bar{t}, b)] \geq W_1[\bar{x}(t, \bar{t}, b)] \geq \mu_1$$

and

$$\int_{t_0}^{t} \dot{V}[s, x(s, \bar{t}, b)] ds \leq \int_{t_0}^{t} (-\mu_1) ds$$

so that for all $t \geq t_0$,

$$V[t, \bar{x}(t, \bar{t}, b)] \leq V[t_0, \bar{x}(t_0, \bar{t}, b)] - \mu_1(t - t_0)$$

Since V is bounded, it follows that if t is sufficiently large,

$$V[t, \bar{x}(t, \bar{t}, b)] < 0$$

This contradicts the hypothesis that $V(t, x)$ is positive definite.

Hence, if $x(t, \bar{t}, b)$ is a solution such that $|b| < L$ and if $\delta > 0$, then there exists $t' > \bar{t}$ such that

$$V[t', x(t', \bar{t}, b)] < \delta \qquad (5.4)$$

Also, if $\tilde{t}, \tilde{t}_0 \in [t_1, \infty)$ and $\tilde{t} < \tilde{t}_0$, then since $\dot{V}(t, x)$ is negative definite,

$$V[\tilde{t}_0, x(\tilde{t}_0, \bar{t}, b)] - V[\tilde{t}, x(\tilde{t}, \bar{t}, b)] = \int_{\tilde{t}}^{\tilde{t}_0} \dot{V}[s, x(s, \bar{t}, b)] \, ds \leq 0$$

That is, $V[t, x(t, \bar{t}, b)]$ is monotonic nonincreasing. This fact, combined with (5.4), shows that

$$\lim_{t \to \infty} V[t, x(t, \bar{t}, b)] = 0$$

and hence also

$$\lim_{t \to \infty} W[x(t, \bar{t}, b)] = 0$$

Now, given $\delta \in (0, A)$, let

$$\mu = \min_{|x| \in [\delta, A]} W(x)$$

There exists T such that if $t > T$, then

$$W[x(t, \bar{t}, b)] < \frac{\mu}{2}$$

Hence, $t > T$ implies

$$|x(t, \bar{t}, b)| < \delta$$

This completes the proof of the Lyapunov asymptotic stability theorem.   ⬜

Lyapunov Instability Theorem. *Suppose there exists a function $V(t, x)$ with domain S which satisfies the following hypotheses:*

(1) $V(t, x)$ *has an infinitesimal upper bound;*

(2) $\dot{V}(t, x)$ *is positive definite on S;*

(3) *there exists $T > t_1$ such that if $\tilde{t} \geq T$ and k is a positive constant, then there exists $c \in R^n$ such that $|c| < k$ and such that $V(\tilde{t}, c) > 0$.*

Then $x(t) \equiv 0$ is a solution of (5.1) which is not stable.

Proof   If $x(t)$ is a solution which is not identically zero, it follows by uniqueness of solution that for all $t > t_1$ for which $x(t)$ is defined

$$|x(t)| \neq 0$$

Hence if $x(t)$ is defined for $t = t_0$, then for all $t \geq t_0$ for which $x(t)$ is defined,

$$V[t, x(t)] - V[t_0, x(t_0)] = \int_{t_0}^{t} \dot{V} ds > 0 \tag{5.5}$$

Let $t_0 > T$ and let $\varepsilon > 0$. By hypothesis, there exists c such that

$$|c| < \min(a, \varepsilon)$$

and such that

$$V(t_0, c) > 0 \tag{5.6}$$

Let $x(t) = x(t, t_0, c)$. We will complete the proof of the theorem by showing that there is a finite value $\tau > t_0$ such that the domain of $x(t, t_0, c)$ does not contain τ. Since V has infinitesimal upper bound, there exists $\lambda \in (0, a)$ such that if $|x| < \lambda$ and $t > t_1$, then

$$|V(t, x)| < V(t_0, c) \tag{5.7}$$

If solution $x(t) = x(t, t_0, c)$ is defined for $t > t_0$, then by (5.5) and (5.6),

$$V[t, x(t)] > V[t_0, x(t_0)] > 0$$

It follows from (5.7) that for $t > t_1$,

$$|x(t)| \geq \lambda$$

By hypothesis, there is a positive definite function $W(x)$ such that for all $(t, x) \in S$, $\dot{V}(t, x) \geq W(x)$. Let

$$\mu = \min_{|x| \in [\lambda, a]} W(x)$$

Then

$$\dot{V}[t, x(t)] \geq W[x(t)] \geq \mu \qquad (5.8)$$

Since $V(t, x)$ is bounded, that is, there exists $L > 0$ such that for all $(t, x) \in S$, $V(t, x) < L$, then by (5.5) and (5.8)

$$L > V[t, x(t)] \geq V[t_0, x(t_0)] + \mu(t - t_0)$$

Thus the set of values $t > t_0$ for which $x(t)$ is defined has a finite upper bound. This completes the proof of the Lyapunov instability theorem. $\square$

Lyapunov functions can also be used to investigate other kinds of stability. We describe its use for two such kinds of stability.

Definition   Given the differential equation

$$x' = f(t, x) \qquad (5.1)$$

and the n-vector function $U(t, x)$ where f and U have continuous first derivatives with respect to t and the components of x for all $(t, x) \in S$ where

$$S = \{(t, x) \mid t > -\delta, \ |x| < a\}$$

with $a > 0$ and $\delta > 0$; suppose also that for all $t > -\delta$,

$$f(t, 0) = 0$$

Then the solution $x(t) \equiv 0$ of equation (5.1) is *stable under persistent disturbances* if $\varepsilon > 0$ implies there exist positive numbers $d_1(\varepsilon)$ and $d_2(\varepsilon)$ such that if for $t > -\delta$ and $|x| < \varepsilon$,

$$|U(t, x)| < d_1(\varepsilon)$$

and if $x(t)$ is a solution of

$$x' = f(t, x) + U(t, x) \qquad (5.9)$$

such that $|x(0)| < d_2(\varepsilon)$, then $x(t)$ is defined for all $t > 0$ and $|x(t)| < \varepsilon$ for all $t > 0$.

Stability Under Persistent Disturbances Theorem.   *Suppose there is a function* $V(t, x)$ *with domain S with the following properties:*

(1) *there exist positive definite functions* $W(x)$, $W_1(x)$ *such that if* $(t, x) \in S$,

$$W(x) \leq V(t, x) \leq W_1(x)$$

(2) *there exists a positive definite function* $W_2(x)$ *such that if* $(t, x) \in S$,

$$\dot{V} = \nabla V \cdot f + V_t \leq -W_2(x)$$

(3) *there exists a positive constant M such that if $(t, x) \in S$, then*

$$\left| \frac{\partial V}{\partial x_i} (t, x) \right| \le M \qquad (i = 1, \ldots, n)$$

Then the solution $x(t) \equiv 0$ of (5.1) is stable under persistent disturbances.

Proof Let $\varepsilon < a$ where a is the number which appears in the definition of the set S. Let

$$m = \min_{|x| = \varepsilon} W(x)$$

Then there exists $r \in (0, 1)$ such that

$$\min_{r\varepsilon \le |x| \le \varepsilon} W(x) \ge \frac{3}{4} m$$

Since $W_1(0) = 0$ and W_1 is continuous, there exists a number $d \in (0, \varepsilon)$ such that if $|x| \le d$, then

$$W_1(x) \le \frac{m}{2}$$

Let μ be a number such that

$$0 < \mu < \min \{W_2(x)/d \le |x| \le \varepsilon\}$$

Let $k \in (0, 1)$ and let

$$d_1(\varepsilon) = \frac{k\mu}{nM}$$

Now consider a solution $x(t)$ of (5.9) and suppose $|x(0)| < d$. If $|x(t)| < d$ for all $t > 0$ for which $x(t)$ is defined, then since $0 < d < a$, it follows (by Extension Theorem 1.1) that $x(t)$ is defined for all $t > 0$ and the proof is complete. Otherwise let t_1 be such that

$$|x(t_1)| = d$$

Then if $\bar{t} > t_1$ and if for $s \in [t_1, \bar{t}]$, $x(s)$ is defined and

$$d \le |x(s)| \le \varepsilon$$

we obtain at any point $(s, x(s))$

$$
\begin{aligned}
V_t + (\nabla V) \cdot (f + U) &= V_t + \nabla V \cdot f + \nabla V \cdot U \\
&= \dot{V} + \nabla V \cdot U \\
&\le -W_2 + \nabla V \cdot U \\
&< -\mu + nM \left(\frac{k\mu}{nM} \right) = -\mu(1 - k) < 0
\end{aligned}
$$

Therefore

$$V[\bar{t}, x(\bar{t})] \le V[t_1, x(t_1)] \le W_1[x(t_1)] \le \frac{m}{2} \tag{5.10}$$

Also

$$|x(\bar{t})| < r\varepsilon \tag{5.11}$$

because if

$$|x(\bar{t})| \geq r\varepsilon$$

then

$$V[\bar{t}, x(\bar{t})] \geq W[x(\bar{t})] \geq \frac{3}{4} m$$

which contradicts (5.10).

Now suppose it is not true that

$$|x(t)| < \varepsilon$$

for all $t > t_1$. Then there exists $\tilde{t} > t_1$ such that

$$|x(\tilde{t})| = \varepsilon$$

Let

$$s_1 = \min\{t/|x(t)| = \varepsilon\}$$
$$s_2 = \max\{t/t < s_1 \text{ and } |x(t)| = d\}$$

Then $s_2 < s_1$ and if $s \in [s_2, s_1]$, then

$$d \leq |x(s)| \leq \varepsilon$$

But then by the same argument used to obtain (5.11),

$$|x(s_1)| < r\varepsilon$$

This contradicts the definition of s_1. Hence for all $t > 0$,

$$|x(t)| < \varepsilon$$

With $d_2(\varepsilon) = d$, the theorem is proved. $\quad\square$

Now we consider equation (5.1) again but now let the set S be defined by:

$$S = \{(t, x)/t_1 < t, \ |x| < \infty\}$$

As before, we assume that $x(t) \equiv 0$ is a solution of (5.1). A function $V(t, x)$ is defined to be positive [negative] semidefinite and positive [negative] definite on the set S exactly as earlier.

Definition   Suppose that for each (t_0, x^0) such that $t_0 > t_1$ and $x^0 \in R^n$, the solution $x(t, t_0, x^0)$ of (5.1) is defined for all $t > t_0$, and that, given $\varepsilon > 0$, there exists $\delta > 0$ such that if

$$|x(\bar{t}, t_0, x^0)| < \delta$$

where $\bar{t} > t_1$, then for all $t > \bar{t}$,

$$|x(t, t_0, x_0)| < \varepsilon$$

and that

$$\lim_{t \to \infty} x(t, t_0, x^0) = 0$$

Then the solution $x(t) \equiv 0$ of (5.1) is *globally asymptotically stable*.

Global Asymptotic Stability Theorem.   *Suppose that there exists a function $V(t, x)$ with domain S such that $V(t, x)$ has the following properties:*

(1) *$V(t, x)$ has an infinitesimal upper bound on each set $S_1 = \{(t, x)/t_1 < t,\ |x| < a\}$ where a is a positive constant;*

(2) *$\dot{V}(t, x)$ is negative definite on S;*

(3) *$V(t, x)$ is positive definite on S, that is, there is a function $W(x)$ such that $W(0) = 0$, $W(x) > 0$ if $|x| \neq 0$ and for all $(t, x) \in S$,*

$$V(t, x) \geq W(x)$$

(4) $\lim_{|x| \to \infty} W(x) = \infty.$

Then $x(t) \equiv 0$ is a globally asymptotically stable solution of (5.1).

Proof   Let $x(t) = x(t, \bar{t}, b)$ be a solution of (5.1) such that $\bar{t} > t_1$ and $b \neq 0$. Since $\dot{V}(t, x)$ is negative semidefinite, then as shown in the proof of the stability theorem, the function $V[t, x(t)]$ is monotonic nonincreasing. Hence for $t \geq \bar{t}$,

$$W[x(t)] \leq V[t, x(t)] \leq V[\bar{t}, b]$$

Hence from condition (4), it follows that the set

$$\{x(t)/t \geq \bar{t}\}$$

is bounded. Hence $x(t)$ is defined for all $t \geq \bar{t}$ (by Exercise 15, Chapter 1). As in the proof of the asymptotic stability theorem, it follows that

$$\lim_{t \to \infty} |x(t)| = 0 \qquad \qquad \Box$$

Applications of the Second Method

As a first application of the Lyapunov second method, we give another proof of the asymptotic stability theorem for nonlinear systems in Chapter 4. For this proof, we need the following lemma.

Lemma 5.1 *If A is a real $n \times n$ matrix and γ is a nonzero real number, there is a real nonsingular matrix P such that*

$$PAP^{-1} = \begin{bmatrix} D_1 & & & \\ & D_2 & & \\ & & \ddots & \\ & & & D_m \end{bmatrix}$$

where each D_j is a real square matrix associated with an eigenvalue λ_j of A. If λ_j is real,

$$D_j = \begin{bmatrix} \lambda_j & & & \\ \gamma & \lambda_j & & \\ & \gamma & & \\ & & \ddots & \\ & & \gamma & \lambda_j \end{bmatrix} \tag{5.12}$$

and if $\lambda = \alpha_j + i\beta_j$,

$$D_j = \begin{bmatrix} \alpha_j & -\beta_j & & & & & \\ \beta_j & \alpha_j & & & & & \\ \gamma & 0 & \alpha_j & -\beta_j & & & \\ 0 & \gamma & \beta_j & \alpha_j & & & \\ & & & & \ddots & & \\ & & & & \gamma & 0 & \alpha_j & -\beta_j \\ & & & & 0 & \gamma & \beta_j & \alpha_j \end{bmatrix} \tag{5.13}$$

Proof From the real canonical form, it follows that it is sufficient to prove the lemma for a $q \times q$ matrix of the form

$$C = \begin{bmatrix} \lambda & 1 & & \\ & \ddots & \ddots & \\ & & & 1 \\ & & & \lambda \end{bmatrix}$$

where λ is real, or a $2q \times 2q$ matrix of the form

$$D = \begin{bmatrix} \alpha & \beta & 1 & 0 & & & \\ -\beta & \alpha & 0 & 1 & & & \\ & & \alpha & \beta & & & \\ & & -\beta & \alpha & & 1 & 0 \\ & & & & & 0 & 1 \\ & & & & \ddots & \alpha & \beta \\ & & & & & -\beta & \alpha \end{bmatrix}$$

where α, β are real.
    First, let

$$R = \begin{bmatrix} 0 & \cdot\ \cdot\ \cdot & 0 & 1 \\ 0 & \cdot\ \cdot\ \cdot & 1 & 0 \\ & & & \vdots \\ 1 & \cdot\ \cdot\ \cdot & & 0 \end{bmatrix}$$

Then $R^{-1} = R$ and

$$RCR^{-1} = \begin{bmatrix} \lambda & 0 & & \cdot\ \cdot\ \cdot & 0 \\ 1 & \lambda & & \cdot\ \cdot\ \cdot & 0 \\ 0 & 1 & \lambda & \cdot\ \cdot\ \cdot & 0 \\ & & \ddots & \ddots & \\ & & & 1 & \lambda \end{bmatrix}$$

Next, let

$$S = \begin{bmatrix} 1 & 0 & \cdot\ \cdot\ \cdot & 0 \\ 0 & \gamma & \cdot\ \cdot\ \cdot & 0 \\ \cdot & & & \\ 0 & \cdot & \cdot\ \cdot\ \cdot & \gamma^{q-1} \end{bmatrix}$$

Then

$$S^{-1} = \begin{bmatrix} 1 & 0 & & \\ 0 & \frac{1}{\gamma} & & \\ & & \ddots & \\ & & & \frac{1}{\gamma^{q-1}} \end{bmatrix}$$

and

$$SRCR^{-1}S^{-1}$$

$$= \begin{bmatrix} 1 & & & \\ & \gamma & & \\ & & \ddots & \\ & & & \gamma^{q-1} \end{bmatrix} \begin{bmatrix} \lambda & 0 & \cdots & 0 \\ 1 & \lambda & & \\ & \ddots & \ddots & \\ & & 1 & \lambda \end{bmatrix} \begin{bmatrix} 1 & \frac{1}{\gamma} & & \\ & & \ddots & \\ & & & \frac{1}{\gamma^{q-1}} \end{bmatrix}$$

$$= \begin{bmatrix} \lambda & 0 & \cdots & & 0 \\ \gamma & \gamma\lambda & 0\cdots & & 0 \\ \gamma^2 & \gamma^2\lambda & \cdots & & 0 \\ & & & & \\ & \gamma^{q-1} & \gamma^{q-1}\lambda \end{bmatrix} \begin{bmatrix} 1 & \frac{1}{\gamma} & & \\ & & \ddots & \\ & & & \frac{1}{\gamma^{q-1}} \end{bmatrix}$$

$$= \begin{bmatrix} \lambda & & & & \\ \gamma & \lambda & & & \\ & & \ddots & & \\ & \gamma & & \lambda & \\ & & & \ddots & \\ & & & \gamma & \lambda \end{bmatrix}$$

To treat the matrix D: if R is a $2q \times 2q$ matrix, then

$$RDR^{-1} = \begin{bmatrix} \alpha & -\beta & & & & & & \\ \beta & \alpha & & & & & & \\ 1 & 0 & \alpha & -\beta & & & & \\ 0 & 1 & \beta & \alpha & & & & \\ & & 1 & 0 & \alpha & -\beta & & \\ & & 0 & 1 & \beta & \alpha & & \\ & & & & \cdot & \cdot & \cdot & \cdot \\ & & & & & 1 & 0 & \alpha & -\beta \\ & & & & & 0 & 1 & \beta & \alpha \end{bmatrix}$$

Let S be the $2q \times 2q$ matrix:

$$S = \begin{bmatrix} I_2 & & & \\ & \gamma I_2 & & \\ & & \ddots & \\ & & & \gamma^{q-1}I_2 \end{bmatrix}$$

where I_2 is the 2×2 identity matrix. □

We let

$$V(t, x) = \sum_{i=1}^{n} x_i^2$$

Then $V(t, x)$ is certainly positive definite on (t, x)-space and has an infinitesimal upper bound. Hence, by the asymptotic stability theorem in this chapter, it is sufficient to show that $\dot{V}(t, x)$ is negative definite for all x such that $|x| < a$, where a is some positive number. We prove this for the case where matrix A is a D_j as described in (5.12) and for the case where D_j is described as in (5.13). Suppose first that D_j has the form (5.12) and suppose $\lambda_j < -\sigma < 0$. Then

$$\begin{aligned}
\dot{V} &= 2x_1 x_1' + \cdots + 2x_n x_n' \\
&= 2x_1 \lambda_j x_1 + 2x_2(\gamma x_1 + \lambda_j x_2) + \cdots + 2x_n(\gamma x_{n-1} + \lambda_j x_n) \\
&\quad + 2x_1 h_1 + \cdots + 2x_n h_n
\end{aligned}$$

where $h(t, x) = (h_1(t, x), \ldots, h_n(t, x))$.

Then

$$\begin{aligned}
\dot{V} &< -2\sigma(x_1^2 + \cdots + x_n^2) + 2\gamma(x_1 x_2 + x_2 x_3 + \cdots + x_{n-2} x_{n-1} + x_{n-1} x_n) \\
&\quad + \sum_{j=1}^{n} 2x_j h_j \\
&< -2\sigma(x_1^2 + \cdots + x_n^2) + 2\gamma[2(x_1^2 + \cdots + x_n^2)] + \sum 2x_j h_j
\end{aligned}$$

Choose $\gamma = \frac{\sigma}{4}$. Then

$$\dot{V} < -\sigma(x_1^2 + \cdots + x_n^2) + \sum 2x_j h_j$$

Since $|h(t, x)| = o(|x|)$, then if $\sum_{j=1}^{n} x_j^2$ is sufficiently small,

$$\dot{V} \le -\frac{\sigma}{2}(x_1^2 + \cdots + x_n^2).$$

If D_j has the form (5.13), let us, for convenience, delete the subscripts and write $\lambda_j = \alpha + i\beta$. Then

$$\begin{aligned}
\dot{V} &= 2x_1(\alpha x_1 - \beta x_2) + 2x_2(\beta x_1 + \alpha x_2) + 2x_3(\gamma x_1 + \alpha x_3 - \beta x_4) \\
&\quad + 2x_4(\gamma x_2 + \beta x_3 + \alpha x_4) + \cdots + 2x_{n-1}(\gamma x_{n-3} + \alpha x_{n-1} - \beta x_n) \\
&\quad + 2x_n(\gamma x_{n-2} + \beta x_{n-1} + \alpha x_n) + 2x_1 h_1 + \cdots + 2x_n h_n \\
&= 2\alpha(x_1^2 + x_2^2 + \cdots + x_n^2) + 2\gamma x_1 x_3 + 2\gamma x_2 x_4 + \cdots + 2\gamma x_{n-3} x_{n-1} \\
&\quad + 2\gamma x_{n-2} x_n + 2x_1 h_1 + \cdots + 2x_n h_n
\end{aligned}$$

If $\alpha < -\sigma < 0$, then

$$\begin{aligned}
\dot{V} &\le -2\sigma(x_1^2 + \cdots + x_n^2) + \gamma(x_1^2 + x_3^2) + \gamma(x_2^2 + x_4^2) \\
&\quad + \cdots + \gamma(x_{n-2}^2 + x_n^2) + 2x_1 h_1 + \cdots + 2x_n h_n
\end{aligned}$$

If we choose $\gamma \in (0, \sigma)$, then

$$\dot{V} \le -\sigma(x_1^2 + \cdots + x_n^2) + 2x_1 h_1 + \cdots + 2x_n h_n$$

Since $|h(t, x)| = o(|x|)$, then if $x_1^2 + \cdots + x_n^2$ is sufficiently small,

$$\dot{V} \leq -\frac{\sigma}{2}(x_1^2 + \cdots + x_n^2). \qquad \square$$

Next, we use the Lyapunov second method to prove the instability theorem for nonlinear systems in Chapter 4. (As might be expected, the instability theorem can be proved without resorting to Lyapunov functions, but such a proof is lengthy and the details are rather oppressive.)

In the interests of brevity and clarity we give a complete proof only for a special case, but it is clear how to carry out the same steps for the general case. We assume that the matrix A has eigenvalues $\alpha + i\beta, \alpha - i\beta, a + ib, a - ib$ where $\alpha > 0, a < 0$. The eigenvalues $\alpha + i\beta, \alpha - i\beta$ each have algebraic multiplicity p and the eigenvalues $a + ib, a - ib$ each have algebraic multiplicity q. Also we assume if Lemma 5.1 is applied to A, we obtain

$$PAP^{-1} =
\begin{bmatrix}
\alpha & -\beta & & & & & & & & & & \\
\beta & \alpha & & & & & & & & & & \\
\gamma & 0 & \alpha & -\beta & & & & & & & & \\
0 & \gamma & \beta & \alpha & & & & & & & & \\
& & & & \ddots & & & & & & & \\
& & & & & \gamma & 0 & \alpha & -\beta & & & \\
& & & & & 0 & \gamma & \beta & \alpha & & & \\
& & & & & & & & & a & -b & & \\
& & & & & & & & & b & a & & \\
& & & & & & & & & \gamma & 0 & a & -b \\
& & & & & & & & & 0 & \gamma & b & a \\
& & & & & & & & & & & & \ddots \\
& & & & & & & & & & & \gamma & 0 & a & -b \\
& & & & & & & & & & & 0 & \gamma & b & a
\end{bmatrix}$$

(5.14)

The upper block in the matrix in (5.14) is a $2p \times 2p$ matrix and the lower block is a $2q \times 2q$ matrix.

Let

$$V(t, x) = \sum_{j=1}^{2p} x_j^2 - \sum_{j=2p+1}^{2p+2q} x_j^2$$

By using the same kind of calculations as in the preceding proof, we obtain: if γ is sufficiently close to zero, then

$$\dot{V}(t, x) \geq \frac{\alpha}{2}(x_1^2 + \cdots + x_{2p}^2) + 2x_1 h_1 + \cdots + 2x_{2p} h_{2p}$$
$$- \left(\frac{a}{2}\right)(x_{2p+1}^2 + \cdots + x_{2p+2q}^2) + 2x_{2p+1} h_{2p+1} + \cdots + 2x_{2p+2q} h_{2p+2q}$$

Since a is negative and $|h(t, x)| = 0(|x|)$, then if $|x|$ is sufficiently small, it follows that there exists $R > 0$ such that if $|x| < R$, then $\dot{V}(t, x)$ is positive definite. Also $V(t, x)$ has an infinitesimal upper bound. Finally if c is any nonzero vector of the form

$$(c_1, \ldots, c_{2p}, 0, \ldots, 0)$$

then

$$V(t, c) = \sum_{j=1}^{2p} c_j^2 > 0$$

Thus the hypotheses of the Lyapunov instability theorem are satisfied and we may apply it to obtain the instability theorem in Chapter 4.

Exercises

1. Show that the function

$$V(t, x) = x_1^2 + (\sin^2 t)x_2^2$$

   is positive semidefinite but not positive definite.

2. Show that

$$V(t, x) = x_1^2 + (1 + \sin^2 t)x_2^2$$

   is positive definite and has an infinitesimal upper bound.

3. Show that

$$V(t, x) = \sin(tx)$$

   is bounded but does not have an infinitesimal upper bound.

4. Show that $(0, 0, 0)$ is a globally asymptotically stable solution of

$$x' = -x - xy^2 - x^3$$
$$y' = -7y + 3x^2y - 2yz^2 - y^3$$
$$z' = -5z + y^2z - z^3$$

   Hint: Use the global asymptotic stability theorem with

$$V(t, x) = 3x^2 + y^2 + 2z^2.$$

5. Given the n-dimensional system

$$x' = Ax + f(x) \tag{5.15}$$

where the matrix A is such that each eigenvalue of A has negative real part, show that there exists $b > 0$ such that if

$$|f(x)| < b|x|$$

for all x with $x \neq 0$, then 0 is a globally asymptotically stable solution of (5.15).

Chapter 6

Periodic Solutions

Periodic Solutions for Autonomous Systems

The study of periodic solutions of ordinary differential equations has been a significant part of mathematics for several centuries. The earliest motivation for the study was classical mechanics, especially celestial mechanics. Poincaré and Lyapunov, motivated by this subject, developed theory which has been extended and applied in the twentieth and twentyfirst centuries to study radically different topics: radio circuits, control theory, chemical oscillations, population theory, epidemics, and numerous problems in physiology.

We have already in Chapter 3 encountered periodic solutions of autonomous systems, most notably the Poincaré-Bendixson theorem, and our goal in this chapter is to describe some extensions of these results, especially those that may be effective in applications. As pointed out earlier, there is no direct generalization of the Poincaré-Bendixson theorem. It is not difficult to give examples of solutions of a three-dimensional system which are bounded and which do not have equilibrium points among their ω-limit points but which are nevertheless neither periodic nor do their orbits approach orbits of periodic solutions (see Exercise 10 in Chapter 3). So one must be resigned to the prospect of imposing more hypotheses. Further, these hypotheses can be expected to be strong for the simple reason that R^2 is very different from R^n ($n > 2$). A number of useful results in this direction are described in detail in Farkas [1994, Section 5.4, pp. 251–268].

There is, however, a different direction that can be taken in applications if the system being modelled with a differential equation is not fully understood, as is often the case with biological systems. To describe this direction, we consider briefly the general question of modeling such imperfectly known systems. We look first at some examples of limit cycles which display various stability properties.

Example 6.1 Consider:

$$x' = -y + \frac{x}{\sqrt{x^2 + y^2}} [1 - (x^2 + y^2)]$$

$$y' = x + \frac{y}{\sqrt{x^2 + y^2}} [1 - (x^2 + y^2)]$$

Using the polar coordinate $r = \sqrt{x^2 + y^2}$, we may write this example as:

$$x' = -y + \frac{x}{r}(1 - r^2) \tag{6.1}$$

$$y' = x + \frac{y}{r}(1 - r^2) \tag{6.2}$$

Since

$$rr' = xx' + yy'$$

then, multiplying (6.1) by x and (6.2) by y and adding, we obtain

$$rr' = r(1 - r^2)$$

or

$$r' = (1 - r^2) \tag{6.3}$$

Since

$$\theta' = \frac{xy' - yx'}{r^2}$$

then, multiplying (6.1) by $-y$ and (6.2) by x and adding and then dividing by r^2, we obtain

$$\theta' = \frac{y^2 + x^2}{r^2} = 1$$

Since (6.3) can be written:

$$\frac{1}{2}\left\{\frac{r'}{1-r} + \frac{r'}{1+r}\right\} = 1$$

or

$$\frac{1}{2}\left\{\frac{dr}{1+r} + \frac{dr}{1-r}\right\} = dt$$

then by integrating we obtain

$$\ell n \left|\frac{1+r}{1-r}\right| = 2t + C = 2t + \ell n \left|\frac{1+r_0}{1-r_0}\right|$$

where $r = r_0$ at $t = 0$. Then if $0 < r < 1$ and $r_0 < 1$

$$\frac{1+r}{1-r} = \left[\frac{1+r_0}{1-r_0}\right]e^{2t}$$

or

$$r = \frac{Ke^{2t} - 1}{Ke^{2t} + 1} \tag{6.4}$$

where

$$K = \left[\frac{1 + r_0}{1 - r_0}\right]$$

If $r > 1$ and $r_0 > 1$

$$r = \frac{Ke^{2t} + 1}{Ke^{2t} - 1} \tag{6.5}$$

where

$$K = \frac{r_0 + 1}{r_0 - 1}$$

Inspection of (6.4) and (6.5) shows that $r \to 1$ as $t \to \infty$. Also, $r = 1$ is the orbit of the periodic solution

$$x(t) = \cos t$$
$$y(t) = \sin t \tag{6.6}$$

Hence (6.4) and (6.5) show that all solutions whose orbits are inside the circle $r = 1$ spiral toward the circle and all solutions whose orbits are outside the circle $r = 1$ spiral toward the circle. It is clear that solution (6.6) is asymptotically orbitally stable, and it is easy to show (Exercise 2) that (6.6) is phase asymptotically stable. Thus (6.6) is a limit cycle and every solution approaches this limit cycle.

If all limit cycles had such strong stability properties, the Poincaré-Bendixson theorem would be far more valuable in applied mathematics. Unfortunately, many limit cycles have no stability properties that have any physical significance, as we show now with examples.

Example 6.2

$$x' = \alpha(\sqrt{x^2 + y^2})x + \beta y$$
$$y' = -\beta x + \alpha(\sqrt{x^2 + y^2})y$$

where $\alpha(r)$ is a monotonic differentiable function for all $r > 0$ and

$$\alpha(r) = 0 \qquad r \leq 1$$
$$\alpha(r) < 0 \qquad r > 1$$

(For an example of such a function, see Exercise 3.) If

$$V(x, y) = x^2 + y^2$$

then

$$\dot{V} = 2\alpha x^2 + 2\beta xy - 2\beta xy + 2\alpha y^2$$
$$= 2\alpha(x^2 + y^2)$$

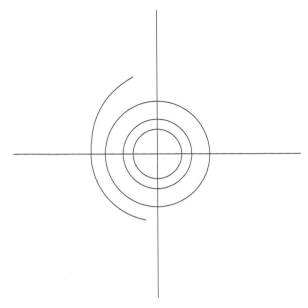

Figure 6.1

where $\alpha = \alpha(r)$. Hence, every orbit which intersects a circle with center 0 and radius > 1 crosses the circle into its interior as t increases. On the other hand, as shown in Chapter 3 (see Case V), every orbit which passes through a point in the interior of the circle with center 0 and radius 1 is a circle with center 0. Hence, the orbits are as sketched in Figure 6.1. The solution

$$x(t) = \cos \beta t$$
$$y(t) = \sin \beta t \tag{6.7}$$

is clearly a limit cycle. It is indeed the Ω-limit set of each solution which passes through a point outside the circle with center 0 and radius 1. This is intuitively quite clear, but the proof is slightly complicated by the fact that $\dot{V} = 0$ if $r = 1$. To give a precise proof, let $(\bar{x}(t), \bar{y}(t))$ be a solution which passes through a point outside the circle with center 0 and radius 1. If

$$V(x, y) = x^2 + y^2$$

then

$$\dot{V} = 2\alpha(\sqrt{\bar{x}^2 + \bar{y}^2})([\bar{x}(t)]^2 + [\bar{y}(t)]^2)$$

and hence $V[\bar{x}(t), \bar{y}(t)]$ is monotonic decreasing. Also, if there exists a strictly increasing sequence $\{t_n\}$ such that $t_n \to \infty$ and such that

$$\lim_{n \to \infty} \{[\bar{x}(t_n)]^2 + [\bar{y}(t_n)]^2\} = B > 1$$

then for all t,

$$V[\bar{x}(t), \bar{y}(t)] = \{[\bar{x}(t)]^2 + [\bar{y}(t)]^2\} \geq B > 1 \tag{6.8}$$

and there exists $M > 0$ such that for all t

$$\dot{V} \leq \alpha(B)\{[x(t)]^2 + [y(t)]^2\} < -M < 0$$

Let $\varepsilon > 0$ and suppose $\bar{t}$ is such that

$$\{[\bar{x}(\bar{t})]^2 + [\bar{y}(\bar{t})]^2\} < B + M\varepsilon$$

Let $\tilde{t}$ be such that $\tilde{t} - \bar{t} > 2\varepsilon$. Then

$$V[\bar{x}(\tilde{t}), \bar{y}(\tilde{t})] - V[\bar{x}(\bar{t}), \bar{y}(\bar{t})] = [\dot{V}(\tau)][\tilde{t} - \bar{t}] < -M(2\varepsilon)$$

where $\tau \in (\bar{t}, \tilde{t})$. Hence

$$V[\bar{x}(\tilde{t}), \bar{y}(\tilde{t})] < B + M\varepsilon - M(2\varepsilon) = B - M\varepsilon$$

This contradicts (6.8). Hence, if $\{t_n\}$ is a sequence such that $t_n \to \infty$, then

$$\lim_{n \to \infty} \{[\bar{x}(t_n)]^2 + [\bar{y}(t_n)]^2\} = 1$$

Thus the solution (6.7) is stable, but is certainly not asymptotically orbitally stable or phase asymptotically stable because of the behavior of the orbits in the interior of the circle with center 0 and radius 1.

If we were to investigate a physical system which was described by Example 6.2, then if there were small disturbances of the system, the solutions of the equations could not be used to make definite predictions about the behavior of the physical system. It would be possible to predict that, after sufficient time had elapsed, both $x(t)$ and $y(t)$ would be less than or equal to one (the orbit would be in the unit circle), but no other predictions about values of $x(t)$ and $y(t)$ could be made. If a small disturbance shifted the physical system from one circular orbit to another, there would be no tendency for the physical system to return to the original orbit. Hence, if small disturbances occurred fairly frequently, the only prediction that could be made would be that $(x(t), y(t))$ would tend to remain in the unit circle. Certainly no prediction of periodicity could be made.

Example 6.3

$$x' = \alpha(\sqrt{x^2 + y^2})x + \beta y$$
$$y' = -\beta x + \alpha(\sqrt{x^2 + y^2})y$$

where $\alpha(r)$ is a differentiable function for $r > 0$ and

$$(r) < 0 \qquad r > 1$$
$$\alpha(r) = 0 \qquad r = 1$$
$$\alpha(r) < 0 \qquad r < 1$$

By the same kind of study as used in Example 6.2, it follows that the circle $r = 1$ is a limit cycle and is the Ω-limit set of every orbit which passes through a point which is outside the circle with center 0 and radius 1. On the other hand, by application of a familiar argument, it follows that every orbit which passes through a point in the interior of the circle with center 0 and radius 1 approaches the origin as $t \to \infty$. The limit cycle $r = 1$ is said in this case to be *semi-stable*.

Stability of the Periodic Solutions

These examples make it clear that it is necessary to investigate not only the existence of periodic solutions but also the stability. If the system (6.1) and (6.2) in Example 6.1 is deemed to be a reliable model or description of a physical system, then the limit cycle in Example 6.1 with its strong stability properties can be considered a reliable predictor of oscillatory or periodic behavior of the physical system. However the limit cycles obtained in Examples 6.2 and 6.3 have unsatisfactory stability properties and could not be deemed reliable predictors of periodic behavior in a physical system.

The reader may well ask what the word "deemed," in the preceding paragraph means. There is no formal answer to this question. Mathematical models or descriptions for the imperfectly known systems we are considering are obtained by using, to some extent, intuition, that is the modeling process has an element of "by guess and by gosh." In many biological systems, the modelmaker (deriver of the ordinary differential equation) can be reasonably confident that there are disturbances or elements in the system which are not described or taken into account in the model. Thus if a solution of the model (differential equation) is to predict real behavior of the biological system then the solution must have the following property: If the biological system is "moving along" the solution and if the biological system is "disturbed" so that the system is "kicked onto" another solution, then as the biological system "moves along" this other solution, we require that this other solution approach the original solution.

Thus we are left with two questions for which there are no entirely satisfactory answers.

(i) How does one know if the model is a valid description of the biological system?

Answer: If the model predicts correctly an experimentally observed result, that is a hopeful sign. But that is only evidence that the model may be reliable. (Put in a cheerless way, that only shows that the model is not entirely wrong.)

(ii) How does one know if a particular solution of the differential equation can be used to predict the actual behavior of the biological system?

Answer: The rather vaguely described property stated at the end of the preceding paragraph should be satisfied. One precise formulation of that property can be made by using the rigorously defined concept of asymptotic stability discussed in Chapter 4. (Recall that the limitations of that concept are discussed at the end of Chapter 4.)

Sell's Theorem

These remarks suggest that if one is concerned with applications, then after establishing the existence of the periodic solution the next step is to determine whether it is asymptotically stable. However, it might be just as reasonable to find the asymptotically stable solutions and then determine if any of these solutions is periodic. For the autonomous case, there is good evidence for using the latter procedure. That evidence is a striking result due to Sell [1966]. Sell's theorem states that the Ω-limit set of a bounded phase asymptotically stable solution is the orbit of a phase asymptotically stable periodic solution (which may, of course, be an equilibrium point). Thus if it is established that there exists a bounded phase asymptotically stable solution S such that $\Omega(S)$ contains no equilibrium points, then $\Omega(S)$ is the orbit of a phase asymptotically stable periodic solution. We have thereby replaced a two-step procedure (finding a periodic solution and then proving that it is asymptotically stable) with a one-step procedure (investigating phase asymptotically solutions). However, the one-step procedure is, by no means, simple in all cases.

In order to prove Sell's theorem (which we will state formally later), we prove first that the Ω-limit set of a bounded uniformly stable solution contains the orbit of an almost periodic uniformly stable solution (the Sell-Deysach theorem).

We consider the n-dimensional autonomous system

$$x' = f(x) \tag{6.9}$$

where f is a continuous n-vector function with domain U, an open set in R^n such that f satisfies a local Lipschitz condition at each point of U.

Definition   Solution $x(t)$ of (6.9) is *bounded* if there exist numbers t_0 and M such that for all $t \geq t_0$ solution $x(t)$ is defined and

$$|x(t)| < M$$

Definition   Let $h(t)$ be a real-valued continuous function from the real t-axis into R^n. Then $h(t)$ is *almost periodic* if and only if: given $\varepsilon > 0$ then there exists a number $L(\varepsilon)$ such that in every open interval $(t_0, t_0 + L(\varepsilon))$, where t_0 is a real number, there is a number T_0 such that for all real t,

$$|h(t + T_0) - h(t)| < \varepsilon$$

The set of numbers $\{T_0\}$ is called a *relatively dense set*. Note that a periodic function $p(t)$ is a trivial example of an almost periodic function: if T is the period of $p(t)$, let $L(\varepsilon) = T + \delta$ where $\delta > 0$. Then if t_0 is any real number, there is a number of the form mT, where m is an integer, in the interval $(t_0, t_0 + T + \delta)$ and for all t,

$$|p(t + mT) - p(t)| = 0$$

Sell-Deysach Theorem. *If $x(t)$ is a bounded uniformly stable solution of (6.9), then there is an almost periodic uniformly stable solution $\bar{x}(t)$ of (6.9) such that $O[\bar{x}(t)] \subset \Omega[x(t)]$.*

Proof   We divide the proof into three lemmas. The first lemma, which is the main result, is a kind of n-dimensional analog of the Poincaré-Bendixson theorem.   ☐

Definition   A solution $x(t)$ of (6.9) defined for all real t is *recurrent* if, given $\varepsilon > 0$, then there exists a positive number $T(\varepsilon)$ such that if t_1 and t_2 are real numbers, then there exists $t_3 \in (t_1, t_1 + T(\varepsilon))$ such that

$$|x(t_2) - x(t_3)| < \varepsilon$$

That is, if t_1 is any fixed real number, every point in $0[x(t)]$ is within ε of the curve

$$[x(t)/t \in (t_1, t_1 + T(\varepsilon))]$$

Lemma 6.1   *If $x(t)$ is a bounded solution of (6.9), there is a recurrent solution $\bar{x}(t)$ of (6.9) such that $O[\bar{x}(t)] \subset \Omega[x(t)]$.*

Proof   By Theorem 3.4, the set $\Omega[x(t)]$ is a nonempty invariant, closed, bounded (and therefore compact) set. Also, by Theorem 3.5, the set $\Omega[x(t)]$ contains a minimal set.

The proof of Lemma 6.1 follows from:

Birkhoff's Theorem.   *Every orbit of a compact minimal set M is recurrent.*

Proof of Birkhoff's Theorem. Let $x(t)$ be a solution such that $O[x(t)] \subset M$, and suppose $x(t)$ is not recurrent. Then there exists a number $r > 0$, a sequence $\{x(\bar{t}_\nu)\}$ and a sequence of pairs $\{t_\nu, T_\nu\}$ such that $T_\nu \to \infty$ and such that if

$$D_\nu = \{x(t)/t \in (t_\nu - T_\nu, t_\nu + T_\nu)\}$$

then for each ν

$$\operatorname*{glb}_{p \in D_\nu} |x(\bar{t}_\nu) - p| \geq r$$

By taking subsequences if necessary, let

$$\bar{x} = \lim_{\nu \to \infty} x(\bar{t}_\nu)$$

and

$$\bar{u} = \lim_{\nu \to \infty} x(t_\nu)$$

Let $u(t)$ be a solution of (6.9) such that

$$u(0) = \bar{u}$$

Let $T > 0$. Since the solution depends continuously on the initial value (Corollary 1.1 to Existence Theorem 1.1), there exists a positive number $\delta = \delta(r/3, T)$ such that if $v(t)$ is a solution of (6.9) and

$$|u(0) - v(0)| < \delta$$

then for all $t \in [-T, T]$,

$$|u(t) - v(t)| < \frac{r}{3}$$

Let v be such that

$$T_v > T$$
$$|\bar{u} - x(t_v)| < \delta$$

and

$$|\bar{x} - x(\bar{t}_v)| < \frac{r}{3}$$

For any fixed $t \in (-T, T)$:

$$|x(\bar{t}_v) - x(t_v + t)| \geq r$$

Hence, for $t \in (-T, T)$,

$$
\begin{aligned}
|u(t) - \bar{x}| &\geq |x(\bar{t}_v) - x(t_v + t)| \\
&\quad - |u(t) - x(t_v + t)| \\
&\quad - |x(\bar{t}_v) - \bar{x}| \\
&\geq \frac{r}{3}
\end{aligned}
$$

Since T was chosen arbitrarily, it follows that for all real t,

$$|u(t) - \bar{x}| \geq \frac{r}{3} \tag{6.10}$$

Since M is closed,

$$\bar{x} \in M$$

and

$$\bar{u} \in M$$

Since M is invariant,

$$O[u(t)] \subset M$$

By (6.10), $\overline{O[u(t)]}$ is a proper closed subset of M. Since $\overline{O[u(t)]}$ is invariant, we have a contradiction to the hypothesis that M is minimal. This completes the proof of Birkhoff's theorem and hence the proof of Lemma 6.1. □

Lemma 6.2   *If $x(t)$ is a bounded uniformly stable solution of (6.9) and if $\tilde{x}(t)$ is a solution of (6.9) such that $O[\tilde{x}(t)] \subset \Omega[x(t)]$, then $\tilde{x}(t)$ is uniformly stable.*

Proof   Since $x(t)$ is bounded, then $\Omega[x(t)]$ is bounded. Hence, since $\Omega[x(t)]$ is invariant, then $\tilde{x}(t)$ is defined for all real t. Let $t_0 \geq K$, where K occurs in the definition of uniform stability. Since $O[\tilde{x}(t)] \subset \Omega[x(t)]$ there exists $\tau \geq K$ such that

$$|x(\tau) - \tilde{x}(t_0)| < \frac{1}{2}\left(\delta\left[\min\left(\frac{\delta\left(\frac{\varepsilon}{2}\right)}{2}, \frac{\varepsilon}{2}\right)\right]\right) \tag{6.11}$$

where $\varepsilon > 0$ is given and $\delta(\varepsilon)$ is the function in the definition of uniform stability. Suppose that $u(t)$ is a solution of (6.9) and there are numbers t_1, t_2 such that $t_2 \geq K$ and

$$|u(t_1) - \tilde{x}(t_2)| < \frac{\delta\left(\frac{\varepsilon}{2}\right)}{2} \tag{6.12}$$

From (6.11) and the fact that $x(t)$ is uniformly stable, if $t \geq 0$,

$$|x(t + \tau) - \tilde{x}(t + t_0)| < \min\left(\frac{\delta\left(\frac{\varepsilon}{2}\right)}{2}, \frac{\varepsilon}{2}\right) \tag{6.13}$$

Assume $t_0 = t_2$. Then it follows from (6.11) and (6.12) that

$$|u(t_1) - x(\tau)| < \delta\left(\frac{\varepsilon}{2}\right) \tag{6.14}$$

Hence, if $t \geq 0$,

$$|u(t + t_1) - x(t + \tau)| < \frac{\varepsilon}{2} \tag{6.15}$$

Adding (6.13) and (6.15), we obtain if $t \geq 0$

$$|u(t + t_1) - \tilde{x}(t + t_0)| < \varepsilon \tag{6.16}$$

But $t_0 = t_2$. So (6.16) is the desired result. This completes the proof of Lemma 6.2. $\Box$

Lemma 6.3   *If $x(t)$ is a recurrent uniformly stable solution of (6.9), then $x(t)$ is almost periodic.*

Proof   Let K be the constant which appears in the uniform stability condition on $x(t)$. Then the function $y(t) = x(t - K)$ is a recurrent uniformly stable solution of (6.9) and the constant K in the uniform stability condition on $y(t)$ is zero. We show that $y(t)$ is almost periodic. First, given $\varepsilon > 0$, then by the uniform stability of $y(t)$, there exists a $\delta = \delta\left(\frac{\varepsilon}{2}\right)$. Also, since $y(t)$ is recurrent, there exists $T(\delta) > 0$ such that in every interval of length $T(\delta)$ there is a number T_ν such that

$$|y(T_\nu) - y(0)| < \frac{\delta}{2} \tag{6.17}$$

By the uniform stability of $y(t)$, it follows that if $t \geq 0$,

$$|y(T_v + t) - y(t)| < \varepsilon$$

To complete the proof, we need to show that this last inequality also holds for all $t < 0$. For this, we proceed as follows. Let T_v be fixed. By the continuity in the initial condition of solutions of (6.9) (Corollary 1.1), there exists $d > 0$ such that if $w(t)$ is a solution of (6.9) and

$$|y(0) - w(0)| < d$$

then

$$|y(T_v) - w(T_v)| < \frac{\delta}{2}$$

If t is a given real number, then by the recurrence of $y(t)$, if $A > 0$, there is a real number t_1 such that $t_1 < -A$ and $t_1 < t$ and such that

$$|y(0) - y(t_1)| < \min[d, \delta] \tag{6.18}$$

Hence by the definition of d,

$$|y(T_v) - y(T_v + t_1)| < \frac{\delta}{2} \tag{6.19}$$

From (6.17) and (6.19), we have:

$$|y(0) - y(T_v + t_1)| < \delta$$

and hence by the uniform stability of $y(t)$, if $t \geq 0$,

$$|y(t) - y(t + T_v + t_1)| < \frac{\varepsilon}{2} \tag{6.20}$$

Since $y(t)$ is uniformly stable with $K = 0$, it follows from (6.18) that if $t \geq 0$

$$|y(t) - y(t + t_1)| < \frac{\varepsilon}{2} \tag{6.21}$$

From (6.20) and (6.21), it follows that if $t \geq 0$,

$$|y(t + t_1) - y(t + T_v + t_1)| < \varepsilon$$

Since t_1 may be chosen so that $t_1 < -A$ where A is a given positive number, this completes the proof of Lemma 6.3, and hence the proof of the Sell-Deysach theorem. □

Sell's Theorem. *If $x(t)$ is a bounded phase asymptotically stable solution of (6.9), then there is a phase asymptotically stable periodic solution $y(t)$ of (6.9) such that $O[y(t)] = \Omega[x(t)]$.*

Proof By the Sell-Deysach theorem, there is an almost periodic uniformly stable solution $\bar{x}(t)$ of (6.9) such that

$$O[\bar{x}(t)] \subset \Omega[x(t)] \qquad\qquad \square$$

Next we need:

Lemma 6.4 *If $x(t)$ is a bounded phase asymptotically stable solution of (6.9) and if a solution $y(t)$ of (6.9) is such that*

$$O[y(t)] \subset \Omega[x(t)]$$

then $y(t)$ is phase asymptotically stable.

Proof From Lemma 6.2, it is sufficient to show that there exists a $\delta > 0$ such that if t_1, t_2 are real numbers such that $t_2 \geq K$ and if a solution $u(t)$ of (6.9) is such that

$$|u(t_1) - y(t_2)| < \delta$$

then there exists a number t_3 such that

$$\lim_{t \to \infty} |u(t) - y(t_3 + t)| = 0$$

Given $\varepsilon > 0$, let $\delta(\varepsilon)$ be the $\delta(\varepsilon)$ given by the uniform stability of solution $x(t)$. Suppose that there exist numbers t_1, t_2 such that

$$|u(t_1) - y(t_2)| < \frac{\delta(\varepsilon)}{2} \tag{6.22}$$

Since $O[y(t)] \subset \Omega[x(t)]$, there exists $t_3 \geq K$ such that

$$|x(t_3) - y(t_2)| < \frac{\delta(\varepsilon)}{2} \tag{6.23}$$

From (6.22) and (6.23), it follows that

$$|u(t_1) - x(t_3)| < \delta(\varepsilon) \tag{6.24}$$

and since $x(t)$ is phase asymptotically stable, it follows from (6.24) that there exists t_4 such that

$$\lim_{t \to \infty} |u(t) - x(t_4 + t)| = 0 \tag{6.25}$$

and it follows from (6.23) that there exists t_5 such that

$$\lim_{t \to \infty} |y(t) - x(t_5 + t)| = 0 \tag{6.26}$$

Let

$$\tau = t - (t_4 - t_5)$$

Then (6.26) becomes

$$\lim_{\tau \to \infty} |y[\tau + (t_4 - t_5)] - x[t_5 + \tau + (t_4 - t_5)]| = 0$$

or

$$\lim_{\tau \to \infty} |y[\tau + (t_4 - t_5)] - x(\tau + t_4)| = 0 \tag{6.27}$$

From (6.25) and (6.27), we obtain

$$\lim_{\tau \to \infty} |u(\tau) - y[\tau + (t_4 - t_5)]| = 0$$

This completes the proof of Lemma 6.4. □

By Lemma 6.4, the almost periodic solution $\bar{x}(t)$ is phase asymptotically stable. Since $\bar{x}(t)$ is almost periodic, $O[\bar{x}(t)] \subset \Omega[\bar{x}(t)]$. Suppose $u(t)$ is a recurrent solution of (6.9) such that

$$O[u(t)] \subset \Omega[\bar{x}(t)]$$

By Lemma 6.4, solution $u(t)$ is phase asymptotically stable and by Lemma 6.3, solution $u(t)$ is almost periodic. Since

$$O[u(t)] \subset \Omega[\bar{x}(t)]$$

and $u(t)$ is phase asymptotically stable, there exists a number t_1 such that

$$\lim_{t \to \infty} |u(t + t_1) - \bar{x}(t)| = 0 \tag{6.28}$$

Since $u(t)$ is almost periodic, solution $u(t + t_1)$ is almost periodic.

Lemma 6.5 *If f and g are almost periodic functions and if c is a real number, then $cf + g$ is almost periodic.*

Proof See, for example, Besicovitch [1954, p. 4].

By Lemma 6.5, the function $u(t + t_1) - \bar{x}(t)$ is almost periodic and hence by (6.28), for all t, $u(t + t_1) = \bar{x}(t)$. Thus the only recurrent solution $u(t)$ such that

$$O[u(t)] \subset \Omega[\bar{x}(t)]$$

is $\bar{x}(t)$ or $\bar{x}(t + k)$ where k is a constant. Since $\Omega[\bar{x}(t)]$ is a nonempty invariant compact set, then by Theorem 3.5 it contains a compact minimal set M. By Birkhoff's theorem, every orbit in M is the orbit of a recurrent solution. Hence

$$M = O[\bar{x}(t)]$$

Since a minimal set is closed, then

$$O[\bar{x}(t)] = \overline{O[\bar{x}(t)]}$$

Next we use:

Lemma 6.6   *If $u(t)$ is a recurrent solution of (6.9) such that $u(t)$ is not periodic, then*

$$\overline{O[u(t)]} - O[u(t)] = \Omega[u(t)]$$

Before proving Lemma 6.6, we show how to complete the proof of Sell's Theorem. Since

$$O[\bar{x}(t)] = \overline{O[\bar{x}(t)]}$$

then by Lemma 6.6, $\bar{x}(t)$ is periodic. Also $O[\bar{x}(t)] \subset \Omega[x(t)]$. Since $\bar{x}(t)$ is phase asymptotically stable, it follows that

$$\Omega[x(t)] \subset O[\bar{x}(t)]$$

This completes the proof of Sell's Theorem.

Proof of Lemma 6.6. First we show that

$$\overline{O[u(t)]} - O[u(t)] \subset \Omega[u(t)]$$

Since $\Omega[u(t)]$ is closed, it is sufficient to show that

$$\overline{O[u(t)]} - O[u(t)] \subset \Omega[u(t)]$$

Suppose

$$p \in \overline{O[u(t)]} - O[u(t)]$$

Then there is a sequence $\{t_n\}$ such that

$$\lim_{n \to \infty} u(t_n) = p$$

Also $\lim_{n \to \infty} t_n = \infty$ because otherwise p would be a point in $O[u(t)]$. Hence $p \in \Omega[u(t)]$.

In order to prove that

$$\Omega[u(t)] \subset \overline{O[u(t)]} - O[u(t)]$$

it is sufficient to prove that

$$O[u(t)] \subset \overline{O[u(t)]} - O[u(t)]$$

because then, since

$$\overline{O[u(t)]} - O[u(t)]$$

is closed, it follows that

$$\Omega[u(t)] \subset \overline{O[u(t)]} - O[u(t)]$$

Let $\bar{u} \in O[u(t)]$. It is sufficient to show that, given $\varepsilon > 0$, there exists $\bar{u}^0 \in \overline{B_\varepsilon(\bar{u})}$, where

$$B_\varepsilon(\bar{u}) = \{x \in R^n / |x - \bar{u}| < \varepsilon\}$$

such that $\bar{u}^0 \in \overline{O[u(t)]} - O[u(t)]$. Assume that $u(t)$ is such that $u(0) = \bar{u}$. (If $u(t_0) = \bar{u}$ where $t_0 \neq 0$, just let $u(t)$ denote the solution $v(t) = u(t+t_0)$.) Since $u(t)$ is recurrent, there exists a monotonic strictly increasing sequence $\{t_n\}$ such that $t_n > 0$ for all n, $\lim\limits_{n\to\infty} t_n = +\infty$ and $\lim\limits_{n\to\infty} u(t_n) = \bar{u}$. Choose $\tau_1 > t_1$ such that

$$v_1 = u(\tau_1) \in B_\varepsilon(\bar{u})$$

Then since $\tau_1 > t_1$ and $u(t)$ is not periodic,

$$v_1 \notin \{u(t)/t \in [-t_1, t_1]\}$$

and

$$\delta_1 = \mathop{\mathrm{glb}}\limits_{t\in[-t_1,t_1]} |v_1 - u(t)| > 0$$

Let

$$\varepsilon_1 = \min\left[\frac{\varepsilon}{2}, \varepsilon - |\bar{u} - v_1|, \frac{1}{2}\delta_1\right]$$

Then if

$$B_{\varepsilon_1}(v_1) = \{x \in R^n / |x - v_1| < \varepsilon_1\}$$

we have

$$B_{\varepsilon_1}(v_1) \subset B_\varepsilon(\bar{u}) \tag{6.29}$$

and

$$\overline{B_{\varepsilon_1}(v_1)} \cap \{u(t)/t \in [-t_1, t_1]\} = \phi$$

Now assume $v_{n-1} = u(\tau_{n-1})$ and ε_{n-1} have been defined. Since $u(t)$ is recurrent, we can choose $\tau_n > t_n$ such that

$$v_n = u(\tau_n) \in B_{\varepsilon_{n-1}}(v_{n-1}) \tag{6.30}$$

Let

$$\delta_n = \mathop{\mathrm{glb}}\limits_{t\in[-t_n,t_n]} |v_n - u(t)|$$

and define

$$\varepsilon_n = \min\left[\frac{\varepsilon_{n-1}}{2}, \varepsilon_{n-1} - |v_n - v_{n-1}|, \frac{1}{2}\delta_n\right]$$

Then from the definition of ε_n, it follows that

$$B_{\varepsilon_n}(v_n) \subset B_{\varepsilon_{n-1}}(v_{n-1}) \tag{6.31}$$

Also from the definition of δ_n

$$\overline{B_{\varepsilon_n}(v_n)} \cap \{u(t)/t \in [-t_n, t_n]\} = \phi \tag{6.32}$$

The sequence $\{v_n\}$ is such that

$$|v_n - v_{n-1}| < \varepsilon_{n-1} \le \frac{\varepsilon}{2^{n-1}}$$

Thus $\{v_n\}$ is a Cauchy sequence and converges to a point v. Since $\{v_n\} \subset O[u(t)]$, then $v \in \overline{O[u(t)]}$ and since by (6.29) and (6.31), $|v_n - \bar{u}| < \varepsilon$ for all n, then $|v - \bar{u}| \le \varepsilon$.

To complete the proof of Lemma 6.6, we show that $v \notin O[u(t)]$. Suppose there exists τ such that

$$v = u(\tau)$$

Take $t_n > |\tau|$. Then

$$v \in \{u(t)/t \in [-t_n, t_n]\} \tag{6.33}$$

But since, by (6.31)

$$v \in \overline{B_{\varepsilon_n}(v_n)}$$

and, by (6.32),

$$\overline{B_{\varepsilon_n}(v_n)} \cap \{u(t)/t \in [-t_n, t_n]\} = \phi$$

then

$$v \notin \{u(t)/t \in [-t_n, t_n]\}$$

which contradicts (6.33).   ∎

Once we pass outside the realm of phase asymptotically stable solutions, the situation is very different. There may or there may not be a periodic solution. As was already realized by Poincaré, the general behavior of solutions if $n > 2$ can be extremely complicated, but this was not fully seen until the advent of computers. Computer analysis shows that even the solutions of simple looking nonlinear systems with $n > 2$ can be very irregular and unpredictable. This observation is part of the motivation for the development of the theory of chaos. For an enlightening introduction to chaos with references, see Farkas [1994, Section 7.6].

Notice that if the solutions of a given equation display such irregularity, then the equation is not very useful in applications because the solutions are unpredictable and hence do not yield much useful information about the future behavior of the system being modelled by the equation. Thus we are confronted by two possible alternatives (both somewhat distasteful). First, the differential equation is not a realistic model and must be replaced with a more accurate model. Second, the system being modelled is too complex to admit a description in terms of an ordinary differential equation.

Periodic Solutions for Nonautonomous Systems

So far we have looked only at the problem of periodic solutions for autonomous systems. One can ask a parallel question for nonautonomous systems: given the system

$$x' = f(t, x)$$

where f has period T as a function of t, that is, for all x and all t,

$$f(t + T, x) = f(t, x)$$

then does the system have a solution of period T?

Notice that this question differs from the question about autonomous systems in that it is more limited. We are given a period T and we seek a solution with that period. In studying the autonomous systems we must search for a periodic solution but we have no idea what the value of the period should be. This might lead one to suspect that the question about periodic solutions for nonautonomous equations would be easier to answer. This is, in fact, the case. We will show now that application of the Brouwer fixed point theorem leads easily to the existence of periodic solutions in rather general circumstances. For cases in which the fixed point theorem cannot be applied, the more general theory of topological degree is sometimes applicable.

First we formulate our problem precisely. We consider the n-dimensional equation

$$x' = f(t, x) \tag{6.34}$$

where f has domain $R \times R^n$ and f has continuous first derivatives in all variables at each point of $R \times R^n$. We assume that f as a function of t has period T, that is, for each $(t, x) \in R \times R^n$,

$$f(t + T, x) = f(t, x)$$

and we ask the question: Does equation (6.34) have a solution $x(t)$ of period T? To answer this question, we prove first another version of Lemma 2.4.

Lemma 6.7   *A solution $x(t)$ of (6.34) has period T if and only if*

$$x(T) - x(0) = 0 \tag{6.35}$$

Proof If $x(t)$ has period T, equation (6.35) obviously holds. Suppose (6.35) is true. The function $y(t) = x(t + T)$ is also a solution of (6.34) because, using the periodicity of f as a function of t, we have:

$$\frac{dy}{dt} = \frac{d}{dt} x(t + T) = f[t + T, x(t + T)] = f[t, x(t + T)] = f[t, y(t)]$$

But by (6.35),

$$y(0) = x(T) = x(0)$$

Hence by the uniqueness of solution, it follows that for all t

$$x(t + T) = y(t) = x(t)$$

This completes the proof of Lemma 6.7. $\quad\square$

Now we assume that there is a bounded open set U in R^n such that $\bar{U}$ is a homeomorphism of a closed ball and if $x(t, c)$ is the solution of (6.34) such that

$$x(0, c) = c \in \bar{U}$$

then

$$x(T, c) \in \bar{U}$$

That is, the mapping M defined by

$$M : c \rightarrow x(T, c)$$

is a mapping which takes $\bar{U}$ into $\bar{U}$. It follows from Corollary 1.1 in Chapter 1 that M is continuous. Hence by the Brouwer fixed point theorem (see Appendix), it follows that M has a fixed point in $\bar{U}$, that is, there exists $c_0 \in \bar{U}$ such that

$$x(0, c_0) = c_0 = x(T, c_0)$$

Hence by Lemma 6.7, $x(t, c_0)$ is a solution of period T of equation (6.34). Summarizing this discussion, we obtain:

Theorem 6.1 *Given the n-dimensional system*

$$x' = f(t, x) \tag{6.34}$$

where f has domain $R \times R^n$, and f has continuous first derivatives in all variables at each point of $R \times R^n$ and there exists a positive number T such that for all $(t, x) \in R \times R^n$

$$f(t + T, x) = f(t, x)$$

Suppose there exists a bounded open set $U \subset R^n$ such that U is a homeomorphism of a closed ball and such that if $x(t, c)$ denotes the solution of (6.34) with

$$x(0, c) = c$$

then if $c \in \bar{U}$, it follows that $x(T, c) \in \bar{U}$. Then equation (6.34) has a solution $x(t, c_0)$, where $c_0 \in \bar{U}$, such that $x(t, c_0)$ has period T.

We add a few remarks about the significance of the conditions in this theorem:

1. From the point of view of pure mathematics, the hypothesis on the bounded open set U seems very strong. However, from the point of view of some applications, it is a rather natural condition, as has already been pointed out earlier for the autonomous case.

2. The periodic solution $x(t, c_0)$ may be trivial, that is, it may happen that $x(t, c_0)$ is a constant solution. This can happen only if for all $t \in [0, T]$,

$$f(t, c_0) = 0 \qquad (6.36)$$

Thus if the theorem is applied and one wishes to conclude that there exists a nontrivial periodic solution, it is sufficient to show that there is no point $c_0 \in \bar{U}$ such that (6.36) holds for all $t \in [0, T]$.

3. The theorem gives no information about the number of periodic solutions or the stability properties of the periodic solutions. There seems little hope of obtaining such information unless stronger hypotheses are imposed. In the next chapter we will study the problem of periodic solutions of quasilinear nonautonomous systems. In this case, by using more refined topological methods, we will be able to get an estimate on the number of periodic solutions and some information about stability.

4. More refined topological methods, especially use of topological degree, can be used in the study of equation (6.34). Notable work is due to Gomory [1956]. For extensive references to other work, see Mawhin and Rouché [1973].

Exercises

1. Prove that if (x_0, y_0) is an equilibrium point of

$$x' = P(x, y)$$
$$y' = Q(x, y)$$

and if the eigenvalues of the matrix

$$\begin{bmatrix} P_x(x_0, y_0) & P_y(x_0, y_0) \\ Q_x(x_0, y_0) & Q_y(x_0, y_0) \end{bmatrix}$$

both have positive real parts, then (x_0, y_0) is contained in the interior of a simple closed curve such that no solution enters that interior. (Hint: make the transformation of variables $t \to -\tau$ and show that (x_0, y_0) is an asymptotically stable equilibrium point of the resulting system.)

2. Show that $r = 1$ is the orbit of a phase asymptotically stable solution of the system

$$x' = -y + \frac{x}{r}(1 - r^2) \tag{6.1}$$

$$y' = x + \frac{y}{r}(1 - r^2) \tag{6.2}$$

3. Prove that the function

$$\alpha(r) = -\exp\left(\frac{-1}{(r-1)^2}\right) \quad r > 1$$

$$\alpha(r) = 0 \qquad\qquad\qquad 0 < r \le 1$$

is a monotonic nonincreasing differentiable function for $r > 0$.

4. Prove the following generalization of Theorem 4.5.

Theorem. *If $x(t)$ is a nontrivial almost periodic solution of the autonomous system*

$$x' = f(x)$$

then $x(t)$ is not asymptotically stable.
(Hint: use the definition of almost periodic and construct a proof parallel to the proof of Theorem 4.5.)

Chapter 7

Perturbation Theory: The Poincaré Method

Introduction

Chapters 7, 8, and 9 are all concerned with the same problem, which can be stated roughly as follows: Given a differential equation

$$\frac{dx}{dt} = F(t, x, \varepsilon) \tag{7.1}$$

where x is an n-vector, F has period $T(\varepsilon)$ in t, where $T(\varepsilon)$ is a differentiable function of ε, and ε is a real parameter such that $|\varepsilon|$ is small; suppose that for $\varepsilon = 0$, system (7.1) has a soluton $\bar{x}(t)$ of period $T(0)$. Then does (7.1) have a solution of period $T(\varepsilon)$ for all $|\varepsilon|$ and is this periodic soluton "near" the given periodic solution $\bar{x}(t)$? Equation (7.1) is said to be *unperturbed* if $\varepsilon = 0$ and *perturbed* if $\varepsilon \neq 0$. Thus our objective is to use knowledge of the unperturbed equation to study the perturbed equation.

We shall be concerned primarily with establishing the existence of periodic solutions and investigating their stability. Students who have already encountered computational perturbation methods may have doubts about all this emphasis on existence. Why not just use such a method and just go ahead and compute? Unfortunately if one "just computes," complications arise even with simple equations. For an enlightening discussion with examples, see Greenberg [1978, Chapter 25].

The problem we will study is very old and very important; old because it has been regarded as a serious problem for centuries in celestial mechanics; very important because it has arisen in disparate subjects, especially since the beginning of the twentieth century. The earlier work, inspired by celestial mechanics, was taken over for use in electrical engineering, in particular radio curcuits (see Andronov and Chaikin [1949]). Later it was used in control theory and most recently in biology (see Murray [2003], Keener and Sneyd [1998]).

We will use a method of Poincaré for dealing with this problem. (See Poincaré [1892–1899, Volume I, Chapter III].) As will be seen, the underlying idea of the method is simple (the problem of finding a periodic solution is reduced to solving a system of n equations in n unknowns) but very effective. In Chapter 7, we use the Poincaré method to study periodic solutions of various classes of nonautonomous equations and in Chapter 8 to study periodic solutions of autonomous equations.

The method of Poincaré yields a unified treatment of the problem and requires little technical apparatus. Other approaches do not have these advantages. For example, dynamical system theory is very effective in studying autonomous equations, but to study, for example, Hopf bifuration requires considerable theory (see Farkas [1994]) whereas the Poincaré method requires only the implicit function theorem to prove the Hopf bifurcation theorem. The averaging method can be used to study periodic solutions of nonautonomous equations but as will be shown in Chapter 9, the results obtained are contained in the results given by the Poincaré method. To apply dynamical system theory to a nonautonomous equation requires that the nonautonomous system be converted to an autonomous equation. However, the conversion requires raising the dimension of the system by one, and that step may increase seriously the complications in the problem. See, for example, Guckenheimer and Holmes ([1983], p. ix).

Our first step is to simplify the problem by introducing the variable

$$s = \left[\frac{T(0)}{T(\varepsilon)}\right] t$$

and letting

$$\bar{F}(s, u, \varepsilon) = \frac{T(\varepsilon)}{T(0)} F\left(\frac{T(\varepsilon)}{T(0)} s, u, \varepsilon\right)$$

Then equation (7.1) becomes

$$\frac{du}{ds} = \bar{F}(s, u, \varepsilon) \tag{7.2}$$

and the problem becomes: given a soluton $u_0(t)$ of period $T(0)$ for (7.2) with $\varepsilon = 0$, then does (7.2) have a solution of period $T(0)$ for small $|\varepsilon|$? (See Exercise 1.)

It is also convenient to write $F(t, x, \varepsilon)$ in the form

$$F(t, x, \varepsilon) = f(t, x) + \varepsilon g(t, x, \varepsilon)$$

Thus our problem becomes to find a solution of period T of

$$\frac{dx}{dt} = f(t, x) + \varepsilon g(t, x, e) \tag{7.3}$$

for $|\varepsilon|$ small enough, where f and g have period T in t. Later we will formulate the problem precisely with the appropriate hypotheses on f and g. It is to this problem that we apply the Poincaré method.

Now we give a brief rough description of the method.

Let $x(t, \varepsilon, c)$ be the solution of (7.3) such that

$$x(0, \varepsilon, c) = c$$

Since f and g have period T in t, it is easy to show (Lemma 7.1 in Section A which follows) that a nasc that solution $x(t, \varepsilon, c)$ have period T is that

$$x(T, \varepsilon, c) = x(0, \varepsilon, c) = 0 \tag{7.4}$$

Equation (7.4) may be regarded as an equation in c and ε. If x in (7.3) is an n-vector, then (7.4) is a system of n scalar equations in $c_1, c_2, \ldots, c_n, \varepsilon$ where $c_1, \ldots, c_n$ are the components of initial value c. If (7.4) can be solved for c as a function $c(\varepsilon)$ of ε, then the solution $x(t, \varepsilon, c(\varepsilon))$ has period T.

Thus, the problem of finding periodic solutions is reduced from an equation involving unknown functions to a problem of solving n equations in n real unknowns.

Before proceeding to study of this last problem we emphasize the importance of the steps that have been taken so far. It was Poincaré who had the vision to realize that studying (7.4) was a good concrete step toward finding periodic solutions and the foresight to perceive that solving (7.4) was a "do-able" problem.

The "catch" in trying to solve (7.4) lies in the fact that although we know that the general solution $x(t, \varepsilon, c)$ exists (we impose enough hypotheses on f and g to guarantee that), we have, in general, very little explicit knowledge about the form of $x(t, \varepsilon, c)$. Poincaré [1892–1899, Volume 1, Chapter 3] pointed out how the implicit function theorem could be applied and also how the existence of multiple solutions could be investigated. Important extensions of the use of the implicit function theorem were made (Coddington and Levinson [1955]) by using the real canonical form of a constant matrix; and multiple solutions can be studied by using topological degree.

In Part A of Chapter 7, we study equation (7.3) for the case in which the unperturbed equation

$$\frac{dx}{dt} = f(t, x)$$

has a single periodic solution. In Part B, we study the case in which the unperturbed equation has a continuous family of periodic solutions. An important example of this case is that in which the unperturbed equation is autonomous; and in Part C, we discuss in detail this example.

In Chapter 8, the case in which (7.3) is autonomous, that is, f and g are both independent of t, will be discussed. As we shall see, the autonomous case presents more complications. We will describe the work of Coddington and Levinson (augmented with the use of topological degree), and we apply one theorem to a phase-locking problem which arises in mechanics and biology. Also we obtain the Hopf bifurcation theorem and show how it is related to the other results in Chapter 8.

In Chapter 9, we describe how the averaging method can be applied to some of the problems discussed in Chapters 7 and 8. We show that although the averaging method is a very powerful general method, its value in studying the perturbation problems considered in Chapters 7 and 8 is limited. However, we will also explain how the power of the averaging method really reveals itself in the case of almost periodic solutions.

Throughout the above discussion, we have emphasized the importance of certain problems and techniques for solving them. It is important for those who use differential equations (other scientists or engineers as well as mathematicians) to be aware of the existence of such material. Otherwise, considerable time and energy may be wasted in attempts to discover known results. (We shall see an example of this in Chapter 8.)

A. The Case in Which the Unperturbed Equation is Nonautonomous and has an Isolated Periodic Solution

Formulating the Problem

We consider the equation

$$\frac{dx}{dt} = f(t, x) + \varepsilon g(t, x, \varepsilon) \tag{7.5}$$

where

$$f : R \times R^n \to R^n$$
$$g : R \times R^n \times I \to R^n$$

and I is an interval with center 0 on the real line. (Note that the equation

$$\frac{dx}{dt} = f(t, x) \tag{7.6}$$

is nonautonomous.) We assume also that:

(i) f has continuous third derivatives in t and x and that g has continuous third derivatives in t, x, ε;

(ii) f and g have period T in t, that is, for all t and x

$$f(t + T, x) = f(t, x)$$

and for all t, x, and ε

$$g(t + T, x, \varepsilon) = g(t, x, \varepsilon).$$

(iii) Equation (7.6) has a solution $\bar{x}(t)$ of period T and $\bar{x}(t)$ is an isolated periodic solution, that is, there exists an open set in R^n which contains the set

$$[\bar{x}(t)/0 \le t \le T]$$

but does not contain the corresponding set for any other periodic solution of period T.

(In the discussions in Chapters 7 and 8 which follow, it will be seen that the domains of f and g could be made more restricted and thus the theorems slightly stronger. We omit such restrictions; although they are easily calculated, they would mainly serve to make the statements of the results lengthier and more awkward.)

We study the following problem:

Problem 1. If $|\varepsilon|$ is sufficiently small, does equation (7.5) have a solution $x(t, \varepsilon)$ of period T such that for all $t \in [0, T]$

$$\lim_{\varepsilon \to 0} x(t, \varepsilon) = \bar{x}(t)$$

We investigate Problem 1 by investigating solutions of (7.5) of the form

$$x(t, \varepsilon) = \bar{x}(t) + \varepsilon \, u(t, \varepsilon) \tag{7.7}$$

Subsituting from (7.7) into (7.5) and using Taylor's expansion with a remainder, we have

$$
\begin{aligned}
\frac{d\bar{x}}{dt} + \varepsilon \frac{du}{dt} &= f(t, \bar{x} + \varepsilon \, u) + \varepsilon \, g(t, \bar{x} + \varepsilon \, u, \varepsilon) \\
&= f(t, \bar{x}) + \{f_x(t, \bar{x})\}\varepsilon u + \varepsilon^2 \gamma(t, u, \varepsilon) + \varepsilon[g(t, \bar{x}, 0) \\
&\quad + \{g_x(t, \bar{x}, 0)\}\varepsilon u + \{g_\varepsilon(t, \bar{x}, 0)\}\varepsilon + \varepsilon^2 \mathfrak{G}(t, \bar{x}, \varepsilon)]
\end{aligned} \tag{7.8}
$$

where the twice differential $n \times n$ matrices $f_x(t, \bar{x})$ and $g_x(t, \bar{x}, 0)$ and the twice differentiable n-vector function $g_\varepsilon(t, \bar{x}, 0)$ all have period T in t, and the remainder functions γ and $\mathfrak{G}$ have continuous first derivatives in all variables and have period T in t.

By hypothesis

$$\frac{d\bar{x}}{dt} = f(t, \bar{x}) \tag{7.9}$$

Subtracting (7.9) from (7.8) and dividing by ε, we obtain

$$\frac{du}{dt} = \{f_x(t, \bar{x})\}u + g(t, \bar{x}, 0) + \varepsilon \mathcal{H}(t, \bar{x}, u, \varepsilon) \tag{7.10}$$

where $\mathcal{H}$ has continuous first derivatives in all variables and period T in t. Since the matrix $f_x(t, \bar{x})$ has period T in t then according to the Floquet theory there exists a transformation of the dependent variable so that the equation

$$\frac{dw}{dt} = \{f_x(t, \bar{x})\}w$$

(this is called the *linear variational system* of (7.5) relative to $\bar{x}(t)$) can be reduced to the form

$$\frac{dw}{dt} = Aw$$

where A is a constant matrix in real canonical form. (See Chapter 2.) Thus the problem of studying equation (7.5) is reduced to the study of equation (7.10) and then, by using the Floquet transformation, to the study of the following equation:

$$\frac{dx}{dt} = Ax + G(t) + \varepsilon F(t, x, \varepsilon) \tag{7.11}$$

where A is a constant matrix in real canonical form, the functions F and G have continuous first derivatives in all variables, and F and G have period T in t. Thus Problem 1 can be rephrased as:

Problem 2. If $|\varepsilon|$ is sufficiently small, does equation (7.11) have solutions of period T?

Equation (7.11) is called a *quasilinear* equation.

Solving the Quasilinear Equation

We make a detailed analysis of Problem 2. The initial steps in the analysis form a classical procedure which was originated by Poincaré and has been widely used ever since.

From Existence Theorem 2.1 for linear systems (Chapter 2) and the existence theorem for equation with a parameter (Chapter 1), it follows that if c is a fixed real n-vector and if $|\varepsilon|$ is sufficiently small, there exists a solution $x(t, \varepsilon, c)$ of equation (7.11) which is defined on an open interval which contains $[0, T]$ and which satisfies the initial condition

$$x(0, \varepsilon, c) = c$$

By the variation of constants formula (Chapter 2), solving equation (7.11) for $x(t, \varepsilon, c)$ is equivalent to solving the following integral equation for $x(t, \varepsilon, c)$.

$$x(t, \varepsilon, c) = e^{tA}c + e^{tA} \int_0^t e^{-sA}\{\varepsilon F[s, x(s, \varepsilon, c)\varepsilon] + G(s)\}\, ds \qquad (7.12)$$

In order to search for solutions $x(t, \varepsilon, c)$ which have period T, we use the following simple but useful lemma.

Lemma 7.1   *A nasc that $x(t, \varepsilon, c)$ have period T is that*

$$x(T, \varepsilon, c) = x(0, \varepsilon, c) = 0 \qquad (7.13)$$

Proof   The condition is obviously necessary. To show that it is sufficient, define the function $y(t, \varepsilon, c)$ by

$$y(t, \varepsilon, c) = x(T + t, \varepsilon, c)$$

Then $y(t, \varepsilon, c)$ is a solution of equation (7.11) because

$$\begin{aligned}
y'(t, \varepsilon, c) &= x'(T + t, \varepsilon, c) \\
&= Ax(T + t, \varepsilon, c) + F[T + t, x(T + t, \varepsilon, c), \varepsilon] + G(T + t) \\
&= Ay(t, \varepsilon, c) + F[t, y(t, \varepsilon, c), \varepsilon] + G(t)
\end{aligned}$$

Thus $y(t, \varepsilon, c)$ is a solution of equation (7.11). Also, since

$$y(0, \varepsilon, c) = x(T, \varepsilon, c) = x(0, \varepsilon, c)$$

then by the uniqueness of solution of (7.11), it follows that for all t,

$$y(t, \varepsilon, c) = x(t, \varepsilon, c)$$

That is,

$$x(T + t, \varepsilon, c) = x(t, \varepsilon, c)$$

for all t. This completes the proof of Lemma 7.1. $\Box$

We use equation (7.13) to search for periodic solutions in this way: Since the general solution $x(t, \varepsilon, c)$ of equation (7.13) is (in some abstract sense if not explicitly) known, then we try to solve equation (7.12) for c as a function of ε. In order to do this, we first write (7.13) in as explicit a form as possible and for this purpose, we use equation (7.12). Taking $t = T$ in (7.12), substituting in (7.13), and using the equality $x(0, \varepsilon, c) = c$, we obtain:

$$(e^{TA} - I)c + e^{TA} \int_0^T e^{-sA}\{\varepsilon F[s, x(s, \varepsilon, c), \varepsilon] + G(s)\}\, ds = 0 \qquad (7.14)$$

In summary, to solve Problem 2, it is sufficient to solve equation (7.14) for c as a function of ε. (Remember that equation (7.14) is an n-vector equation. We will find it convenient sometimes to regard (7.14) as a system of n real equations in $\varepsilon, c_1, \ldots, c_n$ where $c_1, \ldots, c_n$ are the components of c.)

First, if $\varepsilon = 0$, equation (7.14) becomes

$$(e^{TA} - I)c + e^{TA} \int_0^T e^{-sA} G(s)\, ds = 0$$

Thus, if the matrix $(e^{TA} - I)$ is nonsingular, equation (7.14) has the initial solution:

$$\varepsilon = 0, \quad c_0 = -(e^{TA} - I)^{-1} e^{TA} \int_0^T e^{-sA} G(s)\, ds$$

Also, if $e^{TA} - I$ is nonsingular, the implicit function theorem can be applied to solve equation (7.14) uniquely for c as a function of ε in a neighborhood of this initial solution. Since A is in real canonical form (see Chapter 2), it is easy to see that $e^{TA} - I$ is nonsingular if and only if the eigenvalues of the matrix TA are all nonzero and are all different from $\pm i(2n\pi/T)$ $(n = 1, 2, \ldots)$. This last condition is equivalent to the condition that the equation

$$x' = Ax$$

has no nontrivial solutions of period T. (Use Exercise 2.) Thus, we have obtained the following classical result (proved by Poincaré):

Theorem 7.2 *If the equation*

$$\frac{dw}{dt} = \{f_x(t, \bar{x}(t))\}w$$

(i.e., the linear variational system of (7.5) relative to the given periodic solution $\bar{x}(t)$) has no nontrivial solutions of period $2n\pi/T$ where $n = 1, 2, \ldots$ (or, equivalently, if matrix A in equation (7.11) has no eigenvalues of the form $(2n\pi/T)i$ $(n = 1, 2, \ldots)$)

then there exist $\eta_1 > 0$, $\eta_2 > 0$ such that for each ε with $|\varepsilon| < \eta_1$, there is a unique vector $c = c(\varepsilon)$, where $c(\varepsilon)$ is a continuous function of ε, such that

$$|c(\varepsilon) - c_0| < \eta_2$$

and such that $c(\varepsilon)$ is a solution of (7.14) and hence that the solution $x(t, c(\varepsilon))$ of (7.11) has period T.

Later, the stability of this periodic solution will be discussed.

The Resonance Case: An Example

Next we suppose that $(e^{TA} - I)$ is a singular matrix. This is sometimes called the *resonance case*. In applications, the resonance case is often more important than the nonresonance case, that is, the case considered in Theorem 7.1. If $e^{TA} - I$ is singular, then we are forced to look at the structure of equation (7.14) more closely, that is, to discuss some of the scalar equations in (7.14). Much of this discussion is quite straightforward but it looks complicated sometimes because there are n scalar equations. So we will begin by considering a two dimensional case in which $(e^{TA} - I)$ is the zero matrix. This will illustrate the underlying ideas without the clutter of $n \times n$ matrices. After that, we will proceed to the general case where we will also discuss the number of solutions and the stability of the solutions.

We consider a two dimensional example of system (7.11) which often arises in practice. We assume that

$$A = \begin{bmatrix} 0 & \beta \\ -\beta & 0 \end{bmatrix}$$

where, for simplicity of notation, we take $\beta = 2\pi$. Then

$$e^{tA} = \begin{bmatrix} \cos \beta t & \sin \beta t \\ -\sin \beta t & \cos \beta t \end{bmatrix} = \begin{bmatrix} \cos 2\pi t & \sin 2\pi t \\ -\sin 2\pi t & \cos 2\pi t \end{bmatrix}$$

and e^{tA} has period $T = 1$. Also

$$e^{TA} - I = \begin{bmatrix} 0 & 0 \\ 0 & 0 \end{bmatrix}$$

If

$$G(t) = \begin{bmatrix} G_1(t) \\ G_2(t) \end{bmatrix}$$

then if $\varepsilon = 0$, equation (7.14) becomes

$$\int_0^T \begin{bmatrix} \cos 2\pi s & -\sin 2\pi s \\ \sin 2\pi s & \cos 2\pi s \end{bmatrix} \begin{bmatrix} G_1(s) \\ G_2(s) \end{bmatrix} ds = 0 \tag{7.15}$$

Thus equation (7.15) is a necessary condition that there exist a periodic solution for sufficiently small $|\varepsilon|$. Note that equation (7.15) is equivalent to the condition that vector $G(t)$ is orthogonal to the linear space of solutions of the equation

$$x' = Ax$$

It follows from (7.15) that equation (7.14) becomes

$$e^{TA} \int_0^T e^{-sA} \varepsilon \{ F[s, x(s, \varepsilon, c), \varepsilon] \} \, ds = 0 \tag{7.16}$$

and our problem becomes that of solving (7.16) for c as a function of ε. It is thus sufficient to solve

$$\int_0^T e^{-sA} \{ F[s, x(s, \varepsilon, c), \varepsilon] \} \, ds = 0 \tag{7.17}$$

for $c = \begin{bmatrix} c_1 \\ c_2 \end{bmatrix}$ as a function of ε. A reasonable procedure is then to look for conditions on c in (7.17) if $\varepsilon = 0$. That is, we seek solutions c of the equation

$$\int_0^T e^{-sA} \{ F[s, x(s, 0, c), 0] \} \, ds = 0 \tag{7.18}$$

If the components of F are F_1 and F_2, then (7.18) can be written as

$$\int_0^T \begin{bmatrix} \cos \beta s & -\sin \beta s \\ \sin \beta s & \cos \beta s \end{bmatrix} \begin{bmatrix} F_1(\ ) \\ F_2(\ ) \end{bmatrix} ds = 0 \tag{7.19}$$

where for $i = 1, 2$,

$$F_i(\ ) = F_i(s, c_1 \cos \beta s + c_2 \sin \beta s, -c_1 \sin \beta s + c_2 \cos \beta s, 0)$$

Now we impose further conditions on F_1 and F_2 that will make it possible to obtain information about the existence of periodic solutions by using the implicit function theorem or topological degree.

Let us assume that for $i = 1, 2$, the function $F_i(s, \xi_1, \xi_2, 0)$ is a polynomial in ξ_1 and ξ_2 (where $\xi_1 = c_1 \cos \beta s - c_2 \sin \beta s$ and $\xi_2 = -c_1 \sin \beta s + c_2 \cos \beta s$) and that F_i is independent of s, that is,

$$F_i(s, \xi_1, \xi_2, 0) = F_i(\xi_1, \xi_2)$$

From the fact that

$$\int_0^T (\sin^m x)(\cos^n x) \, dx \neq 0$$

if and only if m and n are both even integers, it follows that

$$\int_0^1 \begin{bmatrix} \cos \beta s & -\sin \beta s \\ \sin \beta s & \cos \beta s \end{bmatrix} \begin{bmatrix} F_1(\ ) \\ F_2(\ ) \end{bmatrix} ds = \begin{bmatrix} P_1(c_2, c_2) \\ P_2(c_1, c_2) \end{bmatrix}$$

where $P_1(c_1, c_2)$ and $P_2(c_1, c_2)$ are polynomials in c_1 and c_2 and each term in P_1 and P_2 is of the form

$$K c_1^{q_1} c_2^{q_2}$$

where K is a constant coefficient and $q_1 + q_2$ is an odd number. Suppose that

$$P_1(c_1, c_2) = k_1 c_1 + k_2 c_2 + H_1(c_1, c_2)$$
$$P_2(c_1, c_2) = k_3 c_1 + k_4 c_2 + H_2(c_1, c_2)$$

where

$$\det \begin{bmatrix} k_1 & k_2 \\ k_3 & k_4 \end{bmatrix} \neq 0 \tag{7.20}$$

and for $i = 1, 2$, the polynomial $H_i(c_1, c_2)$ consists of terms of the form $Kc_1^{q_1} c_2^{q_2}$ where $q_1 + q_2$ is odd and

$$q_1 + q_2 \geq 3$$

Then $c = 0$, $\varepsilon = 0$ is an initial solution of (7.17) and by (7.20), we may apply the implicit function theorem and conclude that there exist $\delta_1 > 0$ and $\delta_2 > 0$ such that if $|\varepsilon| < \delta_1$, there exists a unique $c(\varepsilon)$ with $|c(\varepsilon)| < \delta_2$ such that $c(\varepsilon)$ is a solution of (7.18).

By applying topological degree we can make a more extensive investigation. For fixed ε, let M_ε denote the mapping from R^2 into R^2 defined by the left side of (7.17). That is,

$$M_\varepsilon: (c_1, c_2) \rightarrow \int_0^T e^{-sA} \{F[s, x(s, \varepsilon, c), \varepsilon]\} \, ds$$

In the following discussion, we use terminology and notation from the Appendix. Let $\bar{U}$ be the closure of a bounded open set in R^2. If

$$\deg[M_\varepsilon, \bar{U}, 0] \neq 0$$

then there exists at least one solution $c \in U$ of equation (7.17). But if

$$\deg[M_0, \bar{U}, 0] \neq 0$$

and if $|\varepsilon|$ is sufficiently small, then by Property 2 (invariance under homotopy) of the Appendix.

$$\deg[M_\varepsilon, \bar{U}, 0] = \deg[M_0, \bar{U}, 0] \neq 0$$

Hence in order to establish the existence of solutions of (7.17), it is sufficient to show that for some set $\bar{U}$,

$$\deg[M_0, \bar{U}, 0] \neq 0$$

Next we impose the following conditions on $P_1(c_1, c_2)$ and $P_2(c_1, c_2)$

$$P_1(c_1, c_2) = q_1(c_1, c_2) + p_1(c_1, c_2)$$
$$P_2(c_1, c_2) = q_2(c_1, c_2) + p_2(c_1, c_2)$$

where $p_1(c_1, c_2)$ is a polynomial homogeneous of degree s_1 in c_1 and c_2 and $p_2(c_1, c_2)$ is a polynomial homogeneous of degree s_2 in c_1 and c_2 where $s_1 > 1$ and $s_2 > 1$. Suppose $q_1(c_1, c_2)$, $q_2(c_1, c_2)$ consist of sums of terms of the form $kc_1^\alpha c_2^\beta$ where k is a constant and

$$\alpha + \beta < \min(s_1, s_2)$$

Let $\bar{M}$ be the mapping defined by

$$\bar{M}: (c_1, c_2) \rightarrow (p_1(c_1, c_2), p_2(c_1, c_2))$$

As shown in Example 15 of the Appendix, if $p_1(c_1, c_2)$ and $p_2(c_1, c_2)$ have no common real linear factors, and if D is a disc with center 0 and radius R, then $\deg[\bar{M}, D, 0]$ is defined and its value is independent of the magnitude of R. Also, as shown in Example 15, $\deg[\bar{M}, D, 0]$ is easily calculated after the real linear factors of $p_1(c_1, c_2)$ and $p_2(c_1, c_2)$ are determined. Moreover, if R is sufficiently large, then it follows from the generalized Rouché's theorem in the Appendix that

$$\deg[\bar{M}, D, 0] = \deg[M_0, D, 0] \tag{7.21}$$

Let us suppose that

$$|\deg[\bar{M}, D, 0]| \neq 0$$

Then by (7.21), we have

$$|\deg[M_0, D, 0]| \neq 0 \tag{7.22}$$

and it follows that for sufficiently small $|\varepsilon|$, equation (7.17) has at least one solution c in D.

Now let us combine the two sets of hypotheses imposed previously on $P_1(c_1, c_2)$ and $P_2(c_1, c_2)$. That is, suppose

$$P_1(c_1, c_2) = k_1 c_1 + k_2 c_2 + \cdots + p_1(c_1, c_2)$$
$$P_2(c_1, c_2) = k_3 c_1 + k_4 c_2 + \cdots + p_2(c_1, c_2)$$

where

$$\det \begin{bmatrix} k_1 & k_2 \\ k_3 & k_4 \end{bmatrix} \neq 0$$

where $p_1(c_1, c_2)$ and $p_2(c_1, c_2)$ are homogeneous of degree s_1 and s_2, respectively where $s_i > 1$ for $i = 1, 2$ and the dots in the definition of $p_1(c_1, c_2)$ represent terms of the form $k c_1^\alpha c_2^\beta$ where k is a constant and $1 < \alpha + \beta < \min(s_1, s_2)$ and similarly in the definition of $P_2(c_1, c_2)$.

By Example 8 in the Appendix, it follows that if d is a disc with center 0 and sufficiently small radius, then

$$|\deg[M_0, d, 0]| = 1 \tag{7.23}$$

For definiteness, assume $\deg[M_0, d, 0] = 1$. But by Property 3 in the Appendix, we have

$$\deg[M_0, d, 0] + \deg[M_0, \overline{D - d}, 0] = \deg[M_0, D, 0] \tag{7.24}$$

Now suppose that $\deg[\bar{M}, D, 0] \neq 1$. Then if radius R is large enough, it follows from the generalized Rouché's theorem in the Appendix that

$$\deg[M_0, D, 0] = \deg[\bar{M}, D, 0] \neq 1$$

and hence by (7.23) and (7.24), we have

$$\deg[M_0, \overline{D - d}, 0] \neq 0$$

Hence it follows that for each ε with $|\varepsilon|$ sufficiently small, our two dimensional example has at least two distinct periodic solutions. The initial value of one periodic solution approaches an initial condition $\begin{bmatrix} c_1 \\ c_2 \end{bmatrix}$ as $\varepsilon \to 0$ where $\begin{bmatrix} c_1 \\ c_2 \end{bmatrix} \in d$ and the initial value of the other periodic solution approaches an initial condition $\begin{bmatrix} c_1 \\ c_2 \end{bmatrix} \in D - d$.

The Resonance Case: Existence and Number of Periodic Solutions

If q is the dimension of the null space of $e^{TA} - I$ and $q < n$, then the first step is to reduce the system of n scalar equations described by equation (7.14) to a system of q equations. (If $q = n$, then, as seen in the example, no reduction need be made.) In either case, the objective is to make the system of q equations as explicit as possible so that (hopefully) the equations can be solved.

Let E_{n-r} denote the null space of $e^{TA} - I$, and let E_r denote the complement in R^n of E_{n-r}, that is,

$$R^n = E_{n-r} \oplus E_r \quad \text{(direct sum)}$$

Let P_{n-r}, P_r denote the projections of R^n onto E_{n-r} and E_r, respectively. Then $P_r^2 = P_r$, $(P_{n-r})^2 = P_{n-r}$ and $P_{n-r}P_r = P_r P_{n-r} = 0$.

It will be convenient in the computations to use a real nonsingular matrix H such that

$$H(e^{TA} - I) = P_r$$

Since A is in real canonical form, matrix H can be explicitly computed. (See Exercise 3.) (A typical matrix H is computed in the proof of Theorem 7.4.) We notice that since $P_{n-r}P_r = 0$ then

$$P_{n-r}He^{TA} = P_{n-r}H \tag{7.25}$$

Now we are ready to investigate equation (7.14) in the resonance case. First we apply matrix H to equation (7.14) and obtain

$$P_r c + He^{TA} \int_0^T e^{-sA}\{\varepsilon F[s, x(s, \varepsilon, c), \varepsilon] + G(s)\}\, ds = 0 \tag{7.26}$$

Since H is a nonsingular matrix, solving equation (7.14) for c is equivalent to solving equation (7.26) for c. Applying P_{n-r} to (7.26) yields:

$$P_{n-r}He^{TA} \int_0^T e^{-sA}\{\varepsilon F[s, x(s, \varepsilon, c), \varepsilon] + G(s)\}\, ds = 0 \tag{7.27}$$

Setting $\varepsilon = 0$ in (7.27), we obtain

$$P_{n-r}He^{TA} \int_0^T e^{-sA}G(s)\, ds = 0$$

and from (7.25) this equation becomes

$$P_{n-r} H \int_0^T e^{-sA} G(s)\, ds = 0$$

Thus we have obtained a version of another classical result which we state as the following theorem:

Theorem 7.2   *A necessary condition that Problem 2 can be solved in the resonance case is that*

$$P_{n-r} H \int_0^T e^{-sA} G(s)\, ds = 0 \qquad (7.28)$$

Condition (7.28) really stems from the standard orthogonality condition for the solution of nonhomogeneous linear algebraic equations.

In the remainder of the discussion we assume that the necessary condition (7.28) in Theorem 7.2 is satisfied.

Now we proceed to solve equation (7.26) for c in terms of ε. Applying P_r and P_{n-r} to (7.26) and using (7.28) and (7.25), we obtain:

$$P_r c + P_r H e^{TA} \int_0^T e^{-sA} \{\varepsilon F[s, x(s, \varepsilon, c), \varepsilon] + G(s)\}\, ds = 0 \qquad (7.29)$$

$$P_{n-r} H \int_0^T e^{-sA} \varepsilon F[s, x(s, \varepsilon, c), \varepsilon]\, ds = 0 \qquad (7.30)$$

Let $P_r c = \xi$, an r-vector, and let $P_{n-r} c = \eta$, an $(n-r)$-vector. Then (7.29) may be written as:

$$\xi + P_r H e^{TA} \int_0^T e^{-sA} \{\varepsilon F[s, x(s, \varepsilon, \xi \oplus \eta), \varepsilon] + G(s)\}\, ds = 0 \qquad (7.31)$$

Equation (7.31) has the initial solution

$$\varepsilon = 0$$
$$\eta = 0$$

$$\xi = -P_r H e^{TA} \int_0^T e^{-sA} \{G(s)\}\, ds$$

and the Jacobian of the left-hand side of (7.31) with respect to ξ at the initial solution is:

$$\det \begin{bmatrix} 1 & \cdot & \cdot & \cdot & 0 \\ \cdot & 1 & & & \\ \cdot & & & & \\ \cdot & & & \ddots & \\ 0 & & & & 1 \end{bmatrix} = 1 \neq 0$$

Hence by the implicit function theorem we can solve uniquely for ξ in terms of η and ε in a neighborhood of the initial solution to obtain:

$$\xi = \xi(\eta, \varepsilon) \tag{7.32}$$

where $\xi(\eta, \varepsilon)$ is differentiable in η. Substituting from (7.32) into (7.30), we reduce the problem to that of solving

$$P_{n-r} H \int_0^T e^{-sA} \varepsilon F[s, x(s, \varepsilon, \xi(\eta, \varepsilon) \oplus \eta), \varepsilon] \, ds = 0 \tag{7.33}$$

for η in terms of ε. Actually we divide (7.33) by ε and seek to solve the equation

$$P_{n-r} H \int_0^T e^{-sA} F[s, x(s, \varepsilon, \xi(\eta, \varepsilon) \oplus \eta)\varepsilon] \, ds = 0 \tag{7.34}$$

for η in terms of ε. Now if equation (7.34) has an initial soluton η_0 if $\varepsilon = 0$ and if the appropriate Jacobian at $\varepsilon = 0$, $\eta = \eta_0$ is nonzero, then the implicit function theorem can be applied.

Use of topological degree makes possible a wider study of the existence and stability of periodic solutions. For fixed ε, the left-hand side of (7.34) may be regarded as a mapping of η which is an $(n - r)$-vector into an $(n - r)$-vector, that is, a mapping M_ε from $R^{(n-r)}$ into $R^{(n-r)}$. We study the solutions of (7.34) by investigating the topological degree at zero of M_ε. The mapping M_ε is not given very explicitly because it is defined in terms of the general solution $x(s, \varepsilon, c)$. While existence theorems assure us that $x(s, \varepsilon, c)$ exists and has suitable properties of continuity and differentiability, we have, in general, no idea of the explicit form of $x(s, \varepsilon, c)$. So there is little hope of computing or estimating directly the topological degree of M_ε if $\varepsilon \neq 0$. However, it is considerably easier to study the mapping M_0. This mapping is given quite explicitly and in many cases, the degree can be computed (as shown in the example in the preceding section). By the invariance under homotopy of the degree (see the discussion in the Appendix) it follows that if $|\varepsilon|$ is sufficiently small, the degree of M_ε is defined and is equal to the degree of M_0.

Let $c = (c_1, \ldots, c_{n-r}, 0, \ldots, 0)$. Then mapping M_0 is described by

$$M_0 : (c_1, \ldots, c_{n-r}) \rightarrow P_{n-r} H \int_0^T e^{-sA} \{F[s, x(s, 0, c), 0]\} \, ds \tag{7.35}$$

Let $\mathcal{B}$ be a ball with center 0 in $R^{(n-r)}$. We consider $\deg[M_0, \mathcal{B}, 0]$. There is no theoretical reason why we need only study the degree relative to a ball $\mathcal{B}$. We could consider the closure of any suitable bounded open set instead of $\mathcal{B}$. But it turns out that it is convenient and sufficient for our purposes to consider a ball $\mathcal{B}$.

We have already computed $P_{n-r}H$ and consequently we could write mapping M_0 more explicitly than is done in (7.35) above. But a more explicit description of M_0 in the general case offers no particular advantages. We shall illustrate the somewhat oppressive task of finding the explicit form of M_0 later in the proof of Theorem 7.4. We summarize the discussion above as a theorem.

Theorem 7.3 *If*

$$\deg[M_0, \mathcal{B}, 0] \neq 0$$

then if $|\varepsilon|$ is sufficiently small, equation (7.11) has a solution $x(t, \varepsilon)$ of period T and $x(0, \varepsilon) = c$ where $P_{n-r}c \in B$.

We obtain also information about the number of periodic solutions $x(t, \varepsilon)$. That is, we have the following theorem.

Theorem 7.4 *If*

$$\deg[M_0, \mathcal{B}, 0] \neq 0$$

and if $\delta > 0$, then there exists a differentiable function $k(t)$ with the following properties:

(i) $k(t)$ *has period T;*

(ii) $\max_t |k(t)| < \delta$;

(iii) *if the function $F(t, x, \varepsilon)$ in equation (7.11) is replaced by*

$$F(t, x, \varepsilon) + k(t),$$

then if $|\varepsilon|$ is sufficiently small, the number of solutions of (7.11) of period T is greater than or equal to $|\deg[M_0, \mathcal{B}, 0]|$;

(iv) *each periodic solution $x(t, \varepsilon)$ depends continuously on ε.*

Theorem 7.4 says roughly that if function F is varied arbitrarily slightly by adding an arbitrarily small function of t, then the absolute value of the degree is a lower bound for the number of periodic solutions and that the periodic solutions depend continuously on ε. Before giving a detailed proof, we give a brief outline of it. The basic idea of the proof is a straightforward application of Theorem 1 in the Appendix. (Theorem 1 is an immediate consequence of Sard's theorem.) By adding the function $k(t)$ to F, the mapping M_0 is changed so that $M_0^{-1}(0)$ is a finite set of points, at each of which the Jacobian of M_0 is nonzero. Conclusion (iii) then follows from Theorem 1 in the Appendix. Conclusion (iv) follows from the implicit function theorem. As will be seen in the detailed proof, we have a wide choice for the function $k(t)$.

The main complication in giving a detailed proof is the description of the conditions on function $h(t)$. The complication arises from the form of the matrix A. For simplicity, we will describe the conditions on the function $h(t)$ for a specific, fairly typical matrix A. Once the particular description has been given, it is easy to see how to obtain the conditions on function $h(t)$ for other matrices A.

Let us assume that matrix A is the real canonical form

$$A = \begin{bmatrix} 0 & \beta & 1 & 0 & & & & \\ -\beta & 0 & 0 & 1 & & & & \\ & & 0 & \beta & & & & \\ & & -\beta & 0 & & & & \\ & & & & 0 & 1 & & \\ & & & & 0 & & & \\ & & & & & & a & 0 \\ & & & & & & 0 & a \end{bmatrix}$$

where $\beta = \frac{2n\pi}{T}$ and n is a nonzero integer, positive or negative, and T is a positive constant, and a is real and nonzero, and all matrix entries which are not specified are zero. We must calculate the mapping M_0 explicitly. First,

$$e^{tA} = \begin{bmatrix} \cos\beta t & \sin\beta t & t\cos\beta t & t\sin\beta t & & & & \\ -\sin\beta t & \cos\beta t & -t\sin\beta t & t\cos\beta t & & & & \\ & & \cos\beta t & \sin\beta t & & & & \\ & & -\sin\beta t & \cos\beta t & & & & \\ & & & & 1 & t & & \\ & & & & 0 & 1 & & \\ & & & & & & e^{at} & 0 \\ & & & & & & 0 & e^{at} \end{bmatrix}$$

and

$$e^{TA} - I = \begin{bmatrix} 0 & 0 & T & 0 & & & & \\ 0 & 0 & 0 & T & & & & \\ 0 & 0 & 0 & 0 & & & & \\ 0 & 0 & 0 & 0 & & & & \\ & & & & 0 & T & & \\ & & & & 0 & 0 & & \\ & & & & & & e^{aT}-1 & 0 \\ & & & & & & 0 & e^{aT}-1 \end{bmatrix}$$

Now we must compute the matrix H.

We consider submatrices and let

$$B = \begin{bmatrix} 0 & \beta & 1 & 0 \\ -\beta & 0 & 0 & 1 \\ & & 0 & \beta \\ & & -\beta & 0 \end{bmatrix}$$

Then

$$e^{TB} - I = \begin{bmatrix} 0 & 0 & T & 0 \\ & 0 & 0 & T \\ & & 0 & 0 \\ & & & 0 \end{bmatrix}$$

A basis for E_{n-r}, the null space of $e^{TB} - I$, is the set of vectors

$$\begin{bmatrix} 1 \\ 0 \\ 0 \\ 0 \end{bmatrix}, \begin{bmatrix} 0 \\ 1 \\ 0 \\ 0 \end{bmatrix}$$

and a basis for E_r is the set of vectors

$$\begin{bmatrix} 0 \\ 0 \\ 1 \\ 0 \end{bmatrix}, \begin{bmatrix} 0 \\ 0 \\ 0 \\ 1 \end{bmatrix}$$

Projection P_{n-r} is described by

$$\begin{bmatrix} 1 & & & \\ & 1 & & \\ & & 0 & \\ & & & 0 \end{bmatrix}$$

and projection P_r is described by

$$\begin{bmatrix} 0 & & & \\ & 0 & & \\ & & 1 & \\ & & & 1 \end{bmatrix}$$

The matrix H corresponding to B is such that

$$(H)[e^{TB} - I) = P_r$$

That is

$$(H) \begin{bmatrix} 0 & 0 & T & 0 \\ 0 & 0 & 0 & T \\ 0 & 0 & 0 & 0 \\ 0 & 0 & 0 & 0 \end{bmatrix} = \begin{bmatrix} 0 & & & \\ & 0 & & \\ & & 1 & \\ & & & 1 \end{bmatrix}$$

Recalling that H is required to be nonsingular, we take

$$H = \begin{bmatrix} 0 & 0 & 1 & 0 \\ 0 & 0 & 0 & 1 \\ \frac{1}{T} & 0 & 0 & 0 \\ 0 & \frac{1}{T} & 0 & 0 \end{bmatrix}$$

To find the matrix H corresponding to

$$C = \begin{bmatrix} 0 & 1 \\ 0 & 0 \end{bmatrix} \quad \text{and} \quad e^{CT} - I = \begin{bmatrix} 0 & T \\ 0 & 0 \end{bmatrix}$$

we note first that the basis for E_{n-r} is the vector $\begin{bmatrix} 1 \\ 0 \end{bmatrix}$. Therefore

$$P_{n-r} = \begin{bmatrix} 1 & 0 \\ 0 & 0 \end{bmatrix}$$

and

$$P_r = \begin{bmatrix} 0 & 0 \\ 0 & 1 \end{bmatrix}$$

Thus the matrix H must satisfy

$$H \begin{bmatrix} 0 & T \\ 0 & 0 \end{bmatrix} = \begin{bmatrix} 0 & 0 \\ 0 & 1 \end{bmatrix}$$

and hence we may take

$$H = \begin{bmatrix} 0 & \frac{1}{T} \\ \frac{1}{T} & 0 \end{bmatrix}$$

Finally a matrix H corresponding to

$$D = \begin{bmatrix} a & 0 \\ 0 & a \end{bmatrix}$$

and

$$e^{TD} - 1 = \begin{bmatrix} e^{aT} - 1 & 0 \\ 0 & e^{aT} - 1 \end{bmatrix}$$

is easily seen to be

$$H = \begin{bmatrix} \frac{1}{e^{aT}-1} & 0 \\ 0 & \frac{1}{e^{aT}-1} \end{bmatrix}$$

since

$$P_{n-r} = \begin{bmatrix} 0 & 0 \\ 0 & 0 \end{bmatrix}$$

and

$$P_r = \begin{bmatrix} 1 & 0 \\ 0 & 1 \end{bmatrix}$$

Then the H corresponding to A is

$$H = \begin{bmatrix} 0 & 0 & 1 & 0 & & & & \\ 0 & 0 & 0 & 1 & & & & \\ \frac{1}{T} & 0 & 0 & 0 & & & & \\ 0 & \frac{1}{T} & 0 & 0 & & & & \\ & & & & 0 & \frac{1}{T} & & \\ & & & & \frac{1}{T} & 0 & & \\ & & & & & & \frac{1}{e^{aT}-1} & 0 \\ & & & & & & 0 & \frac{1}{e^{aT}-1} \end{bmatrix} \qquad (7.36)$$

Now we compute $P_{n-r}H$ when H is given by (7.36), and P_{n-r} is obtained from the P_{n-r} associated with matrices $B, C,$ and D. That is

$$P_{n-r} = \begin{bmatrix} 1 & & & & & & & \\ & 1 & & & & & & \\ & & 0 & & & & & \\ & & & 0 & & & & \\ & & & & 1 & & & \\ & & & & & 0 & & \\ & & & & & & 0 & \\ & & & & & & & 0 \end{bmatrix} \qquad (7.37)$$

$$P_{n-r}H = \begin{bmatrix} 0 & 0 & 1 & 0 & & & & \\ 0 & 0 & 0 & 1 & & & & \\ 0 & 0 & 0 & 0 & & & & \\ 0 & 0 & 0 & 0 & & & & \\ & & & & 0 & \frac{1}{T} & & \\ & & & & 0 & 0 & & \\ & & & & & & 0 & \\ & & & & & & & 0 \end{bmatrix}$$

$$P_{n-r}He^{-sA} = P_{n-r}H$$

$$\times \begin{bmatrix} \cos\beta s & -\sin\beta s & -s\cos\beta s & s\sin\beta s & & & & \\ \sin\beta s & \cos\beta t & -s\sin\beta s & -s\cos\beta s & & & & \\ & & \cos\beta s & -\sin\beta s & & & & \\ & & \sin\beta s & \cos\beta s & & & & \\ & & & & 1 & -s & & \\ & & & & 0 & 1 & & \\ & & & & & & e^{-as} & \\ & & & & & & & e^{-as} \end{bmatrix}$$

$$= \begin{bmatrix} 0 & 0 & \cos\beta s & -\sin\beta s & & & & \\ 0 & 0 & \sin\beta s & \cos\beta s & & & & \\ 0 & 0 & 0 & 0 & & & & \\ 0 & 0 & 0 & 0 & & & & \\ & & & & 0 & \frac{1}{T} & & \\ & & & & 0 & 0 & & \\ & & & & & & 0 & 0 \\ & & & & & & 0 & 0 \end{bmatrix}$$

Thus

$$P_{n-r}H \int_0^T e^{-sA} F[s, x(s, 0, c), 0]\, ds$$

$$= \int_0^T \begin{bmatrix} (\cos\beta s)F_3 - (\sin\beta s)F_4 \\ (\sin\beta s)F_3 + (\cos\beta s)F_4 \\ 0 \\ 0 \\ \frac{1}{T}F_6 \\ 0 \\ 0 \\ 0 \end{bmatrix} ds$$

In the expression $x(s, 0, c)$, which is an argument of F, the c denotes the vector $(c_1, c_2, 0, 0, c_5, 0, 0, 0)$ (where, for simplicity of notation, we assume that $\xi(0,0) = 0$) and thus the mapping M_0 is defined by:

$$M_0:\ (c_1, c_2, c_5) \rightarrow \left(\int_0^T [(\cos\beta s)F_3 - (\sin\beta s)F_4]\, ds,\ \int_0^T [(\sin\beta s)F_3 \right.$$

$$\left. + (\cos\beta s)F_4]\, ds,\ \int_0^T \frac{1}{T}F_6\, ds \right)$$

where the argument $x(s, 0, c)$ of F is such that

$$c = (c_1, c_2, 0, 0, c_5, 0, 0, 0).$$

Let $k_3(t)$, $k_4(t)$, $k_6(t)$ be differentiable functions of period T such that

$$\int_0^T [(\cos \beta s)k_3(s) + (-\sin \beta s)k_4(s)] \, ds \neq 0$$

$$\int_0^T [(\sin \beta s)k_3(s) + (\cos \beta s)k_4(s)] \, ds \neq 0$$

$$\int_0^T k_6(t) \, ds \neq 0$$

For example, suppose $k_3(s) = \cos \beta s$, $k_4(s) = \cos \beta s$, and $k_6(s) = \cos^2 \beta s$. (Note that we have a wide latitude in our choice of $k_3(s)$, $k_4(s)$, and $k_6(s)$.) If δ_3, δ_4, δ_6 are sufficiently small in absolute value, then

$$\max_s |\delta_3 k_3(s) + \delta_4 k_4(s) + \delta_6 k_6(s)| < \delta$$

Also if F_3, F_4, F_6 are replaced by $F_3 + \delta_3 k_3$, $F_4 + \delta_4 k_4$, $F_6 + \delta_6 k_6$, respectively, the mapping M_0 is replaced $M_0 + d$ where d is the constant 3-vector

$$\left[\begin{array}{c} \int_0^T [\delta_3(\cos(\beta s)k_3(s) + \delta_4(-\sin \beta s k_4(s)] \, ds \\ \int_0^T [\delta_3(\sin \beta s)k_3(s) + \delta_4(\cos \beta s)k_4(s)] \, ds \\ \int_0^T \frac{1}{T}\delta_6 k_6(s) \, ds \end{array} \right]$$

Since $\deg[M_0, \mathcal{B}, 0] \neq 0$, then by Theorem 1 in the Appendix, it follows that for most sufficiently small $|\delta_3|$, $|\delta_4|$, and $|\delta_6|$, the equation

$$M_0(c_1, c_2, c_5) + d = 0$$

has a finite set of solutions, that there are at least $|\deg[M_0, \mathcal{B}, 0]|$ such solutions and also that the jacobian of M_0 at each solution is nonzero. This completes the proof of Theorem 7.4.

Stability of the Periodic Solutions

Next, we show that the implicit function theorem and topological degree yield some information about the stability properties of the periodic solutions that have been obtained for equation (7.11).

We seek criteria for the asymptotic stability of the periodic solutions obtained in the preceding section. Suppose that (c_0, ε_0) is a solution of the equation (7.14) which expresses the periodicity condition, that is, suppose that

$$\tilde{F}(c_0, \varepsilon_0) \overset{\text{def}}{=} (e^{TA} - I)c_0 + e^{TA} \int_0^T e^{-sA} \{\varepsilon_0 F[s, x(s, \varepsilon_0, c_0), \varepsilon_0] + G(s)\} \, ds = 0$$

$$(7.38)$$

Then the solution $x(t, \varepsilon_0, c_0)$ of equation (7.38) has period T. Since our object is to study the stability properties of the periodic solution $x(t, \varepsilon_0, c_0)$, the first step is to compare $x(t, \varepsilon_0, c_0)$ with another solution, say $x(t, \varepsilon_0, \bar{c})$, of equation (7.11) with $\varepsilon = \varepsilon_0$. Substituting $x(t, \varepsilon_0, c_0)$ and $x(t, \varepsilon_0, \bar{c})$ into equation (7.11) and subtracting the first equation from the second, we obtain, by using Taylor's expansion:

$$
\begin{aligned}
x'(t, \varepsilon_0, \bar{c}) - x'(t, \varepsilon_0, c_0) &= A[x(t, \varepsilon_0, \bar{c}) - x(t, \varepsilon_0, c_0)] \\
&\quad + \varepsilon_0\{F[t, x(t, \varepsilon_0, \bar{c}), \varepsilon_0] - F[t, x(t, \varepsilon, c_0), \varepsilon_0]\} \\
&= A[x(t, \varepsilon_0, \bar{c}) - x(t, \varepsilon_0, c_0)] \\
&\quad + \varepsilon_0\{B(t, \varepsilon_0)[x(t, \varepsilon_0, \bar{c}) - x(t, \varepsilon_0, c_0)] \\
&\quad + H[x(t, \varepsilon_0, \bar{c}) - x(t, \varepsilon_0, c_0), t, \varepsilon_0]\} \qquad (7.39)
\end{aligned}
$$

where $B(t, \varepsilon_0)$ is the matrix

$$
F_x[t, x(t, \varepsilon_0, c_0), \varepsilon_0]
$$

the elements of which are functions of t which have period T, and if $\xi \in R^n$,

$$
|H(\xi, t, \varepsilon_0)| = o(|\xi|) \qquad (7.40)
$$

for all $t \in [0, T]$, that is,

$$
\lim_{|\xi| \to 0} \frac{|H(\xi, t, \varepsilon_0)|}{|\xi|} = 0
$$

and this limit is uniform in $t \in [0, T]$.

Let us first suppose that the eigenvalues of matrix A all have negative real parts. Then by Theorem 7.1, equation (7.11) has a unique periodic solution $x(t, \varepsilon)$ for each ε such that $|\varepsilon|$ is sufficiently small. Since the eigenvalues of A have negative real parts, then the solution $u(t) \equiv 0$ of

$$
u' = Au
$$

is asymptotically stable. Hence, if $|\varepsilon_0|$ is sufficiently small, then by Theorem 4.3, the solution $u(t) \equiv 0$ of

$$
u' = [A + \varepsilon_0 B(t, \varepsilon_0)]u
$$

is asymptotically stable. By condition (7.40) on H, it follows from the stability theorem for linear homogeneous systems and the corollary to stability and instability theorems for nonlinear systems (both in Chapter 4) that 0 is an asymptotically stable solution of

$$
u' = [A + \varepsilon_0 B(t, \varepsilon_0)]u + \varepsilon_0 H[u, t, \varepsilon_0]
$$

and hence that $x(t, \varepsilon_0, c_0)$ is asymptotically stable. Thus we have proved:

Theorem 7.5   *If the eigenvalues of matrix A all have negative real parts and if $|\varepsilon|$ is sufficiently small, then equation (7.11) has a unique solution $x(t, \varepsilon)$ of period T and $x(t, \varepsilon)$ is asymptotically stable.*

Similar arguments yield the following theorem (see Exercise 4).

Theorem 7.6 *If all the eigenvalues of A are different from $(2n\pi/T)i$ ($n = 0, \pm 1$, $\pm 2, \dots$) and if A has an eigenvalue with positive real part, then equation (7.11) has a unique solution $x(t, \varepsilon)$ of period T and $x(t, \varepsilon)$ is unstable.*

Now we turn to stability of periodic solutions in the resonance case. From (7.40) and the fact that the matrix $B(t, \varepsilon_0)$ has period T, it follows (by the stability theorem for periodic solutions in Chapter 4) that in order to study the stability of $x(t, \varepsilon_0, c_0)$, it is sufficient to study the characteristic multipliers of the matrix

$$A + \varepsilon_0 B(t, \varepsilon_0)$$

Let $U(t)$ be the fundamental matrix of the equation

$$w' = [A + \varepsilon_0 B(t, \varepsilon_0)]w$$

such that $U(0)$ is the identity matrix. Then the characteristic multipliers to be studied are the eigenvalues of $U(T)$. The analysis we make consists of three parts. First we show (Lemma 7.2) that if $D_{c_0} \tilde{F}$ denotes the differential of $\tilde{F}(\cdot, \varepsilon_0)$ at c_0, then

$$U(T) = D_{c_0} \tilde{F} + I$$

where I is the $n \times n$ identity matrix and that hence if $\lambda_1, \dots, \lambda_n$ are the eigenvalues of $U(T)$, then $\lambda_1 - 1, \dots, \lambda_n - 1$ are the eigenvalues of $D_{c_0} \tilde{F}$. Then we impose the assumption that A is such that

$$e^{TA} - I = \begin{bmatrix} 0 & 0 & \\ 0 & 0 & \\ & & C_2 \end{bmatrix}$$

where C_2 is an $(n - 2) \times (n - 2)$ nonsingular matrix all of whose eigenvalues have negative real parts; and we compare the eigenvalues of $D_{c_0} \tilde{F}$ with the eigenvalues of the matrix

$$\begin{bmatrix} \varepsilon_0 D_{c_0} M_{\varepsilon_0} & \\ & C_2 \end{bmatrix}$$

where M_{ε_0} is the mapping defined in equation (7.34) and $D_{c_0} M_{\varepsilon_0}$ is the differential of M_{ε_0} at c_0. Finally, from this comparison, we draw conclusions about the eigenvalues of $U(T)$.

Lemma 7.2 *If the eigenvalues of $U(T)$ are $\lambda_1, \dots, \lambda_n$, then the eigenvalues of the differential of $\tilde{F}(\cdot, \varepsilon_0)$ at c_0 (see equation (7.38)) are $\lambda_1 - 1, \dots, \lambda_n - 1$.*

Proof If we let $u(t) = x(t, \varepsilon_0, \bar{c}) - x(t, \varepsilon_0, c_0)$ and rewrite equation (7.39) as

$$u'(t) = Au(t) + \varepsilon_0 B(t, \varepsilon_0)u(t) + \varepsilon_0 H[u(t), t, \varepsilon_0] \tag{7.41}$$

then by the variation of constants formula, we obtain

$$u(T) = U(T)u(0) + U(T) \int_0^T [U(s)]^{-1} \varepsilon_0 H[u(s), s, \varepsilon_0] \, ds \tag{7.42}$$

If $u(0) = \bar{c} - c_0 = K$, a constant n-vector, then (7.42) may be written

$$u(T) = U(T)K + U(T) \int_0^T [U(s)]^{-1} \varepsilon_0 H[u(s, K), s, \varepsilon_0] \, ds \qquad (7.43)$$

where we now denote $u(s)$ by $u(s, K)$ so that we can indicate the initial condition

$$u(0, K) = K$$

By uniqueness of solution, for all s,

$$u(s, 0) = 0$$

and therefore

$$u(s, K) = u(s, K) - u(s, 0) = \left[\frac{\partial u}{\partial K} (s, 0) \right] K + R(K)$$

where

$$|R(K)| = o(|K|) \qquad (7.44)$$

From condition (7.40) on H and from (7.44), it follows that the differential at 0 of the mapping

$$\tilde{M} : K \to u(T, K)$$

that is, the mapping

$$\tilde{M} : K \to U(T)K + U(T) \int_0^T [U(s)]^{-1} \varepsilon_0 H[u(s, K), s, \varepsilon_0] \, ds$$

is the matrix $U(T)$. But we have also

$$u(T) = x(T, \varepsilon_0, \bar{c}) - x(T, \varepsilon_0, c_0)$$

$$= e^{TA}\bar{c} + e^{TA} \int_0^T e^{-sA} \{\varepsilon_0 F[s, x(s, \varepsilon_0, \bar{c}), \varepsilon_0] + G(s)\} \, ds$$

$$- e^{TA}c_0 - e^{TA} \int_0^T e^{-sA} \{\varepsilon_0 F[s, x(s, \varepsilon_0, c_0), \varepsilon_0] + G(s)\} \, ds$$

$$= e^{TA}(\bar{c} - c_0) + e^{TA} \int_0^T e^{-sA} \{\varepsilon_0 B(s, \varepsilon_0) u(s, \bar{c} - c_0) \qquad (7.45)$$

$$+ \varepsilon_0 H[u(s, \bar{c} - c_0), s, \varepsilon_0]\} \, ds$$

Also by equation (7.33), we have

$$\tilde{F}(\bar{c}, \varepsilon_0) - \tilde{F}(c_0, \varepsilon_0) = (e^{TA} - I)(\bar{c} - c_0)$$

$$+ e^{TA} \int_0^T e^{-sA} \{\varepsilon_0 B(s, \varepsilon_0) u(s, \bar{c} - c_0) \qquad (7.46)$$

$$+ \varepsilon_0 H[u(s, \bar{c} - c_0), s, \varepsilon_0]\} \, ds$$

Equations (7.45) and (7.46) show that the differential at 0 of the mapping $\tilde{M}$, which we know already is $U(T)$, equals the differential at c_0 of $\tilde{F}(c, \varepsilon_0)$ plus the identity map. That is, if $D_{c_0}\tilde{F}$ denotes the differential of $\tilde{F}(\cdot, \varepsilon_0)$ at c_0 and if $D_0\tilde{M}$ denotes the differential of $\tilde{M}$ at 0, then

$$U(T) = D_0\tilde{M} = D_{c_0}\tilde{F} + I$$

Thus if $\lambda_1, \ldots, \lambda_n$ are the eigenvalues of $U(T)$, then $\lambda_1 - 1, \ldots, \lambda_n - 1$ are the eigenvalues of $D_{c_0}\tilde{F}$. This completes the proof of Lemma 7.2. □

Our next object is to use the value of $\deg[M_0, \mathcal{B}, 0]$ to obtain information about the eigenvalues of $D_{c_0}\tilde{F}$. Since

$$e^{TA} - I = \begin{bmatrix} 0 & & \\ & 0 & \\ & & C_2 \end{bmatrix}$$

a basis for E_{n-r} is the set of vectors

$$\begin{bmatrix} 1 \\ 0 \\ \vdots \\ 0 \end{bmatrix}, \begin{bmatrix} 0 \\ 1 \\ 0 \\ \vdots \\ 0 \end{bmatrix}$$

Thus

$$P_{n-r}: (c_1, c_2, \ldots, c_n) \to (c_1, c_2)$$
$$P_r: (c_1, c_2, \ldots, c_n) \to (c_3, \ldots, c_n)$$

Also we may choose

$$H = \begin{bmatrix} 1 & & \\ & 1 & \\ & & (C_2)^{-1} \end{bmatrix}$$

By (7.38),

$$\tilde{F}(c_0, \varepsilon_0) = (e^{TA} - I)c_0 + e^{TA} \int_0^T e^{-sA} \{\varepsilon_0 F[s, x(s, \varepsilon_0, c_0), \varepsilon_0] + G(s)\} \, ds$$

In this case,

$$P_{n-r}(e^{TA} - I) = \begin{bmatrix} 1 & & & \\ & 1 & & \\ & & 0 & \\ & & & \ddots & \\ & & & & 0 \end{bmatrix} \begin{bmatrix} 0 & & \\ & 0 & \\ & & C_2 \end{bmatrix} = 0$$

Also

$$P_{n-r}H = \begin{bmatrix} 1 & & & \\ & 1 & & \\ & & 0 & \\ & & & \ddots & \\ & & & & 0 \end{bmatrix} \begin{bmatrix} 1 & & \\ & 1 & \\ & & (C_2)^{-1} \end{bmatrix} = \begin{bmatrix} 1 & & & \\ & 1 & & \\ & & 0 & \\ & & & \ddots & \\ & & & & 0 \end{bmatrix} = P_{n-r}$$

By Theorem 7.2, we require that

$$P_{n-r}H \int_0^T e^{-sA}G(s)\,ds = 0$$

and hence, in this case, since $P_{n-r}H = P_{n-r}$, the necessary condition is

$$P_{n-r} \int_0^T e^{-sA}G(s)\,ds = 0$$

But letting I_r denote the $(n-2) \times (n-2)$ identity matrix, we have

$$P_{n-r}e^{TA} = \begin{bmatrix} 1 & & & \\ & 1 & & \\ & & 0 & \\ & & & \ddots & \\ & & & & 0 \end{bmatrix} \begin{bmatrix} 1 & & \\ & 1 & \\ & & C_2 + I_r \end{bmatrix} = \begin{bmatrix} 1 & & & \\ & 1 & & \\ & & 0 & \\ & & & \ddots & \\ & & & & 0 \end{bmatrix} = P_{n-r}$$

Hence

$$P_{n-r}e^{TA} \int_0^T e^{-sA}G(s)\,ds = 0$$

and

$$P_{n-r}\tilde{F}(c_0, \varepsilon_0) = P_{n-r} \int_0^T e^{-sA}\{\varepsilon_0 F[s, x(s, \varepsilon_0, c_0), \varepsilon_0]\,ds \qquad (7.47)$$

Also

$$P_r(e^{TA} - I)c_0 = \begin{bmatrix} 0 & & & \\ & 0 & & \\ & & 1 & \\ & & & \ddots & \\ & & & & 1 \end{bmatrix} \begin{bmatrix} 0 & & \\ & 0 & \\ & & C_2 \end{bmatrix} c_0 = C_2 P_r c_0$$

and hence

$$P_r \tilde{F}(c_0, \varepsilon_0) = C_2(P_r c_0) + P_r e^{TA} \int_0^T e^{-sA}\{\varepsilon_0 F[s, x(s, \varepsilon_0, c_0), \varepsilon_0] + G(s)\}\, ds$$

(7.48)

Since C_2 is nonsingular, then the equation

$$P_r \tilde{F}(c_0, \varepsilon_0) = 0$$

can be solved for $P_r c_0$ in terms of $P_{n-r}\, c_0$ and ε_0 by applying the implicit function theorem. (The initial solution is: $P_{n-r} c_0$ fixed, $\varepsilon_0 = 0$, and $P_r c_0 = -C_2^{-1} P_r e^{TA} \int_0^T e^{-sA} G(s)\, ds$.) Denoting that solution by $P_r c_0(P_{n-r} c_0, \varepsilon)$ and substituting it in (7.47), we obtain:

$$P_{n-r} \tilde{F}(c_0, \varepsilon_0) = \varepsilon_0 P_{n-r} \int_0^T e^{-sA}\{F[s, x(s, \varepsilon_0, P_{n-r} c_0 + P_r c_0(P_{n-r} c_0, \varepsilon_0)), \varepsilon_0]\, ds$$

(7.49)

But the expression on the right-hand side of (7.49) is, except for the factor ε_0, the expression in equation (7.34) which defines the mapping M_ε (described after equation (7.34)). Now suppose that the eigenvalues of $D_{c_0} M_{\varepsilon_0}$ have nonzero real parts. Then we conclude from (7.47) and (7.48) that if $|\varepsilon_0|$ is sufficiently small, the eigenvalues of $D_{c_0} \tilde{F}$ have real parts with the same sign as the real parts of the eigenvalues of the matrix

$$\begin{bmatrix} \varepsilon_0 D_{c_0} M_{\varepsilon_0} & \\ & C_2 \end{bmatrix}$$

Hence we are reduced to considering the eigenvalues of $D_{c_0} M_{\varepsilon_0}$. At this point, we use information about the value of $\deg[M_0, \mathcal{B}, 0]$. Suppose that

$$\deg[M_0, \mathcal{B}, 0] > 0$$

If $F(t, x, \varepsilon)$ is changed by adding a "small" function $h(t)$ (i.e., if Theorem 7.4 is applied), then the Jacobian of M_0 is nonzero at each solution c of the equation

$$M_0(c) = 0$$

For convenience in this discussion we use the following notation: If f is a differentiable mapping from R^n into R^n and x is a point in the domain of f, let $J_x f$ denote the Jacobian of f at the point x, that is, $J_x f$ denotes the determinant of $D_x f$, the differential of f at x.

Since $\deg[M_0, \mathcal{B}, 0] > 0$, then there is at least one point c_0 such that

$$J_{c_0} M_0 > 0$$

Suppose the eigenvalues of $D_{c_0} M_0$ are real. Since $J_{c_0} M_0$ is the product of the eigenvalues, then either both eigenvalues are positive or both are negative. If both eigenvalues are positive, and if ε_0 is positive, then both eigenvalues of

$$\varepsilon_0 D_{c_0} M_{\varepsilon_0}$$

are positive. Hence, $D_{c_0}\tilde{F}$ has two eigenvalues with positive real parts and hence by Lemma 7.2, $U(T)$ has two real eigenvalues with real parts greater than one. Hence $x(t, \varepsilon_0, c_0)$ is unstable.

A similar argument shows that if both the eigenvalues of $D_{c_0}M_0$ are negative, then if ε_0 is positive, and sufficiently small, solution $x(t, \varepsilon_0, c_0)$ is asymptotically stable.

If

$$J_{c_0}M_{\varepsilon_0} < 0$$

then both eigenvalues of $D_{c_0}M_{\varepsilon_0}$ are real and one eigenvalue is positive and one is negative. The same sort of arguments as before show that $x(t, \varepsilon_0, c_0)$ is unstable.

If $J_{c_0}M_0 > 0$ and the eigenvalues of $D_{c_0}M_0$ are complex conjugates, that is, the eigenvalues are $\alpha + i\beta$ and $\alpha - i\beta$, suppose first that $\alpha > 0$. Then if $\varepsilon_0 > 0$, both eigenvalues of $\varepsilon_0 D_{c_0}M$ have positive real parts and $D_{c_0}\tilde{F}$ has two eigenvalues with positive real parts. Hence, $U(T)$ has two eigenvalues whose real parts are greater than one. Hence, $x(t, \varepsilon_0, c_0)$ is unstable. If $\alpha < 0$, similar arguments show that if $\varepsilon_0 > 0$, then $x(t, \varepsilon_0, c_0)$ is asymptotically stable.

Similar arguments may be used if $\varepsilon_0 < 0$. As pointed out before, we shall not consider the case in which the eigenvalues of $D_{c_0}M$ are $i\beta$ and $-i\beta$, that is, we require that the trace of $D_{c_0}M$ be nonzero.

We may summarize our results in the following theorem.

Theorem 7.7 *If the matrix A satisfies Assumption 1 and*

$$e^{TA} - I = \begin{bmatrix} 0 & 0 \\ & C_2 \end{bmatrix},$$

if $\deg[M_0, \mathcal{B}, 0] \neq 0$, *if the function* $F(t, x, \varepsilon)$ *in equation* (7.12) *is replaced by* $F(t, x, \varepsilon) + k(t)$ *as described in the statement of Theorem 7.4 and if* $\operatorname{tr} D_{c_0}M_0 \neq 0$ *at each point* c_0 *such that* $M_0(c_0) = 0$, *then the following conclusions hold:*

(1) *If* $\deg[M_0, \mathcal{B}, 0] < 0$, *then if* $\varepsilon > 0$ *and* ε *is sufficiently small, equation* (7.12) *has at least* $|\deg[M_0, \mathcal{B}, 0]|$ *distinct unstable periodic solutions.*

(2) *If* $\deg[M_0, \mathcal{B}, 0] > 0$, *then if* $|\varepsilon|$ *is sufficiently small, equation* (7.12) *has at least* $|\deg[M_0, \mathcal{B}, 0]|$ *distinct periodic solutions* $x(t, \varepsilon)$ *and each of these solutions* $x(t, \varepsilon)$ *has the following property: If* $\varepsilon > 0$ *and sufficiently small and both eigenvalues of* $D_{c_0}M_0$ *are positive [negative] solution* $x(t, \varepsilon)$ *is unstable [asymptotically stable]. If* $\varepsilon > 0$ *and sufficiently small and the real part of each eigenvalue of* $D_{c_0}M_0$ *is positive [negative] solution* $x(t, \varepsilon)$ *is unstable [asymptotically stable].*

We leave the case

$$e^{TA} - I = \begin{bmatrix} 0 & \\ & C_1 \end{bmatrix}$$

as an exercise (Exercise 5).

B. The Case in Which the Unperturbed Equation has a Family of Periodic Solutions: The Malkin-Roseau Theory

In Part A, we studied, among others, the quasilinear equation

$$\frac{dx}{dt} = Ax + \varepsilon F(t, x, \varepsilon)$$

The majority of the study concerned the resonance case, that is, the case in which

$$\det[e^{TA} - I] = 0$$

Another way of describing the resonance case is to impose the following hypothesis: the unperturbed equation

$$\frac{dx}{dt} = Ax \tag{7.50}$$

has a family of periodic solutions, that is, the linear space of solutions of period T of (7.50) has dimension $n \geq 1$. In Part B, we will impose this kind of hypothesis on the general equation

$$\frac{dx}{dt} = f(t, x) + \varepsilon g(t, x, \varepsilon)$$

that is, we assume that the equation

$$\frac{dx}{dt} = f(t, x) \tag{7.51}$$

has a family of solutions of period T. This problem was studied by Malkin [1959]. We shall follow the later more extensive treatment given by Roseau [1966, Chapter 18]. (Roseau points out that the problem was studied earlier by J. Haag.) We add a description of how the results can be augmented by using topological degree. We also indicate how a further study of the stability of the solutions can be carried out.

An important special case of (7.51) is the case in which $f(t, x)$ is independent of t, that is, equation (7.51) is autonomous. (Then if $x(t)$ is periodic solution of (7.51) and if r is an arbitrary fixed real value, the function $x(t + r)$ is also a periodic solution, and thus we obtain a one-parameter family of periodic solutions.) Since this special case is important in applications, we give it a detailed treatment in Part C.

We consider the n-dimensional equation

$$\frac{dx}{dt} = f(t, x) + \varepsilon g(t, x, \varepsilon) \tag{7.5}$$

where $(t, x, \varepsilon) \in R \times R^n \times I$, where I is an interval on the real line with midpoint zero, and f, g have period T in t. Also f has continuous third derivatives in t and x and g has continuous second derivatives in t, x and ε. We impose:

Assumption 1 The equation

$$\frac{dx}{dt} = f(t, x) \tag{7.51}$$

has a set of solutions each with period T which depend on a set of m real parameters

$$\gamma_1, \ldots, \gamma_m$$

where $m < n$. We denote these periodic solutions by

$$z(t, \gamma_1, \ldots, \gamma_m) \qquad \text{or} \qquad z(t, \gamma)$$

and we require that the function z have continuous second derivatives in $\gamma_1, \ldots, \gamma_m$ for all real $\gamma_j (j = 1, \ldots, m)$.

We study:

Problem 3: If $|\varepsilon|$ is sufficiently small, does equation (7.5) has a solution $x(t, \varepsilon)$ of period T such that as $\varepsilon \to 0$, this solution approaches one of the solutions $z(t, \gamma_1, \ldots, \gamma_m)$ of (7.51).

First, let $x(t, \varepsilon)$ denote a solution of (7.5) and consider the expression:

$$x(t, \varepsilon) - z(t, \gamma) = \varepsilon u(t, \varepsilon)$$

Then

$$\frac{d}{dt}(\varepsilon u(t, \varepsilon)) = \frac{dx}{dt} - \frac{dz}{dt}$$

and

$$\varepsilon \frac{du}{dt} = f(t, z(t\gamma) + \varepsilon u) + \varepsilon g(t, z(t, \gamma) + \varepsilon u, \varepsilon) - f(t, z(t, \gamma)) \qquad (7.52)$$

Let

$$A(t, \gamma) = \left[\frac{\partial f_i}{\partial x_j} (z(t, \gamma)) \right]$$

and

$$h(t, \gamma) = g(t, z(t, \gamma), 0)$$

Then (7.52) becomes

$$\varepsilon \frac{du}{dt} = A(t, \gamma)\varepsilon u + \varepsilon h(t, \gamma) + f(t, z(t, \gamma) + \varepsilon u) - f(t, z(t, \gamma))$$
$$-A(t, \gamma)\varepsilon u + \varepsilon g(t, z(t, \gamma) + \varepsilon u, \varepsilon) - \varepsilon g(t, z(t, \gamma), 0)$$

$$\frac{du}{dt} = A(t, \gamma)u + h(t, \gamma) + \frac{1}{\varepsilon}[f(t, z + \varepsilon u) - f(t, z)]$$
$$-\varepsilon A(t, \gamma)u + \varepsilon g(t, z + \varepsilon u, \varepsilon) - \varepsilon g(t, z, 0)] \qquad (7.53)$$

Let

$$G(t, u, \gamma, \varepsilon) = \frac{1}{\varepsilon^2}[f(t, z(t, \gamma) + \varepsilon u) - f(t, z(t, \gamma)) - \varepsilon A(t, \gamma)u$$
$$+\varepsilon g(t, z(t, \gamma) + \varepsilon u, \varepsilon) - \varepsilon g(t, z(t, \gamma, 0))]$$

Then $G(t, u, \gamma, \varepsilon)$ has a continuous derivative with respect to ε at $\varepsilon = 0$ and equation (7.53) may be written as

$$\frac{du}{dt} = A(t, \gamma)u + h(t, \gamma) + \varepsilon G(t, u, \gamma, \varepsilon) \qquad (7.54)$$

In order to solve Problem 3, it is sufficient to study the periodic solutions of equation (7.54). For this study, we impose an additional hypothesis. Since

$$\frac{\partial z}{\partial t}(t, \gamma) = f[t, z(t, \gamma)] \tag{7.55}$$

then differentiating (7.56) with respect to γ_j ($j = 1, \ldots, m$) we obtain

$$\frac{\partial}{\partial \gamma_j} \frac{\partial z}{\partial t}(t, \gamma) = \left[\frac{\partial f_i}{\partial x_j}(t, z(t, \gamma))\right] \frac{\partial z}{\partial \gamma_j}(t, \gamma)$$

or

$$\frac{\partial}{\partial t}\left[\frac{\partial}{\partial \gamma_j} z(t, \gamma)\right] = A(t, \gamma)\frac{\partial z}{\partial \gamma_j}(t, \gamma)$$

Thus $\frac{\partial z}{\partial \gamma_j}$ is a solution of period T of the linear homogeneous equation

$$\frac{du}{dt} = A(t, \gamma)u \tag{7.56}$$

We impose:
Assumption 2 For each value of γ, the periodic solutions

$$\frac{\partial z}{\partial \gamma_1}(t, \gamma), \ldots, \frac{\partial z}{\partial \gamma_m}(t, \gamma)$$

of (7.56) are linearly independent. Also these periodic solutions are a basis for the linear space of periodic solutions (of period T) of equation (7.56).

Remark It follows from Assumption 2 that the set of functions $z(t, \gamma)$ describes a manifold because for each value of t the rank of the matrix

$$\left[\frac{\partial z}{\partial \gamma_j}(t, \gamma)\right]$$

is m.

Now it follows that the space $\mathcal{L}$ of periodic solutions (of period T) of the adjoint equation of (7.56), that is, the equation

$$\frac{du}{dt} + [A(t, \gamma)]^*u = 0 \tag{7.57}$$

has dimension m. Let $v^1(t), \ldots, v^m(t)$ be a basis for $\mathcal{L}$. Applying Theorem 2.16 in Chapter 2, we conclude that if $\gamma^0 = (\gamma_1^0, \ldots, \gamma_m^0)$ is a solution of the system of equations

$$U_k(\gamma_1, \ldots, \gamma_m) = \int_0^T \left\{\sum_{j=1}^n [v_{jk}(s, \gamma_1, \ldots, \gamma_m)][h_j(s, \gamma_1, \ldots, \gamma_m)]\right\} ds = 0$$

$$\tag{7.58}$$

or

$$U_k(\gamma_1, \ldots, \gamma_m) = \int_0^T v^k \cdot h(s, \gamma_1, \ldots, \gamma_m)] \, ds = 0 \qquad (k = 1, \ldots, m)$$

where

$$\begin{bmatrix} v_{1k} \\ v_{2k} \\ \vdots \\ v_{nk} \end{bmatrix} = v^k$$

then the equation

$$\frac{du}{dt} = A(t, \gamma^0)u + h(t, \gamma^0) \tag{7.59}$$

has a solution $u(t)$ of period T. Also all such solutions $u(t)$ of (7.59) can be described by

$$u(t) = a_1 \frac{\partial z}{\partial \gamma_1} + \cdots + a_m \frac{\partial z}{\partial \gamma_m} + \bar{u}(t) \tag{7.60}$$

where $\bar{u}(t)$ is a particular periodic solution (of period T) of (7.59) and $a_1, \ldots, a_m$ are arbitrary constants.

Now suppose that the system (7.58), $k = 1, \ldots, m$, has such a solution

$$\gamma^0 = (\gamma_1^0, \ldots, c\gamma_m^0)$$

We set $\gamma = \gamma^0$ in equation (7.54) and consider the question of whether the resulting equation

$$\frac{du}{dt} = A(t, \gamma^0)u + h(t, \gamma^0) + \varepsilon G(t, u, \gamma^0, \varepsilon) \tag{7.61}$$

has, for sufficiently small $|\varepsilon|$, a solution $u(t, \varepsilon)$ of period T such that

$$\lim_{\varepsilon \to 0} u(t, \varepsilon) = a_1 \frac{\partial z}{\partial \gamma_1} + \cdots + a_m \frac{\partial z}{\partial \gamma_m} + \bar{u}(t)$$

where $a_1, \ldots, a_m$ are constants and $\bar{u}(t)$ is a particular solution of period T of equation (7.59).

By Theorem 2.16 and the fact that γ^0 is a solution of system (7.58), $k = 1 \ldots, m$, it follows that there exists such a solution $u(t, \varepsilon)$ if the system of equations

$$W_k(a_1, \ldots, a_m) = \int_0^T [v^k(s, \gamma^0)] \cdot$$

$$\left[G\left(s, a_1 \frac{\partial z}{\partial \gamma_1}(s, \gamma^0) + \cdots + a_m \frac{\partial z}{\partial \gamma_m}(s, \gamma^0) + \bar{u}(s), \gamma^0, 0 \right) \right] ds = 0$$

$$(k = 1, \ldots, m) \tag{7.62}$$

has a solution $(a_1^0, \ldots, a_m^0)$ and the determinant of the matrix

$$\left[\frac{\partial W_i}{\partial a_j} \right]$$

in which the entries are evaluated at $(a_1^0, \ldots, a_m^0)$, is nonzero. That is, by applying the implicit function theorem to (7.62), it follows that there exists a solution $u(t, \varepsilon)$ with period T as a function of t and such that

$$\lim_{\varepsilon \to 0} u(t, \varepsilon) = a_1^0 \frac{\partial z}{\partial \gamma_1}(t, \gamma^0) + \cdots + a_m^0 \frac{\partial z}{\partial \gamma}(t, \gamma^0) + \bar{u}(t)$$

Thus we are reduced to studying the solutions of (7.62).

To study the solutions of (7.62), we need two calculational steps:

Step 1. Equations (7.62) are linear in $a_1, \ldots, a_m$.

Step 2. The coefficient matrix of (7.62), that is,

$$\left[\frac{\partial W_i}{\partial a_j} \right]$$

is equal to the coefficient matrix of (7.58) evaluated at γ^0, that is, the matrix

$$\left[\frac{\partial U_i}{\partial \gamma_j} \right]_{\gamma = \gamma^0}$$

Since the proofs of these steps require some rather oppressive arguments, we shall first show how to obtain the desired results by using the steps and then go back and justify the steps.

Theorem 7.8 *Suppose (7.58) has a solution γ^0 and suppose that*

$$\left[\frac{\partial U_i}{\partial \gamma_j} \right]_{\gamma = \gamma^0}$$

is nonsingular. Then if $|\varepsilon|$ is sufficiently small there exists a solution $u(t, \varepsilon)$ with period T of equation (7.54).

Proof   Since

$$\left[\frac{\partial U_i}{\partial \gamma_j} \right]_{\gamma = \gamma^0}$$

is nonsingular, then by Step 2, the matrix

$$\left[\frac{\partial W_i}{\partial a_j} \right]$$

is nonsingular. This, together with Step 1, completes the proof of Theorem 7.8.   ☐

Just as in the earlier part of the chapter (Theorem 7.4) we can extend the results of Theorem 7.8 by using topological degree instead of the implicit function theorem. We have:

Theorem 7.9   *Let M be the mapping corresponding to (7.54), that is,*

$$M: (\gamma_1, \ldots, \gamma_m) \to (U_1(\gamma_1, \ldots, \gamma_m), \ldots, U_m(\gamma_1, \ldots, \gamma_m))$$

and suppose

$$\deg[M, B^m, 0] \neq 0$$

Then given $\delta > 0$, there exists $k(s) = (k_1(s), \ldots, k_m(s))$ such that

$$|k(s)| < \delta$$

and $k(s)$ has period T and such that if $g(t, x, \varepsilon)$ in equation (7.5) is replaced by

$$g(t, x, \varepsilon) + k(t)$$

then if $|\varepsilon|$ is sufficiently small, the number of solutions $u(t, \varepsilon)$ of (7.54) with period T is greater than or equal to $|\deg[M, B^m, 0]|$.

We may also apply the stability analysis developed earlier (Theorem 7.7) to the solutions of equation (7.54). We obtain somewhat wider stability results than Roseau's results. Roseau requires that $A(t, \gamma^0)$ have exactly one characteristic multiplier equal to one and of algebraic multiplicity one. Our analysis allows the occurrence of other characteristic multipliers equal to one.

In order to prove Steps 1 and 2, we first rewrite equations (7.62) more explicitly. First, we note that since f has continuous third derivatives in t and x and g has continuous second derivatives in t, x, and ε then $G(t, u, \gamma, \varepsilon)$ is defined at $\varepsilon = 0$ and has a continuous derivative with respect to ε at $\varepsilon = 0$. From equation (7.60), it follows that if u is a solution of (7.59), then the kth component of u is

$$u_k = a_1 \frac{\partial z_k}{\partial \gamma_1} + \cdots + a_m \frac{\partial z_k}{\partial \gamma_m} + \bar{u}_k$$

Hence if G_j is the jth component of G, we have

$$G_j(s, u, \gamma^0, 0) = \left[\frac{1}{2} \frac{\partial^2 f_j}{\partial z_k \partial z_\ell}\right] u_k u_\ell$$

$$+ \frac{\partial g_j}{\partial z_k}(z(s, \gamma^0), s, 0)u_k + \frac{\partial g_j}{\partial \varepsilon}(z(s, \gamma^0), s, 0)$$

where we let $\frac{\partial f_j}{\partial z_k}$ denote $\frac{\partial f_j}{\partial x_k}[t, z(t, \gamma^0)]$.

(Here and later we use the convention of indicating that summation takes place if an index is used twice. Thus in the first term on the right, summation over $k = 1, \ldots, n$ and $\ell = 1, \ldots, n$ takes place.)

Hence we may rewrite (7.62) as

$$W_k(a_1, \ldots, a_m) = \frac{1}{2} a_p a_q \int_0^T v_{jk} \left(\frac{\partial^2 f_j}{\partial z_\ell \partial z_u} \right) \frac{\partial z_\ell}{\partial \gamma_p} \frac{\partial z_\mu}{\partial \gamma_q} \, ds$$

$$+ a_q \int_0^T v_{jk} \left[\frac{1}{2} \left(\frac{\partial^2 f_j}{\partial z_\ell \partial z_\mu} \right) \left(\frac{\partial z_\ell}{\partial \gamma_q} \bar{u}_\mu + \frac{\partial z_\mu}{\partial \gamma_q} \bar{u}_\ell \right) \right.$$

$$\left. + \frac{\partial g_j}{\partial z_\ell} \frac{\partial z_\ell}{\partial \gamma_q} \right] ds + \mathcal{R}(k = 1, \ldots, m) \tag{7.63}$$

where $\mathcal{R}$ consists of terms which are independent of $a_1, \ldots, a_m$.

Now we are ready to prove Steps 1 and 2.

Proof of Step 1: First, we prove

$$\frac{d}{dt} \left(\frac{\partial^2 z_j}{\partial \gamma_p \partial \gamma_q} \right) = \frac{\partial^2 f_j}{\partial z_\ell \partial z_m} \frac{\partial z_\ell}{\partial \gamma_p} \frac{\partial z_m}{\partial \gamma_q} + \frac{\partial f_j}{\partial z_\ell} \frac{\partial^2 z_\ell}{\partial \gamma_p \partial \gamma_q} \tag{7.64}$$

Integrating the equation

$$\frac{dz(t, \gamma)}{dt} = f[t, z(t, \gamma)]$$

we obtain

$$z(t, \gamma) = z(0, \gamma) + \int_0^t f[s, z(s, \gamma)] \, ds \tag{7.65}$$

Differentiating the jth component of (7.65) successively with respect to γ_q, γ_p and t, we get:

$$\frac{\partial z_j}{\partial \gamma_q} = \frac{\partial z_j(0, \gamma)}{\partial \gamma_q} + \int_0^t \frac{\partial f_j}{\partial z_\mu} \frac{\partial z_\mu}{\partial \gamma_q} \, ds$$

$$\frac{\partial^2 z_j}{\partial \gamma_p \partial \gamma_q} = \frac{\partial^2 z_j(0, \gamma)}{\partial \gamma_p \partial \gamma_q} + \int_0^t \left[\frac{\partial^2 f_j}{\partial z_\ell \partial z_\mu} \frac{\partial z_\mu}{\partial \gamma_q} \frac{\partial z_\ell}{\partial \gamma_p} + \frac{\partial f_j}{\partial z_\mu} \frac{\partial^2 z_\mu}{\partial \gamma_p \partial \gamma_q} \right] ds$$

$$\frac{d}{dt} \left(\frac{\partial^2 z_j}{\partial \gamma_p \partial \gamma_q} \right) = \frac{\partial^2 f_j}{\partial z_\ell \partial z_\mu} \frac{\partial z_\mu}{\partial \gamma_q} \frac{\partial z_\ell}{\partial \gamma_p} + \frac{\partial f_j}{\partial z_\mu} \frac{\partial^2 z_\mu}{\partial \gamma_p \partial \gamma_q}$$

This completes the proof of (7.64). Applying (7.64) to (7.63), we may write the coefficient of $a_p a_q$ as

$$\frac{1}{2} \int_0^T v_{jk} \left[\frac{d}{ds} \left(\frac{\partial^2 z_j}{\partial \gamma_p \partial \gamma_q} \right) - \frac{\partial f_j}{\partial z_\mu} \frac{\partial^2 z_\mu}{\partial \gamma_p \partial \gamma_q} \right] ds$$

Integrating the first term in the integrand by parts, we have

$$\int_0^T v_{jk} \left[\frac{d}{ds} \left(\frac{\partial^2 z_j}{\partial \gamma_p \partial \gamma_q} \right) \right] ds = \int_0^T \frac{d}{ds} \left\{ v_{jk} \frac{\partial^2 z_j}{\partial \gamma_p \partial \gamma_q} \right\} ds - \int_0^T \left(\frac{dv_{jk}}{ds} \right) \frac{\partial^2 z_j}{\partial \gamma_p \partial \gamma_q} \, ds$$

Since

$$v_{jk} \frac{\partial^2 z_j}{\partial \gamma_p \partial \gamma_q}$$

has period T, then

$$\int_0^T \frac{d}{ds} \left[v_{jk} \frac{\partial^2 z_j}{\partial \gamma_p \partial \gamma_q} \right] ds = 0$$

hence the coefficient of $a_p a_q$ becomes

$$\frac{1}{2} \int_0^T \left\{ -\left(\frac{d v_{jk}}{ds} \right) \frac{\partial^2 z_j}{\partial \gamma_p \partial \gamma_q} - v_{jk} \frac{\partial f_j}{\partial z_\mu} \frac{\partial^2 z_\mu}{\partial \gamma_p \partial \gamma_q} \right\} ds$$

In the second term of this integrand, that is,

$$v_{jk} \frac{\partial f_j}{\partial z_\mu} \frac{\partial^2 z_\mu}{\partial \gamma_p \partial \gamma_q}$$

we sum over μ and over j. Since the value of the resulting sum is independent of the notation used, we may write the second integrand as

$$v_{\mu k} \frac{\partial f_\mu}{\partial z_j} \frac{\partial^2 z_j}{\partial \gamma_p \partial \gamma_q}$$

that is, we reverse the roles of μ and j. Then the coefficient of $a_p a_q$ becomes

$$\frac{1}{2} \int_0^T \left\{ -\left(\frac{d v_{jk}}{ds} \right) \frac{\partial^2 z_j}{\partial \gamma_p \partial \gamma_q} - v_{\mu k} \frac{\partial f_\mu}{\partial z_j} \frac{\partial^2 z_j}{\partial \gamma_p \partial \gamma_q} \right\} ds$$

$$= -\frac{1}{2} \int_0^T \left\{ \frac{d v_{jk}}{ds} + v_{\mu k} \frac{\partial f_\mu}{\partial z_j} \right\} \frac{\partial^2 z_j}{\partial \gamma_p \partial \gamma_q} ds$$

But the vector

$$v^k = \begin{bmatrix} v_{1k} \\ \vdots \\ v_{nk} \end{bmatrix}$$

satisfies the equation

$$\frac{d v^k}{dt} + \left[\frac{\partial f_i}{\partial x_j} \right]^* v^k = 0$$

where

$$\left[\frac{\partial f_i}{\partial x_j} \right]^* = \left[\frac{\partial f_i}{\partial x_j} (t, z(t, \gamma^0)) \right]^* = \left[\frac{\partial f_i}{\partial z_j} \right]^*$$

Thus

$$\frac{d}{dt} \begin{bmatrix} v_{1k} \\ \vdots \\ v_{nk} \end{bmatrix} + \begin{bmatrix} \frac{\partial f_1}{\partial x_1} & \frac{\partial f_2}{\partial x_1} & \cdots & \frac{\partial f_n}{\partial x_1} \\ \vdots & & & \\ \frac{\partial f_1}{\partial x_n} & \cdot & \cdot & \frac{\partial f_n}{\partial x_n} \end{bmatrix} \begin{bmatrix} v_{1k} \\ \vdots \\ v_{nk} \end{bmatrix}$$

or

$$\frac{dv_{jk}}{dt} = -\frac{\partial f_\mu}{\partial x_j} v_{\mu k} \qquad (j = 1, \ldots, n)$$

and hence the coefficient of $a_p a_q$ is zero. This completes the proof of Step 1.

Proof of Step 2: The (k, q) element of the coefficient matrix

$$\left[\frac{\partial W_i}{\partial a_j}\right]$$

is the coefficient of a_q in the equation (7.63). Inspection shows that this coefficient is

$$\int_0^T v_{jk}\left[\frac{\partial^2 f_j}{\partial z_v \partial z_\ell}\bar{u}_\ell + \frac{\partial g_j}{\partial z_v}\right]\frac{\partial z_v}{\partial \gamma_q}\, ds \qquad (7.66)$$

Now we turn to calculations of the entries in the matrix

$$\left[\frac{\partial U_i}{\partial \gamma_j}\right]_{\gamma=\gamma^0}$$

Since by (7.58)

$$U_k(\gamma_1, \ldots, \gamma_m) = \int_0^T [v^k \cdot h(s, \gamma_1, \ldots, \gamma_m)]\, ds$$

then by definition of $h(s, \gamma_1, \ldots, \gamma_m)$, we have

$$\begin{aligned} U_k(\gamma_1, \ldots, \gamma_m) &= \int_0^T v^k \cdot g(s, z(s, \gamma), 0)\, ds \\ &= \int_0^T (v_{jk}(g_j(s, z(s, \gamma), 0)\, ds \end{aligned} \qquad (7.67)$$

Differentiation of (7.67) with respect to γ_q yields

$$\frac{\partial U_k}{\partial \gamma_q} = \int_0^T \left\{\left(\frac{\partial v_{jk}}{\partial \gamma_q}\right)(g_j(s, z(s, \gamma), 0)) + (v_{jk})\left(\frac{\partial g_j}{\partial z_\mu}\frac{\partial z_\mu}{\partial \gamma_q}\right)\right\}\, ds \qquad (7.68)$$

If $\gamma = \gamma^0$, then by definition of $\bar{u}(t)$

$$\frac{d\bar{u}_j}{dt} = \left[\frac{\partial f_j}{\partial z_\ell}\right]\bar{u}_\ell + g_j[t, z(t, \gamma^0), 0] \qquad (7.69)$$

Substituting from (7.69) into (7.68), we have

$$\frac{\partial U_k}{\partial \gamma_q}\Bigg|_{\gamma=\gamma^0} = \int_0^T \left\{\left(\frac{\partial v_{jk}}{\partial \gamma_q}\right)\left[\frac{d\bar{u}_j}{ds} - \frac{\partial f_j}{\partial z_\ell}\bar{u}_\ell\right] + (v_{jk})\left(\frac{\partial g_j}{\partial z_\mu}\frac{\partial z_\mu}{\partial \gamma_q}\right)\right\}\, ds \qquad (7.70)$$

Now we apply integration by parts to the first term on the right-hand side of (7.70) and obtain

$$\int_0^T \frac{\partial v_{jk}}{\partial \gamma_q} \frac{d\bar{u}_j}{ds} ds + \int_0^T \left\{ \frac{d}{ds} \left(\frac{\partial v_{jk}}{\partial \gamma_q} \right) \right\} \bar{u}_j \, ds$$

$$= \int_0^T \frac{d}{ds} \left[\left(\frac{\partial v_{jk}}{\partial \gamma_q} \right) \bar{u}_j \right] ds$$

$$= \left(\frac{\partial v_{jk}}{\partial \gamma_q} \bar{u}_j \right) \Big]_0^T$$

$$= 0$$

Hence we may rewrite (7.70) as

$$\frac{\partial U_k}{\partial \gamma_q} \Big]_{\gamma=\gamma^0} = - \int_0^T \left\{ \frac{d}{ds} \left(\frac{\partial v_{jk}}{\partial \gamma_q} \right) \bar{u}_j + \frac{\partial v_{jk}}{\partial \gamma_q} \frac{\partial f_j}{\partial z_\ell} \bar{u}_\ell \right\} ds + \int_0^T \left\{ v_{jk} \frac{\partial g_j}{\partial z_\mu} \frac{\partial z_\mu}{\partial \gamma_q} \right\} ds$$

$$(7.71)$$

Since

$$\frac{dv_{jk}}{dt} + \frac{\partial f_\ell}{\partial z_j} v_{\ell k} = 0$$

then

$$\frac{d}{dt} \left(\frac{\partial v_{jk}}{\partial \gamma_q} \right) = \frac{\partial}{\partial \gamma_q} \left(\frac{dv_{jk}}{dt} \right) = - \frac{\partial f_\ell}{\partial z_j} \frac{\partial v_{\ell k}}{\partial \gamma_q} - \frac{\partial^2 f_\ell}{\partial z_\nu \partial z_j} v_{\ell k} \frac{\partial z_\nu}{\partial \gamma_q} \qquad (7.72)$$

Substituting from (7.72) into the first integrand on the right-hand side of (7.71), we obtain for the first integrand

$$- \frac{\partial f_\ell}{\partial z_j} \frac{\partial v_{\ell k}}{\partial \gamma_q} \bar{u}_j - \frac{\partial^2 f_\ell}{\partial z_\nu \partial z_j} v_{\ell k} \frac{\partial z_\nu}{\partial \gamma_q} \bar{u}_j + \frac{\partial v_{jk}}{\partial \gamma_q} \frac{\partial f_j}{\partial z_\ell} \bar{u}_\ell$$

Hence (7.71) becomes

$$\frac{\partial U_k}{\partial \gamma_q} \Big]_{\gamma=\gamma_0} = - \int_0^T \left\{ - \frac{\partial f_\ell}{\partial z_j} \frac{\partial v_{\ell k}}{\partial \gamma_q} \bar{u}_j - \frac{\partial^2 f_\ell}{\partial z_\nu \partial z_j} v_{\ell k} \frac{\partial z_\nu}{\partial \gamma_q} \bar{u}_j \right.$$

$$\left. + \frac{\partial v_{jk}}{\partial \gamma_q} \frac{\partial f_j}{\partial z_\ell} \bar{u}_\ell \right\} ds + \int_0^T \left\{ v_{jk} \frac{\partial g_j}{\partial z_\mu} \frac{\partial z_\mu}{\partial \gamma_q} \right\} ds \qquad (7.73)$$

By a change in subscript notation, we have

$$\frac{\partial f_\ell}{\partial z_j} \frac{\partial v_{\ell k}}{\partial \gamma_q} \bar{u}_j = \frac{\partial f_j}{\partial z_\ell} \frac{\partial v_{jk}}{\partial \gamma_q} \bar{u}_\ell$$

and hence

$$\frac{\partial U_k}{\partial \gamma_q}\Bigg]_{\gamma=\gamma_0} = \int_0^T \left\{ \frac{\partial^2 f_\ell}{\partial z_\nu \partial z_j} v_{\ell k} \frac{\partial z_\nu}{\partial \gamma_q} \bar{u}_j + v_{jk} \frac{\partial g_j}{\partial z_\mu} \frac{\partial z_\mu}{\partial \gamma_q} \right\} ds$$

But except for subscript notation, this is the same as (7.66). This completes the proof of Step 2.

C. The Case in Which the Unperturbed Equation is Autonomous

As pointed out earlier, an important special case of the equation studied in Part B is the equation

$$\frac{dx}{dt} = f(x) + \varepsilon g(t, x, \varepsilon) \tag{7.74}$$

where the unperturbed equation is autonomous. Now we study that special case in some detail.

We will impose the following assumptions on equation (7.74).

1. Let $(t, x, \varepsilon) \in R \times R^n \times I$ where I is an interval in R with midpoint zero, and let f have continuous third derivatives in x and suppose that the equation

$$\frac{dx}{dt} = f(x) \tag{7.75}$$

has a solution $\bar{x}(t)$ of period T such that the matrix

$$\left[\frac{\partial f_i}{\partial x_j}(\bar{x}(t)) \right]$$

has $(n - 1)$ characteristic multipliers with absolute value less than 1 (i.e., $(n - 1)$ characteristic exponents with negative real part).

2. Let g have continuous second derivatives in t, x, and ε and suppose $g(t, x, \varepsilon)$ is periodic in t with period $T(1 + \varepsilon m(\varepsilon))$ where m is a continuous function of ε which is to be determined.

First, a few remarks about the significance of these assumptions. Since (7.75) is autonomous then it follows that $\bar{x}(t + \gamma)$, where γ is any real number, is also a solution of (7.74), and thus we have a 1-parameter family of solutions of period T of (7.75). The hypothesis on the matrix

$$\left[\frac{\partial f_i}{\partial x_j}(\bar{x}(t)) \right]$$

guarantees that the linear space of solutions of period T of

$$\frac{du}{dt} = \left[\frac{\partial f_i}{\partial x_j}(\bar{x}(t)) \right] u \tag{7.76}$$

has dimension one (see Theorem 2.14 in Chapter 2). Hence the linear space of solutions of period T of (7.76) has as a basis the function

$$\frac{\partial \bar{x}}{\partial \gamma}(t)$$

In Part B, this last statement was a hypothesis (Assumption 2 in Part B). Here we replace Assumption 2 in Part B with a more concrete sufficient condition, that is, the condition that the matrix

$$\left[\frac{\partial f_i}{\partial x_j}(\bar{x}(t)) \right]$$

has $(n-1)$ characteristic multipliers with absolute value less than one.

In the second assumption we have assumed that $g(t, x, \varepsilon)$ has period $T(1+\varepsilon m(\varepsilon))$. The natural question is why not assume that g has period T as in Parts A and B. The answer lies in the fact that (7.75) is autonomous. We shall discuss this more at length at the beginning of Chapter 8.

In order to apply the theory in Part B to equation (7.74), we first make the change of variable

$$t = \tau(1 + \varepsilon m)$$

where $m = m(\varepsilon)$, a differentiable function of ε, in equation (7.74) and obtain

$$\frac{dx}{d\tau} = \frac{dx}{dt}\frac{dt}{d\tau} = (1 + \varepsilon m)[f(x) + \varepsilon g(\tau(1 + \varepsilon m), x, \varepsilon)]$$

$$\frac{dx}{d\tau} = f(x) + \varepsilon\{mf(x) + (1 + \varepsilon m)g(\tau(1 + \varepsilon m), x, \varepsilon)\}$$

or, if $G(\tau, x, \varepsilon)$ denotes the expression in curly brackets,

$$\frac{dx}{d\tau} = f(x) + \varepsilon G(\tau, x, \varepsilon) \tag{7.77}$$

where $G(\tau, x, \varepsilon)$ has period T in τ. By applying the theory in Part B to equation (7.77), we will show that under certain conditions equation (7.77) has a solution of period T and hence that equation (7.74) has a solution of period $T(1 + \varepsilon m)$.

The equation

$$\frac{dx}{d\tau} = f(x) \tag{7.78}$$

has the 1-parameter family of periodic solutions $\bar{x}(\tau + \gamma)$ and for each fixed γ the linear variational equation is

$$\frac{dx}{d\tau} = \left[\frac{\partial f_i}{\partial x_j}[\bar{x}(\tau + \gamma)] \right] x \tag{7.79}$$

with adjoint system

$$\frac{dx}{d\tau} + \left[\frac{\partial f_i}{\partial x_j}[\bar{x}(\tau + \gamma)] \right]^* x = 0 \tag{7.80}$$

Also, it follows that $\frac{\partial \bar{x}}{\partial \gamma}(\tau + \gamma)$ is a basis for the linear space of solutions of period T of (7.79). Hence equation (7.80) has exactly one linearly independent periodic (of period T) solution $w(t + \gamma)$. Applying Theorem 7.8, we conclude: if the equation

$$U(\gamma) = \int_0^T [w(s + \gamma)] \cdot G(s, \bar{x}(s + \gamma), 0) \, ds = 0 \qquad (7.81)$$

has a solution γ_0 and $U'(\gamma_0) \neq 0$, then if $|\varepsilon|$ is sufficiently small there exists a solution $x(t, \varepsilon)$ of period T of equation (7.77). Since

$$G(\tau, \bar{x}(\tau + \gamma), \varepsilon) = m(\varepsilon) f(\bar{x}(\tau + \gamma)) + (1 + \varepsilon m) g[\tau (1 + \varepsilon m(x)), \bar{x}(\tau + \gamma), \varepsilon]$$

then

$$G(\tau, \bar{x}(\tau + \gamma), 0) = m(0) f(\bar{x}(\tau + \gamma)) + g[\tau, \bar{x}(\tau + \gamma), 0]$$

and hence (7.81) may be written

$$U(\gamma) = \int_0^T [w(s + \gamma)] \cdot \{m(0) f[\bar{x}(s + \gamma)] + g[s, \bar{x}(s + \gamma), 0]\} \, ds$$

Summarizing, we have:

Theorem 7.10   *If the equation*

$$U(\gamma) = \int_0^T [w(s + \gamma)] \cdot \{m(0) f[\bar{x}(s + \gamma)] + g[s, \bar{x}(s + \gamma), 0]\} \, ds = 0 \quad (7.82)$$

has a solution γ_0 such that $U'(\gamma_0) \neq 0$, then if $|\varepsilon|$ is sufficiently small, there exists a solution $x(t, \varepsilon)$ of equation (7.74) such that $x(t, \varepsilon)$ has solution $T(1 + \varepsilon m(\varepsilon))$.

Proof   It follows from Theorem 7.8 that equation (7.77) has a solution $x(\tau, \varepsilon)$ of period T and

$$\lim_{\varepsilon \to 0} x(\tau, \varepsilon) = a_1^0 \frac{\partial \bar{x}}{\partial \tau}(\tau + \gamma^0) + \tilde{x}(\tau) \qquad (7.83)$$

where $\tilde{x}(\tau)$ is a particular periodic solution (chosen in advance) of the linear equation

$$\frac{dx}{d\tau} = \left\{ \frac{\partial f_i}{\partial x_j} [\bar{x}(\tau + \gamma^0)] \right\} x + G(\tau, \bar{x}(\tau + \gamma^0), 0) \qquad (7.84)$$

Use of the change of variables

$$\tau = \frac{t}{1 + \varepsilon m}$$

in equation (7.77) yields

$$\frac{dx}{d\tau} = \frac{dx}{dt}(1 + \varepsilon m) = f(x) + \varepsilon \{m f(x) + (1 + \varepsilon m) g(t, x, \varepsilon)\}$$

$$= (1 + \varepsilon m) f(x) + (1 + \varepsilon m) \varepsilon g(t, x, \varepsilon)$$

This last equation is equation (7.74). That is

$$\frac{dx}{dt} = f(x) + \varepsilon g(t, x, \varepsilon) \qquad (7.85)$$

and $x(\frac{t}{1+\varepsilon m}, \varepsilon)$ is a solution of period $T(1 + \varepsilon m)$ of (7.74) such that

$$\lim_{\varepsilon \to 0} x\left(\frac{t}{1 + \varepsilon m}, \varepsilon\right) = a_1^0 \frac{\partial}{\partial t} \bar{x}(t + \gamma^0) + \tilde{x}(t)$$

This completes the proof of Theorem 7.10.   ⬜

Since $\omega(s + \gamma)$ and $\bar{x}(s + \gamma)$ each have period T in γ (because $\omega(t + \gamma)$ and $\bar{x}(s + \gamma)$ have period T in t), then $U(\gamma)$ has period T in γ, and it follows that in applying Theorem 7.10 it is sufficient to investigate solutions γ of equation (7.82) such that $0 \leq \gamma \leq T$. Notice also that since

$$U(0) = U(T)$$

then $U(\gamma)$ has an even number of changes of sign in the interval $[0, T]$.

Exercises

1. Show that if the equation (7.2) has a solution of period $T(0)$, then the equation (7.1) has a solution of period $T(\varepsilon)$.

2. Prove: the equation

$$\frac{dx}{dt} = Ax$$

   has no nontrivial solutions of period T if and only if the eigenvalues of matrix TA are different from $\pm 2n\pi i (n = 0, 1, 2, \ldots)$.

3. Find matrix H such that

$$H(e^{TA} - I) = P_r$$

4. Prove Theorem 7.6.

5. Prove an analog of Theorem 7.7 for the case

$$e^{TA-I} = \begin{bmatrix} 0 & \\ & C_1 \end{bmatrix}$$

Chapter 8

Perturbation Theory: Autonomous Systems and Bifurcation Problems

Introduction

This is a continuation of Chapter 7 because we are again considering the existence of periodic solutions in a perturbation problem and the underlying approach is again the Poincaré method. But now the unperturbed and perturbed equations are both autonomous. That is, we consider an equation of the form

$$\frac{dx}{dt} = f(x, \varepsilon)$$

Following a common usage, we term such problems bifurcation problems.

The fact that the unperturbed and perturbed equations are both autonomous introduces two complications: First, we have few hints about the period of any such desired periodic solution; second, as observed before, if $x(t)$ is a solution of period T of an autonomous equation then $x(t + k)$, where k is an arbitrary real number, is also a solution of period T. Thus the unperturbed equation has either no periodic solution or a continuous family of periodic solutions.

In this chapter, we describe first a classical result that is analogous to the Poincaré theorem for nonautonomous equations (Theorem 7.1). The hypothesis for this analogous case is that the given periodic solution $\bar{x}(t)$ for the unperturbed equations is such that the matrix

$$f_x[\bar{x}(t), 0]$$

has the number one as a simple characteristic multiplier. Then we describe an application of this result to the problem of phaselocking. Next, we look at the case in which the characteristic multiplier one of

$$f_x[\bar{x}(t), 0]$$

is not simple. Using the Poincaré method leads to complications in this case, and the analysis required is lengthier. For this discussion, we depend upon the work of Coddington and Levinson [1955].

Finally we discuss the Hopf bifurcation which differs from the previous bifurcation problems we have discussed, in that the parameter ε plays a different role. Nevertheless to obtain the Hopf bifurcation theorem, we use an adaptation of the Coddington-Levinson approach. For a different approach using the center manifold theorem, see Farkas [1994, Chapter 7].

The Classical Problem

We consider the equation

$$\frac{dx}{dt} = f(x, \varepsilon) \tag{8.1}$$

where the n-vector function f has continuous third derivatives at each point

$$(x, \varepsilon) \in R^n \times I$$

where I is an open interval on the real line with midpoint zero and we assume that if $\varepsilon = 0$, equation (8.1) has a nontrivial solution $\bar{x}(t)$ of period T, that is, a periodic solution which is not an equilibrium point. If equation (8.1) with $\varepsilon = 0$ describes a physical system, then the existence of periodic solution $x(t)$ implies that the physical system has a "natural" period of oscillation equal to T. But if the system is perturbed by an influence which is described in equation (8.1) by ε, then the resulting system may not have the same oscillating behavior, that is, if $\varepsilon \neq 0$, equation (8.1) may not have a solution of period T. However, the physical system may oscillate with a different period, that is, equation (8.1) may have a solution of period $T_1 \neq T$. A simple example of this is the pendulum, the period of which depends on the length of the pendulum.

This suggests that we reformulate the problem in this way: We seek a function $T(\varepsilon)$ with domain an open interval J with midpoint 0 and with $J \subset I$ such that $T(0) = T$ and such that if $|\varepsilon|$ is sufficiently small, equation (8.1) has a solution $\bar{x}(t, \varepsilon)$ of period $T(\varepsilon)$ with

$$\lim_{\varepsilon \to 0} x(t, \varepsilon) = \bar{x}(t)$$

We assume for the present that

$$T(\varepsilon) = T(0) + \tau(\varepsilon)$$

where $\tau(0) = 0$. Then if $x(t, c, \varepsilon)$ is a solution of (8.1) such that

$$x(0, c, \varepsilon) = c$$

the condition that $x(t, c, \varepsilon)$ have period $T(\varepsilon)$ is

$$x(T(0) + \tau(\varepsilon), c, \varepsilon) - x(0, c, \varepsilon) = 0 \tag{8.2}$$

As in Chapter 7, we seek a periodic solution by solving equation (8.2) for c as a function of ε. The problem is more complicated than the problem for nonautonomous equations in Chapter 7 because we must also solve for the function $\tau(\varepsilon)$. Since equation (8.2) consists of n scalar equations, then at first it seems unlikely that we can solve for the $n + 1$ unknowns consisting of τ and the n components of vector c. Fortunately, as we shall shortly see, the fact that equation (8.1) is autonomous saves us from this dismaying situation.

Our first objective is to obtain a result analogous to the classical result of Poincaré for nonautonomous equations (Theorem 7.1). For this purpose, we impose the following hypothesis on the given periodic solution $\bar{x}(t)$ of the equation

$$\frac{dx}{dt} = f(x, 0)$$

We assume that the linear variational equation

$$\frac{dy}{dt} = \{f_x[\bar{x}(t), 0]\}y \tag{8.3}$$

is such that its linear space of solutions of period T has dimension one, that is, the matrix

$$f_x[\bar{x}(t), 0]$$

has the number one as a simple characteristic multiplier. (The corresponding assumption in Theorem 7.1 is that the linear variational equation has no solutions of period T. But we cannot hope to use that hypothesis here because equation (8.3) has the nontrivial periodic solution $\frac{d\bar{x}}{dt}$.)

Next, by translation of axes, we may assume that the given periodic solution $\bar{x}(t)$ is such that

$$\bar{x}_1(0) = 0.$$

By rotation of axes, we may assume that

$$\bar{x}_1'(0) \neq 0$$
$$\bar{x}_2'(0) = \bar{x}_3'(0) = \cdots = \bar{x}_n'(0) = 0$$

Now we want to show that if a periodic solution $x(t, c, \varepsilon)$ of (8.1) gets close enough to $\bar{x}(t)$, that is, if there exist t_1 and t_2 so that

$$|x(t_2, c, \varepsilon) - \bar{x}(t_1)|$$

is small enough, then the orbit of $x(t, c, \varepsilon)$ crosses the plane $x_1 = 0$. First, if

$$\tilde{x}(t, c, \varepsilon) = x(t + t_2 - t_1, c, \varepsilon)$$

then

$$|\tilde{x}(t_1, c, \varepsilon) - \bar{x}(t_1)| = |x(t_2, c, \varepsilon) - \bar{x}(t_1)|$$

Since we are concerned with orbits, we can work with $\tilde{x}(t, c, \varepsilon)$ and the hypothesis becomes: there exists t_1 so that

$$|\tilde{x}(t_1, c, \varepsilon) - \bar{x}(t_1)| \tag{8.4}$$

is small enough.

As assumed earlier,

$$\frac{d\bar{x}_1}{dt} = M \neq 0$$

Thus $\bar{x}(t)$ crosses the plane $x_1 = 0$ at $t = 0$ and hence there exist t_3 and t_4 such that

$$\bar{x}_1(t_3) = m$$
$$\bar{x}_1(t_4) = -m$$

where $m > 0$ and m may be chosen as small as desired. For definiteness, assume $M > 0$ and $t_4 < 0 < t_3$. By the existence theorem for equations with a parameter, Chapter 1, Exercise 13, if (8.4) is small enough, that is, if $|c - \bar{x}(0)|$ and $|\varepsilon|$ are small enough then

$$|\tilde{x}(t, c, \varepsilon) - \bar{x}(t)| < D \quad \text{for} \quad t \in I$$

where D is a given positive number such that

$$D < m$$

and I is a closed interval which contains $[t_4, t_3]$ and $[-m, m]$. Thus

$$|\tilde{x}_1(t_3, c, \varepsilon) - \bar{x}(t_3)| < D < m = \bar{x}_1(t_1)$$

and hence

$$\tilde{x}_1(t_3, c, \varepsilon) > 0$$

Also

$$|\tilde{x}_1(t_4, c, \varepsilon) - \bar{x}(t_4)| < D < m$$

Since $\bar{x}(t_4) = -m$, then

$$\tilde{x}_1(t_4, c, \varepsilon) < 0$$

Therefore $\tilde{x}_1$ changes sign between t_3 and t_4, say at t_0.

It remains to show that

$$\frac{d\bar{x}_1}{dt}(t_0, c, \varepsilon) \neq 0$$

If $|c - \bar{x}(0)|$ and $|\varepsilon|$ are sufficiently small and K is a positive constant, then for $|t| \leq K$

$$|\tilde{x}(t, c, \varepsilon) - \bar{x}(t)|$$

is sufficiently small so that

$$|f[\tilde{x}(t, c, \varepsilon)] - f[\bar{x}(t)]| < \frac{M}{2}$$

or, since

$$\frac{dx}{dt} = f(x, \varepsilon)$$

we have

$$\left| \frac{d\tilde{x}}{dt} - \frac{d\bar{x}}{dt} \right| < \frac{M}{2}$$

and a fortiori

$$\left| \frac{d\tilde{x}_1}{dt} - \frac{d\bar{x}_1}{dt} \right| < \frac{M}{2} \tag{8.5}$$

But if m is sufficiently small then (8.5) implies that

$$\frac{d\tilde{x}_1}{dt}(t_0, c, \varepsilon) \neq 0$$

Thus, we have obtained:

Lemma 8.1   *If there exist t_1, t_2 and a periodic solution $x(t, c, \varepsilon)$ of equation (8.1) such that*

$$|x(t_2, c, \varepsilon) - \bar{x}(t_1)|$$

is sufficiently small, then the orbit of $x(t, c, \varepsilon)$ crosses the plane $x_1 = 0$, that is, $x_1(t, c, \varepsilon)$ intersects the plane $x_1 = 0$ and $\frac{dx_1}{dt} \neq 0$ at the point of intersection.

Lemma 8.1 shows that there exists a value t_0 such that the periodic solution $x(t, c, \varepsilon)$ is such that

$$x_1(t_0, c, \varepsilon) = 0$$

where $x_1(t, c, \varepsilon)$ denotes the first component of $x(t, c, \varepsilon)$. Hence, we may "reparameterize" the solution $x(t, c, \varepsilon)$ as $x(t - t_0, c, \varepsilon)$ which has zero as its first component at $t = 0$. Thus, in studying periodic solutions we need only seek periodic solutions with first components equal to zero at $t = 0$.

Now we are ready to investigate equation (8.2). First we note that (8.2) may be written as:

$$x[T(0) + \tau(\varepsilon), c, \varepsilon] - c = 0 \tag{8.6}$$

where, as follows from Lemma 8.1, we may assume that

$$c = \begin{bmatrix} 0 \\ c_2 \\ \vdots \\ c_n \end{bmatrix}$$

We solve (8.6) for $\tau, c_2 \ldots, c_n$ as functions of ε by using the implicit function theorem. First, equation (8.6) has the initial solution

$$\varepsilon = 0, \tau = 0, \bar{c}_2 = x_2(0), \ldots, \bar{c}_n = x_n(0)$$

It is therefore sufficient to prove that the appropriate Jacobian at that initial solution is nonzero. The appropriate Jacobian is the determinant of the matrix:

$$
\begin{bmatrix}
x_1'(T(0)) & \frac{\partial x_1}{\partial c_2}(T(0)) & \cdots & \frac{\partial x_1}{\partial c_n}(T(0)) \\
x_2'(T(0)) & \frac{\partial x_2}{\partial c_2} - 1 & \cdots & \frac{\partial x_2}{\partial c_2} \\
\vdots & \vdots & & \vdots \\
x_n'(T(0)) & \frac{\partial x_n}{\partial c_2} & \cdots & \frac{\partial x_n}{\partial c_n} - 1
\end{bmatrix}
$$

Since

$$
x_1'(T(0)) = x_1'(0) \neq 0
$$

and, for $j = 2, \ldots, n$,

$$
x_j'(T(0)) = x_j'(0) = 0
$$

it is sufficient to prove that

$$
\det
\begin{bmatrix}
\frac{\partial x_2}{\partial c_2} - 1 & \cdots & \frac{\partial x_2}{\partial c_n} \\
& \cdots & \\
\frac{\partial x_n}{\partial c_2} & & \frac{\partial x_n}{\partial c_n} - 1
\end{bmatrix}
\neq 0 \tag{8.7}
$$

To prove (8.7), we argue as follows. Let k be the vector

$$
k =
\begin{bmatrix}
k_1 \\
c_2 \\
\vdots \\
c_n
\end{bmatrix}
$$

where $|k_1|$ is small and

$$
|c_j - \bar{c}_j| \qquad (j - 2, \ldots, n)
$$

is small, and consider the solution $x(t, k, 0)$ of (8.1). A fundamental matrix of the linear variational equation

$$
y' = [f_x[x(t, k, 0)]y \tag{8.8}
$$

is

$$
X(t) =
\begin{bmatrix}
\frac{\partial x_1}{\partial k_1} & \frac{\partial x_1}{\partial c_2} & \cdots & \frac{\partial x_1}{\partial c_n} \\
\vdots & \vdots & & \vdots \\
\frac{\partial x_n}{\partial k_1} & \frac{\partial x_n}{\partial c_2} & \cdots & \frac{\partial x_n}{\partial c_n}
\end{bmatrix}
$$

But the components of the solution $x(t, k, 0)$ may be written as

$$x_1(t, k, 0) = k_1 + t\mu_1(t, k)$$
$$x_j(t, k, 0) = c_j + t\mu_j(t, k) \qquad (j = 2, \ldots, n)$$

where $\mu_j(t, k)$, with $j = 1, \ldots, n$, is a differentiable function of t and the components of k. Thus

$$\frac{\partial x_1}{\partial k_1}(0, k, 0) = 1$$

$$\frac{\partial x_j}{\partial k_j}(0, k, 0) = 0 \qquad (j = 2, \ldots,)$$

Now this holds true in particular for $k_1 = 0$. That is, if $k_1 = 0$, the first column of $X(t)$ is a solution of (8.8) with the initial value (at $t = 0$)

$$\begin{bmatrix} 1 \\ 0 \\ \vdots \\ 0 \end{bmatrix}$$

Next let

$$\tilde{c} = \begin{bmatrix} 0 \\ c_2 \\ \vdots \\ c_n \end{bmatrix}$$

Then the column vector

$$\begin{bmatrix} \frac{\partial x_1}{dt}(t, \tilde{c}, 0) \\ \vdots \\ \frac{\partial x_n}{dt}(t, \tilde{c}, 0) \end{bmatrix}$$

is a solution of (8.8) with the initial value

$$\begin{bmatrix} x_1'(0) \\ 0 \\ \vdots \\ 0 \end{bmatrix}$$

It follows by the uniqueness of solution of (8.8) that

$$\frac{1}{x_1'(0)} \begin{bmatrix} \frac{\partial x_1}{\partial t}(t, \tilde{c}, 0) \\ \vdots \\ \frac{\partial x_n}{\partial t}(t, \tilde{c}, 0) \end{bmatrix} = \begin{bmatrix} \frac{\partial x_1}{\partial k_1}(t, \tilde{c}, 0) \\ \vdots \\ \frac{\partial x_n}{\partial k_1}(t, \tilde{c}, 0) \end{bmatrix}$$

Since $\frac{dx}{dt}(t, \tilde{c}, 0)$ has period $T = T(0)$, it follows that

$$\frac{\partial x_j}{\partial k_1}(0, \tilde{c}, 0) = \frac{\partial x_j}{\partial k_1}(T, \tilde{c}, 0) \qquad (j = 1, \ldots, n)$$

Thus

$$X(T) = \begin{bmatrix} 1 & \frac{\partial x_1}{\partial c_2} & \cdots & \frac{\partial x_1}{\partial c_n} \\ 0 & & & \\ \vdots & \vdots & & \vdots \\ 0 & \frac{\partial x_n}{\partial c_2} & \cdots & \frac{\partial x_n}{\partial c_n} \end{bmatrix} \tag{8.9}$$

where

$$\frac{\partial x_i}{\partial c_j} = \frac{\partial x_j}{\partial c_j}(T, \tilde{c}, 0) \qquad \begin{cases} i = 1, \ldots, n \\ j = 1, \ldots, n \end{cases}$$

But by hypothesis the number 1 is a simple root of the equation

$$\det[X(T) - I] = 0$$

Hence from (8.9) it follows that

$$\det \begin{bmatrix} \frac{\partial x_2}{\partial c_2}(T, \tilde{c}, 0) - 1 & \cdots & \frac{\partial x_2}{\partial c_n} \\ 0 & & \\ \vdots & & \vdots \\ \frac{\partial x_n}{\partial c_2} & \cdots & \frac{\partial x_n}{\partial c_n} - 1 \end{bmatrix} \neq 0$$

That is, we have obtained (8.7) and hence our proof is complete. Summarizing, we write:

Theorem 8.1   *Suppose the equation*

$$\frac{dx}{dt} = f(x, \varepsilon)$$

satisfies the following hypotheses:

(1) *The n-vector function f has continuous third derivatives at each point*

$$(x, \varepsilon) \in R^n \times I$$

where I is an open interval on the real line with midpoint 0.

(2) *The equation*

$$x' = f(x, \varepsilon)$$

has a nontrivial solution x(t) of period T.

(3) *The matrix*

$$f_x[x(t), 0]$$

which has period T has the number one as a simple characteristic multiplier.

Conclusion: There exists δ > 0 and a differentiable function τ(ε) with τ(0) = 0 such that if |ε| < δ then there exists a unique solution x(t, ε) of

$$x' = f(x, \varepsilon)$$

such that x(t, ε) has period T + τ(ε) and

$$\lim_{\varepsilon \to 0} x(t, \varepsilon) = x(t) \tag{8.10}$$

Corollary If all the characteristic multipliers of $f_x[x(t), 0]$, other than the simple characteristic multiplier one, have absolute value less than one, then $x(t, \varepsilon)$ is phase asymptotically stable.

Proof This follows from (8.10) and the phase asymptotic stability theorem for periodic solutions (Chapter 4). □

An Application: Phaselocking

Suppose that we have given a set of structures all of which have the same design and all of which are ocillating systems of roughly the same frequency. Suppose also that these structures are such that each can affect the frequencies of some of the others, that is, coupling occurs. If all of the systems then oscillate with the same frequency, we say that phaselocking occurs.

Phaselocking arises in mechanical, electrical, and biological systems. The problems range from interactions among precisely tuned pendulum clocks, because of coupling through very small vibrations, to interactions among cardiac fibers which must oscillate (electrically) at the same frequency in order for the heart to function. (See Winfree [1967] for a number of examples and an interesting discussion.)

Here we consider the problem of weak coupling, that is, the case in which the influences of coupling are small. We assume that the set of N structures can be described by the equations

$$\frac{dx^{(1)}}{dt} = f(x^{(1)}) + \varepsilon F_1(x^{(1)}, \ldots, x^{(N)}, \varepsilon)$$

$$\frac{dx^{(N)}}{dt} = f(x^{(N)}) + \varepsilon F_N(x^{(1)}, \ldots, x^{(N)}, \varepsilon) \tag{8.11}$$

where $x^{(1)}, \ldots, x^{(N)}$ are n-vectors, the function f is an n-vector, that is,

$$f: R^n \to R^n$$

and $F_1, \ldots, F_N$ are mappings into R^n from

$$\underbrace{R^n \times \cdots \times R^n}_{N \text{ times}} \times I = R^{Nn} \times I$$

where I is an open interval on the real line with midpoint 0. Further, we assume that the equation

$$\frac{dx}{dt} = f(x) \tag{8.12}$$

has a solution $x(t)$ of period T and that the linear variational system of (8.12) relative to $x(t)$ has the number one as a simple characteristic multiplier. Then if one system (say the jth system) is isolated (from the other systems) it would be described by

$$\frac{dx^{(j)}}{dt} = f(x^{(j)}) + \varepsilon F_j(0, \ldots, x^{(j)}, 0, \ldots, 0, \varepsilon) \tag{8.13}$$

and if $|\varepsilon|$ is sufficiently small, then by Theorem 8.1, equation (8.13) has a periodic solution $\bar{x}^{(j)}(t)$ with period $T + \tau_j(\varepsilon)$ such that

$$\lim_{\varepsilon \to 0} \bar{x}^{(j)}(t) = x(t)$$

where $x(t)$ is the given solution of period T of equation (8.12) and such that

$$\lim_{\varepsilon \to 0} \tau_j(\varepsilon) = 0$$

Thus if the individual systems were all isolated from one another, all the individual systems would oscillate with periods close to T, but the periods would not, in general, be the same.

Now we want to find conditions under which phaselocking occurs. The procedure consists essentially in setting up the usual equations and then studying them by using the techniques in the proof of Theorem 8.1. We give the proof for the case $N = 2$. Following the proof, we indicate (in Remark 1) how to deal with the case $N > 2$.

We consider the system

$$\frac{dx}{dt} = f(x) + \varepsilon F(x, y, \varepsilon)$$

$$\frac{dy}{dt} = f(y) + \varepsilon G(x, y, \varepsilon) \tag{8.14}$$

where

$$f: \ R^n \to R^n$$
$$F: \ R^n \times R^n \times I \to R^n$$
$$G: \ R^n \times R^n \times I \to R^n$$

Then since $x(t)$ is a solution of period T of equation (8.12) it follows that $(x(t), y(t))$ where $y(t) = x(t)$, is a solution of period T of system (8.14) with $\varepsilon = 0$. By translation of the $x_1 - $ axis, $\ldots, x_n - $ axis, we may assume that

$$x_1(0) = 0 \tag{8.15}$$

and by rotation of the same axes, we may assume that

$$\begin{cases} x_1'(0) \neq 0 \\ x_2'(0) = x_3'(0) = \cdots = x_n'(0) = 0 \end{cases} \tag{8.16}$$

Since $x(t) = y(t)$, we have also

$$\begin{cases} y_1(0) = 0 \\ y_1'(0) \neq 0 \\ y_2'(0) = y_3'(0) = \cdots = y_n'(0) = 0 \end{cases} \tag{8.17}$$

Now let $(x(t, c, d, \varepsilon), y(t, c, d, \varepsilon))$ denote the solution of (8.14) with initial value (c, d) at $t = 0$, that is,

$$x(0, c, d, \varepsilon) = c$$
$$y(0, c, d, \varepsilon) = d$$

Then the periodicity condition given in equation (8.14) of the previous section becomes

$$x(T + \tau, c, d, \varepsilon) - c = 0$$
$$y(T + \tau, c, d, \varepsilon) - d = 0 \tag{8.18}$$

From Lemma 8.1, it follows that the components of the initial condition (c, d) can be chosen as

$$c = \begin{bmatrix} 0 \\ c_2 \\ \vdots \\ c_n \end{bmatrix}$$

and

$$d = \begin{bmatrix} d_1 \\ d_2 \\ \vdots \\ d_n \end{bmatrix}$$

We want to solve (8.18) for $\tau, c_2, \ldots, c_n, d_1, \ldots, d_n$ as functions of ε.

(The optimistic among us might hope that we could take $d_1 = 0$. Unfortunately, as is indicated in the proof of Lemma 8.1, setting $c_1 = 0$ is made possible by reparameterizing the possible periodic solutions. But the reparameterizing depends on using $\varepsilon \neq 0$. So we would in general obtain different reparameterizings for the first equation and the second equation.)

First, equation (8.18) has the initial solution: $\varepsilon = 0$, $\tau = 0$, $\bar{c}_2 = x_2(0), \ldots, \bar{c}_n = x_n(0), \bar{d}_1 = y_1(0) = 0, \bar{d}_2 = y(0) = x_2(0), \ldots, \bar{d}_n = y_n(0) = x_n(0)$ or using (8.15) to (8.17), we have for the initial solution

$$\varepsilon = 0, \tau = 0, \bar{c}_2 = x_2(0), \ldots, \bar{c}_n = x_n(0)$$
$$\bar{d}_1 = 0, \bar{d}_2 = x_2(0), \ldots, \bar{d}_n = x_n(0)$$

From the fact that $x(t) = y(t)$ and from (8.15) to (8.17), the differential at the initial solution of (8.18) is the matrix

$$\mathcal{M} = \begin{bmatrix} \mathcal{A}_1 & \\ & \mathcal{A}_2 \end{bmatrix}$$

where $\mathcal{A}_j$ $(j = 1, 2)$ is an $n \times n$ matrix and all other entries of $\mathcal{M}$ are zero, and

$$\mathcal{A}_1 = \begin{bmatrix} x_1'(T(0)) & \frac{\partial x_1}{\partial c_2}(T(0)) & \cdots & \frac{\partial x_1}{\partial c_n}(T(0)) \\ x_2'(T(0)) & \frac{\partial x_2}{\partial c_2} - 1 & \cdots & \frac{\partial x_2}{\partial c_n} \\ \vdots & & & \vdots \\ x_n'(T(0)) & \frac{\partial x_n}{\partial c_2} & \cdots & \frac{\partial x_n}{\partial c_n} - 1 \end{bmatrix}$$

and

$$\mathcal{A}_2 = \begin{bmatrix} \frac{\partial x_1}{\partial c_1}(T(0)) - 1 & \frac{\partial x_1}{\partial c_2}(T(0)) & \cdots & \frac{\partial x_1}{\partial c_n}(T(0)) \\ \frac{\partial x_2}{\partial c_1} & \frac{\partial x_2}{\partial c_2} - 1 & & \vdots \\ \vdots & \vdots & & \\ \frac{\partial x_n}{\partial c_1} & \frac{\partial x_n}{\partial c_2} & & \frac{\partial x_n}{\partial c_n} - 1 \end{bmatrix}$$

We showed in the proof of Theorem 8.1 that $x_1'(T(0)) = 1$ and $x_2'(T(0)) = \cdots = x_n'(T(0)) = 0$ and that

$$\det \mathcal{A}_1 \neq 0$$

We also showed that

$$\frac{\partial x_1}{\partial c_1}(T(0)) = 1$$

$$\frac{\partial x_j}{\partial c_1}(T(0)) = 1 \qquad j = 2, \ldots, n$$

and that at $T(0)$

$$\det \begin{bmatrix} \frac{\partial x_2}{\partial c_2} - 1 & & \frac{\partial x_2}{\partial c_n} \\ \vdots & & \\ \frac{\partial x_n}{\partial c_2} & & \frac{\partial x_n}{\partial c_n} - 1 \end{bmatrix} \neq 0 \qquad (8.19)$$

Thus

$$\det A_2 = 0$$

Hence we cannot apply the implicit function directly to (8.18). However, suppose we consider the system S consisting of the first n equations in (8.18) and the last $(n-1)$ equations in (8.18), that is, the system consisting of all the equations in (8.18) except the $(n+1)$th equation. Since $\det A_1 \neq 0$ and by (8.19), it follows that if we regard S as $(2n-1)$ equations in the $(2n-1)$ unknowns $\tau, c_2, \ldots, c_n, d_2, d_3, \ldots, d_n$ then we can solve S for these unknowns in terms of d_1 and ε. That is, we obtain

$$\tau = \tau(\varepsilon, d_1) \quad \text{where} \quad \tau(0, \bar{d}_1) = 0$$
$$c_2 = c_2(\varepsilon, d_1) \quad \text{where} \quad c_2(0, \bar{d}_1) = \bar{c}_2$$

$$\cdot \ \cdot \ \cdot$$

$$c_n = c_n(\varepsilon, d_1) \quad \text{where} \quad c_n(0, \bar{d}_1) = \bar{c}_n$$
$$d_2 = d_2(\varepsilon, d_1) \quad \text{where} \quad d_2(0, \bar{d}_1) = \bar{d}_2$$
$$d_3 = d_3(\varepsilon, d_1) \quad \text{where} \quad d_3(0, \bar{d}_1) = \bar{d}_3$$

$$\cdot \ \cdot \ \cdot$$

$$d_n = d_n(\varepsilon, d_1) \quad \text{where} \quad d_n(0, \bar{d}_1) = \bar{d}_n$$

If we substitute these expressions in the $(n+1)$th equation, we obtain then a scalar equation in d_1 and ε_1. This scalar equation, call it $\mathcal{E}$, has the initial solution $d_1 = \bar{d}_1 = 0$ and $\varepsilon = 0$. If we can solve this equation $\mathcal{E}$ near this initial solution for d_1 as a function of ε, then we obtain a solution of (8.18) and hence a periodic solution of (8.1).

Theorem 8.2 *Suppose that equation (8.1) satisfies the following conditions:*

(i) *Functions f, $F_1, \ldots, F_N$ have continuous third derivatives in all variables.*

(ii) *The equation*

$$\frac{dy}{dt} = f(y)$$

has a nontrivial solution $y(t)$ of period T such that the matrix $f_y[y(t)]$ has $(n-1)$ characteristic multipliers with absolute value different from one.

(iii) *The scalar equation $\mathcal{E}$ can be solved for d_1 as a function of ε.*

Conclusion: If $|\varepsilon|$ is sufficiently small, then system (8.14) has a solution of period $T + \tau(\varepsilon)$, say $(x(t, \varepsilon), (y(t, \varepsilon)))$. Also

$$\lim_{\varepsilon \to 0} T + \tau(\varepsilon) = T$$

and

$$\lim_{\varepsilon \to 0}(x(t, \varepsilon), y(t, \varepsilon)) = (x(t), y(t))$$

Remarks.

1. If $N > 2$, then using the same procedure, the problem is reduced to solving a system of $(N - 1)$ equations for $(N - 1)$ variables as functions of ε.

2. Anyone concerned with a real-world example of phaselocking would find the condition that $|\varepsilon|$ be sufficiently small a serious limitation of Theorem 8.2. The theorem can be applied only to a case in which the influences of the various systems upon one another are very small. (This condition might hold for certain mechanical or electrical systems but not for physiological systems.)

3. Theorem 8.1 is a classical theorem. It is termed "well-known" by Coddington and Levinson [1952] and it is proved in their book (Coddington and Levinson [1955]). Theorem 8.2 is scarcely more than a corollary to Theorem 8.1. Kopell and Ermentrout have studied special cases of these questions in a series of articles ([1986] and succeeding papers). Their discussion would probably have been different if they had been aware of the classical theorem.

The Case in Which the Classical Hypothesis is Violated

In the previous section we have assumed that the given equation

$$\frac{dx}{dt} = f(x, \varepsilon)$$

has, for $\varepsilon = 0$, a periodic solution $x(t)$ such that

$$f_x[x(t), 0]$$

has one characteristic multiplier equal to one and that characteristic multiplier has multiplicity one. As a result of this and because we must allow for the period of the periodic solution changing as ε changes, we are able to reduce the problem to one for which the implicit function theorem can be used. Now we want to look at the case in which $f_x[x(t)]$ has a characteristic multiplier equal to one but the characteristic multiplier is not simple (the characteristic multiplier has multiplicity greater than one). The problems which arise correspond to the problems for the resonance case in the study of nonautonomous equation (Chapter 7).

Thus, as might be expected, the problems are more complicated than those studied in the previous sections of this chapter. Because of these complications, we will study a special case in which the characteristic multiplier has multiplicity two.

We study the equation

$$\frac{dx}{dt} = Ax + \varepsilon f(x, \varepsilon) \tag{8.20}$$

where A is the constant $n \times n$ matrix

$$\begin{bmatrix} 0 & 1 & \\ -1 & 0 & \\ & & D \end{bmatrix}$$

in which D is a constant $(n - 2) \times (n - 2)$ matrix such that the eigenvalues of D all have nonzero real parts, the function f has continuous second derivatives at each point of $R^n \times I$ and I is an open interval on the real line with midpoint 0.

We observe first that each nonzero solution of

$$\frac{dx}{dt} = Ax \tag{8.21}$$

has as its first component

$$c_1 \cos t + c_2 \sin t$$

where c_1, c_2 are constants not both zero. For any such pair c_1, c_2 there exists t_0 such that

$$c_1 \cos t_0 + c_2 \sin t_0 = 0 \tag{8.22}$$

Also if the derivative of the first component is zero at t_0, that is, if

$$-c_1 \sin t_0 + c_2 \cos t_0 = 0 \tag{8.23}$$

then it follows at once from (8.22) and (8.23) that $c_1 = c_2 = 0$, a contradiction to our assumption. Hence if the first component is zero at t_0, its derivative is nonzero at t_0 and thus the first component changes sign at t_0.

Using the variation of constants formula, we may represent the solution $x(t, c, \varepsilon)$ of equation (8.20) as:

$$x(t, c, \varepsilon) = e^{tA}c + \varepsilon \int_0^t e^{(t-\sigma)A} f[x(\sigma, c, \varepsilon), \varepsilon] d\sigma \tag{8.24}$$

We seek a solution $x(t, c, \varepsilon)$ of period

$$2\pi + \varepsilon m(\varepsilon)$$

where $m(\varepsilon)$ is a differentiable function of ε which is to be determined. (As will be seen a little later in the calculations (equations (8.30) to (8.32), it is a good strategy to use the expresson $\varepsilon m(\varepsilon)$ rather than using simply a function of ε.)

A nasc that $x(t, c, \varepsilon)$ have period $2\pi + \varepsilon m(\varepsilon)$ is

$$x(2\pi + \varepsilon m(\varepsilon), c, \varepsilon) - x(0, c, \varepsilon) = 0 \tag{8.25}$$

or, using (8.24), we may write (8.25) as:

$$e^{(2\pi + \varepsilon m)A}c - c + \varepsilon \int_0^{2\pi + \varepsilon m} e^{(2\pi + \varepsilon m - \sigma)A} f[x(\sigma, c, \varepsilon), \varepsilon] d\sigma = 0 \tag{8.26}$$

We have already remarked that given a solution $x(t, c, 0)$ then the first component of $x(t, c, 0)$ changes sign at some value t_0. Hence we conclude from equation (8.24) that if $|\varepsilon|$ is sufficiently small, the first component of $x(t, c, \varepsilon)$ must change sign at a value t_1 close to t_0. That is, we have: for each ε such that $|\varepsilon|$ is sufficiently small there is a t_1 such that

$$x_1(t_1, c, \varepsilon) = 0$$

Since $x(t, c, \varepsilon)$ is a solution of (8.20) which is autonomous, then

$$\tilde{x}(t, c, \varepsilon) = x(t_1 + t, c, \varepsilon)$$

is a solution of (8.20) such that

$$\tilde{x}_1(0, c, \varepsilon) = x_1(t_1, c, \varepsilon) = 0$$

Thus we have shown that if $x(t, c, \varepsilon)$ is a periodic solution of (8.20) and if $|\varepsilon|$ is sufficiently small, then it can be reparameterized as a periodic solution $x(t, c, \varepsilon)$ such that

$$x_1(0, c, \varepsilon) = 0$$

Hence it is sufficient to search for periodic solutions $x(t, c, \varepsilon)$ such that

$$x(0, c, \varepsilon) = \begin{bmatrix} 0 \\ c_2 \\ \vdots \\ c_n \end{bmatrix}$$

or

$$c = \begin{bmatrix} 0 \\ c_2 \\ \vdots \\ c_n \end{bmatrix}$$

That is, we will set $c_1 = 0$ in (8.26) and then solve for $m, c_2, \ldots, c_n$ as functions of ε. But first we rewrite (8.26) as

$$[e^{(2\pi + \varepsilon m)A} c - e^{2\pi A} c] + [e^{2\pi A} c - c]$$

$$+ \varepsilon \int_0^{2\pi + \varepsilon m} e^{(2\pi + \varepsilon m - \sigma)A} f[x(\sigma, c, \varepsilon), \varepsilon] d\sigma = 0 \qquad (8.27)$$

Using the same notation introduced in Chapter 7 to treat the resonance case, we proceed as follows.

The null space E_{n-2} of $[e^{2\pi A} - I]$ consists of vectors of the form

$$\begin{bmatrix} c_1 \\ c_2 \\ 0 \\ \vdots \\ 0 \end{bmatrix}$$

and E_r consists of vectors of the form

$$\begin{bmatrix} 0 \\ 0 \\ c_3 \\ \vdots \\ c_n \end{bmatrix}$$

Applying P_r to (8.27) yields:

$$P_r[e^{(2\pi+\varepsilon m)A}c - e^{2\pi A}c] + (e^{2\pi D} - I)\begin{bmatrix} c_3 \\ \vdots \\ c_n \end{bmatrix}$$

$$+ P_r\left\{ \varepsilon \int_0^{2\pi+\varepsilon m} e^{(2\pi+\varepsilon m-\sigma)A} f[x(\sigma, c, \varepsilon), \varepsilon]\, d\sigma \right\} = 0 \qquad (8.28)$$

and applying P_{n-r} to (8.27) yields

$$P_{n-r}[e^{(2\pi+\varepsilon m)A} - e^{2\pi A}]c$$

$$+ P_{n-r}\varepsilon \int_0^{2\pi+\varepsilon m} e^{(2\pi+\varepsilon m-\sigma)A} f[x(\sigma, c, \varepsilon), \varepsilon]\, d\sigma = 0 \qquad (8.29)$$

or assuming m and ε are nonzero and dividing by ε,

$$P_{n-r}\left[\left(\frac{e^{(2\pi+\varepsilon m)A} - e^{2\pi A}}{\varepsilon m} \right) \left(\frac{\varepsilon m}{\varepsilon} \right) \right] c$$

$$+ P_{n-r} \int_0^{2\pi+\varepsilon m} e^{(2\pi+\varepsilon m-\sigma)A} f[x(\sigma, c, \varepsilon), \varepsilon]\, d\sigma = 0 \qquad (8.30)$$

Equation (8.28) may be regarded as a system of $(n-2)$ equations in the $(n-2)$ variables $c_3, \ldots, c_n$ to be solved in terms of ε, m, and c_2. If c_2 and m are regarded as fixed, then (8.28) has the initial solution

$$\varepsilon = 0, c_3 = 0, \ldots, c_n = 0$$

The Jacobian at this solution, that is, $\det[e^{2\pi D} - D]$ is nonzero by the hypothesis imposed on D. Hence by the implicit function theorem if $|\varepsilon|$ is sufficiently small, there exists a unique solution

$$c_3(\varepsilon, c_2, m, \ldots, c_n, \varepsilon, c_2, m) \qquad (8.31)$$

of (8.28) and such that

$$c_j(0, c_2, m) = 0, \qquad\qquad j = 3, \ldots, n$$

Substituting (8.31) into (8.29) we obtain a system of two equations in the two variables c_2 and m. One objective is then to solve this system for c_2 and m in terms of ε. To do this, we replace (8.29) by (8.30) if ε and m are nonzero and observe that taking the limit as $\varepsilon \to 0$ in equation (8.30) we have

$$P_{n-r}[e^{2\pi A}Am]c + P_{n-r} \int_0^{2\pi} e^{(2\pi-\sigma)A} f[x(\sigma, c, 0), 0]\, d\sigma = 0 \qquad (8.32)$$

where

$$c = \begin{bmatrix} 0 \\ c_2 \\ c_3 \\ \vdots \\ c_n \end{bmatrix}$$

and

$$c_j = c_j(0, c_2, m) = 0, \qquad j = 3, \ldots, n$$

From the definition of A, we may write (8.32) as

$$m \begin{bmatrix} 0 & 1 \\ -1 & 0 \end{bmatrix} \begin{bmatrix} 0 \\ c_2 \end{bmatrix} + \int_0^{2\pi} \begin{bmatrix} \cos(2\pi - \sigma) & \sin(2\pi - \sigma) \\ -\sin(2\pi - \sigma) & \cos(2\pi - \sigma) \end{bmatrix} \begin{bmatrix} f_1 \\ f_2 \end{bmatrix} d\sigma = 0$$

where f_1, f_2 are components of f and

$$f_k = f_k \left(\begin{bmatrix} \cos \sigma & \sin \sigma \\ -\sin \sigma & \cos \sigma \end{bmatrix} \begin{bmatrix} 0 \\ c_2 \end{bmatrix}, 0 \right) \qquad k = 1, 2$$

or

$$mc_2 = - \int_0^{2\pi} \{ [\cos(2\pi - \sigma)] f_1 + [\sin(2\pi - \sigma)] f_2 \} d\sigma$$

$$0 = - \int_0^{2\pi} \{ [-\sin(2\pi - \sigma)] f_1 + [\cos(2\pi - \sigma)] f_2 \} d\sigma \qquad (8.33)$$

Observing that the right-hand sides of (8.33) are functions of c_2 only, we may rewrite (8.33) as

$$mc_2 = \gamma_1(c_2)$$
$$0 = \gamma_2(c_2) \qquad (8.34)$$

Since

$$\begin{bmatrix} \cos(2\pi - \sigma) & \sin(2\pi - \sigma) \\ -\sin(2\pi - \sigma) & \cos(2\pi - \sigma) \end{bmatrix} = \begin{bmatrix} \cos \sigma & -\sin \sigma \\ \sin \sigma & \cos \sigma \end{bmatrix}$$

and since

$$\int_0^{2\pi} \cos^m t \sin^n t \, dt \neq 0$$

if and only if m and n are even, it follows that if f_1 and f_2 are power series, then γ_1 and γ_2 are power series in c_2 and each term in each power series has an odd exponent. That is

$$\gamma_1(c_2) = c_2^{a_1} P_1(c_2)$$
$$\gamma_2(c_2) = c_2^{a_2} P_2(c_2) \qquad (8.35)$$

where exponents a_1, a_2 are odd, and each term in the power series $P_j(c_2)$, where $j = 1, 2$, has the form

$$k(c_2)^q$$

where k is a constant and q is even, and

$$P_j(0) \neq 0, \qquad j = 1, 2.$$

Hence system (8.34) becomes

$$mc_2 = c_2^{a_1} P_1(c_2)$$
$$0 = c_2^{a_2} P_2(c_2) \tag{8.36}$$

Now let us assume that

$$P_2(c_2) = 0$$

has a solution $\bar{c}_2 \neq 0$ such that $P_2'(\bar{c}_2) \neq 0$. Let

$$\bar{m} = (\bar{c}_2)^{a_1-1} P_1(\bar{c}_2)$$

Then $(\bar{m}, \bar{c}_2)$ is a solution of (8.36) and the Jacobian at $(\bar{m}, \bar{c}_2)$ is the determinant

$$\begin{bmatrix} -\bar{c}_2 & \frac{\partial}{\partial c_2}[c_2^{a_1} P_1(c_2)]\Big|_{c_2=\bar{c}_2} \\ 0 & (\bar{c}_2)^{a_2} P_2'(\bar{c}_2) + a_2(\bar{c}_2)^{a_2-1} P_2(\bar{c}_2) \end{bmatrix}$$

Since $P_2(\bar{c}_2) = 0$, the value of the determinant is

$$-(\bar{c}_2)^{a_2+1} P_2'(\bar{c}_2) \neq 0$$

Hence from the derivation of (8.36) it follows that if $|\varepsilon|$ is sufficiently small there exists a solution

$$c_2 = c_2(\varepsilon)$$
$$m = m(\varepsilon)$$

of equation (8.29) and hence a periodic solution of period $(2\pi + \varepsilon m(\varepsilon))$ of equation (8.20).

Summarizing, we have

Theorem 8.3 *Suppose the equation*

$$x' = Ax + \varepsilon F(x, \varepsilon) \tag{8.37}$$

satisfies the following conditions: (i)

$$A = \begin{bmatrix} 0 & 1 & 0 & \cdots & 0 \\ -1 & 0 & 0 & \cdots & 0 \\ 0 & 0 & & & \\ \vdots & \vdots & & D & \\ 0 & 0 & & & \end{bmatrix}$$

where C is an $(n-2) \times (n-2)$ matrix such that the eigenvalues of D all have nonzero real parts;

(ii) the function f has continuous second derivatives at each point of $R^n \times I$ and I is an open interval on the real line with midpoint 0;

(iii) *the components f_1 and f_2 of f are power series in their variables.*

Then if $P_2(c_2)$, defined in equation (8.35), has a zero of multiplicity one, and if $|\varepsilon|$ is sufficiently small, there exists a differentiable function $m(\varepsilon)$ such that equation (8.37) has a periodic solution of period $2\pi_\varepsilon m(\varepsilon)$.

Remark.

It is clear from the proof that the periodic solution obtained in Theorem 8.3 is not unique. The function $P_2(c_2)$ could have more than one zero of multiplicity one.

Hopf Bifurcation

So far in Chapter 7 and this chapter, we have studied the problem of finding periodic solutions if a parameter ε is such that $|\varepsilon|$ is small enough, that is, ε assumes positive and negative values. This problem arises in many applications, but there are also applications in which it is necessary to determine if there exist periodic solutions just for small positive ε or just for negative ε close to zero.

To see this we consider a physical system which is described by a differential equation in which a parameter ε occurs. Suppose that if $\varepsilon = 0$, the differential equation has an asymptotically stable equilibrium point and thus if $\varepsilon = 0$, the physical system is stable. Now it may happen that for $\varepsilon < 0$ the differential equation continues to have an asymptotically stable equilibrium point (and the physical system is stable). On the other hand, if the parameter ε is increased above zero, the physical system may begin to oscillate with increasing amplitude (the differential equation displays periodic solutions of increasing amplitude) until the system collapses or, if the parameter ceases to increase, just settles down to an oscillation of constant amplitude. This kind of behavior is observed in many real situations: mechanical, acoustical, electrical, biological, and economic systems. Consequently, it is important to study solutions of the corresponding differential equation near the value of the parameter where the system begins to oscillate. Such a study was made by Andronov and Leontovich [1937] for the two-dimensional case, and the n-dimensional case was studied by Hopf [1942]. Such a result (the appearance of an oscillation as a parameter is varied) is frequently called a *Hopf bifurcation*, although a fairer and more accurate term would be an Andronov-Leontovich-Hopf bifurcation.

Mathematically the problems in Hopf bifurcation theory differ from the perturbation problems considered in previous sections of this chapter in the following way: the problems studied earlier were typified by an equation of the form

$$\frac{dx}{dt} = Ax + \varepsilon f(x, \varepsilon)$$

But the Hopf bifurcation problems are typified by an equation of the form

$$\frac{dx}{dt} = [A + \varepsilon L(\varepsilon)]x + f(x, \varepsilon)$$

where f is higher order in x uniformly in ε and, in general, $f(x, 0) \neq 0$. The crucial role played by the parameter ε is how it affects the linear part of the equation.

We will use the Coddington-Levinson approach to study the Hopf bifurcation. In order to allow for the occurrence of cases in which there are periodic solutions only for a limited set of values of ε, for example, there are periodic solutions for $\varepsilon > 0$ and there are no periodic solutions for $\varepsilon < 0$, we will use a somewhat different procedure from before. Earlier in this chapter, the typical procedure was to solve for the initial value $(0, c_2, \ldots, c_n)$ and the period T as functions of ε. In the Hopf bifurcation problem, the procedure is roughly as follows: one of the numbers in the initial value $(0, c_2, \ldots, c_n)$, say c_2, acts as the independent variable and the values $c_3, \ldots, c_n, T$, and ε are determined as functions of c_2. The resulting $\varepsilon(c_2)$ may then be restricted to positive values or to negative values.

The other difference in the technical part of the treatment stems from the fact that unlike the previous work, the ε does not play an important part in the estimates of the nonlinear term. Instead of using the fact that $|\varepsilon|$ is small, we use the fact that f is higher-order and hence if $|x|$ is small, the magnitude of the nonlinear term can be controlled.

We consider the equation

$$\frac{dx}{dt} = Ax + \varepsilon L(\varepsilon)x + f(x, \varepsilon) \tag{8.38}$$

where

$$A = \begin{bmatrix} \begin{matrix} 0 & 1 \\ -1 & 0 \end{matrix} & \\ & D \end{bmatrix}$$

where D is an $(n - 2) \times (n - 2)$ matrix each of whose eigenvalues has nonzero real part, and $L(\varepsilon)$ is an $n \times n$ matrix each of whose entries is a differentiable function of ε for $\varepsilon \in I$, an interval on R with midpoint 0, such that

$$L(\varepsilon) = \begin{bmatrix} \begin{matrix} h(\varepsilon) & 0 \\ 0 & h(\varepsilon) \end{matrix} & \\ & \mathcal{D}(\varepsilon) \end{bmatrix}$$

where $h(0) = 1$ and $\mathcal{D}(\varepsilon)$ is a differentiable $(n - 2) \times (n - 2)$ matrix. For later use, let

$$C = \begin{bmatrix} 0 & 1 \\ -1 & 0 \end{bmatrix}$$

and

$$H(\varepsilon) = \begin{bmatrix} h(\varepsilon) & 0 \\ 0 & h(\varepsilon) \end{bmatrix}$$

Also $f(x, \varepsilon)$ has continuous second derivatives in x and ε for $(x, \varepsilon) \in R^n \times I$ and

$$|f(x, \varepsilon)| = o(|x|)$$

uniformly in ε, that is, for $\varepsilon \in I$ and $x \in R^n$

$$|f(x, \varepsilon)| < \{\eta(|x|)\}|x|$$

where η is a continuous monotonic increasing function with domain $\{r \in R/r \geq 0\}$ and $\eta(0) = 0$.

Since

$$\frac{dx}{dt} = Ax$$

has solutions of period 2π, we seek solutions of period $2\pi\lambda$, where λ is close to 1, for equation (8.38). Let $x(t, c, \varepsilon)$ be a solution of (8.38) such that $x(0, c, \varepsilon) = c$. Then a necessary and sufficient condition that $x(t, c, \varepsilon)$ have period $2\pi\lambda$ is

$$x(2\pi\lambda, c, \varepsilon) = x(0, c, \varepsilon) = c \tag{8.39}$$

As before, we apply the variation of constants formula to (8.39) and obtain

$$[e^{[A+\varepsilon L(\varepsilon)]2\pi\lambda} - I]c + \int_0^{2\pi\lambda} e^{(2\pi\lambda-\sigma)A+\varepsilon L(\varepsilon))} f[x(\sigma, c, \varepsilon), \varepsilon]\,d\sigma = 0 \tag{8.40}$$

Following the procedure used earlier, we seek solutions of (8.38) with initial value c such that

$$c = \begin{bmatrix} 0 \\ c_2 \\ \vdots \\ c_n \end{bmatrix}$$

Then (8.40) may be written as

$$[e^{[A+\varepsilon L(\varepsilon)]2\pi\lambda} + I] \begin{bmatrix} 0 \\ c_2 \\ \vdots \\ c_n \end{bmatrix} + \int_0^{2\pi\lambda} e^{[A+\varepsilon L(\varepsilon)](2\pi\lambda-\sigma)} f[x(\sigma, c_2, \ldots, c_n, \varepsilon), \varepsilon]\,d\sigma = 0$$

$$\tag{8.41}$$

Using again the notation introduced in Chapter 7, we apply P_r to equation (8.41) and obtain

$$[e^{2\pi\lambda[D+\varepsilon \mathcal{D}(\varepsilon)]} - I] \begin{bmatrix} c_3 \\ \vdots \\ c_n \end{bmatrix} + \int_0^{2\pi\lambda} e^{(2\pi\lambda-\sigma)[D+\varepsilon \mathcal{D}(\varepsilon)]} f[x(\sigma, c_2, \ldots, c_n, \varepsilon), \varepsilon]\,d\sigma = 0$$

$$\tag{8.42}$$

where f denotes $\begin{bmatrix} f_3 \\ \vdots \\ f_n \end{bmatrix}$.

To solve (8.42) for $c_3, \ldots, c_n$ in terms of c_2, ε, and λ, we apply the implicit function theorem. Taking $c_2 = 0, \varepsilon = 0, \lambda = 1$ in (8.42), we have

$$[e^{2\pi - D} - I] \begin{bmatrix} c_3 \\ \vdots \\ c_n \end{bmatrix} + \int_0^{2\pi} e^{(2\pi - \sigma)D} f[x(\sigma, 0, c_3, \ldots, c_n, \varepsilon), \varepsilon] \, d\sigma = 0$$

But

$$x(\sigma, 0, 0, \ldots, 0, 0) \equiv 0$$

and $f(0, \varepsilon) = 0$. Hence an initial solution of (8.42) is $c_3 = \cdots = c_n = 0$ at $c_2 = 0$, $\varepsilon = 0, \lambda = 1$ and the appropriate Jacobian is

$$\det[e^{2\pi D} - I]$$

To prove that this is the appropriate Jacobian, we observe first that by the differentiability theorem in Chapter 1, the derivatives $\frac{\partial x}{\partial c_j}$ $(j = 2, \ldots, n)$ exist and hence

$$|x(\sigma, 0, c_3, \ldots, c_n, \varepsilon)| = |x(\sigma, 0, c_3, \ldots, c_n, \varepsilon) - x(\sigma, 0, \ldots, 0, \varepsilon)|$$

$$= \left| \left(\frac{\partial x}{\partial c_2} \right) c_3 + \cdots + \left(\frac{\partial x}{\partial c_n} \right) c_n \right|$$

Thus there exists a constant $M_1 > 0$ such that if $|c_2|, \ldots, |c_n|$ are sufficiently small then

$$|x(\sigma, 0, c_3, \ldots, c_n, \varepsilon)| < M_1 \sum_{j=3}^{n} |c_j| \tag{8.43}$$

Then the proof follows from the hypothesis that

$$|f(x, \varepsilon)| = o(|x|)$$

By the hypothesis on matrix D, it follows that

$$\det[e^{2\pi D} - I] \neq 0$$

Thus we conclude: if $|c_2|, \ldots, |c_n|, |\varepsilon|$, and $|\lambda - 1|$ are small enough we can solve uniquely for $c_3, \ldots, c_n$ in terms of $c_2, \varepsilon, \lambda$, that is, we have

$$c_j = c_j(c_2, \varepsilon, \lambda) \qquad (j = 3, \ldots, n) \tag{8.44}$$

where $c_j(c_2, \varepsilon, \lambda)$ is a differentiable function and

$$c_j(0, 0, 1) = 0$$

Lemma 8.2 *For $j = 3, \ldots, n$, there exists $k_j(c_2, \varepsilon, \lambda)$ such that*

$$c_j(c_2, \varepsilon, \lambda) = c_2 k_j(c_2, \varepsilon, \lambda) \tag{8.45}$$

Proof　As noted before

$$x(\sigma, 0, \varepsilon) = 0$$

From (8.42), we have for sufficiently small $|\varepsilon|$ and $|\lambda - 1|$:

$$\begin{bmatrix} c_3 \\ \vdots \\ c_n \end{bmatrix} = -[e^{2\pi\lambda[D+\varepsilon\mathcal{D}(\varepsilon)]} - I]^{-1}$$

$$\times \left[\int_0^{2\pi\lambda} e^{(2\pi\lambda-\sigma)[D+\varepsilon\mathcal{D}(\varepsilon)]} f[x(\sigma, c_2, c_3 \ldots, c_n, \varepsilon), \varepsilon] \, d\sigma \right] = 0$$

Taking $c_2 = 0$ and using the second-order condition on f, we have: there exist constants $M_2 > 0$ and $M_3 > 0$ such that

$$\sum_{j=3}^n |c_j(0, \varepsilon, \lambda)| < M_2 \left[\int_0^{2\pi\lambda} f[x(\sigma, 0, c_3, \ldots, c_n, \varepsilon)] \, d\sigma \right]$$

$$< M_3 \max_\sigma \eta(|x(\sigma, 0, c_3, \ldots, \varepsilon)|) \times |x(\sigma, 0, c_3, \ldots, c_n, \varepsilon)|$$

Applying (8.43), we have

$$\sum_{j=3}^n |c_j(0, \varepsilon, \lambda)| < M_3 \left[\eta \left(M_1 \sum_{j=3}^n |c_j| \right) \right] \left[M_1 \sum_{j=3}^n |c_j| \right] \tag{8.46}$$

If $\sum_{j=3}^n |c_j|$ is sufficiently small, then

$$M_3 M_1 \eta \left(M_1 \sum_{j=3}^n |c_j| \right) < 1 \tag{8.47}$$

If $|c_j(0, \varepsilon, \lambda)| \neq 0$ for some $j = 3, \ldots, n$, then (8.46) and (8.47) yield a contradiction. Hence for $j = 3, \ldots, n$,

$$c_j(0, \varepsilon, \lambda) = 0$$

Thus

$$c_j(c_2, \varepsilon, \lambda) = c_j(c_2, \varepsilon, \lambda) - c_j(0, \varepsilon, \lambda)$$
$$= c_2 k_j(c_2, \varepsilon, \lambda)$$

where

$$k_j(c_2, \varepsilon, \lambda) = \frac{\partial}{\partial c_2} c_j(\theta c_2, \varepsilon, \lambda)$$

where $0 < \theta < 1$. This completes the proof of Lemma 8.2.　　　　　$\Box$

Now we apply P_{n-r} to (8.41), substitute (8.44) into the result and obtain

$$[e^{2\pi\lambda(c+\varepsilon H(\varepsilon))} - I]\begin{bmatrix} 0 \\ c_2 \end{bmatrix} + \int_0^{2\pi\lambda} [e^{(2\pi\lambda-\sigma)(c+\varepsilon H(\varepsilon))}]f$$

$$[x(\sigma, c_2, c_3(c_2, \varepsilon, \lambda), \dots, c_n(c_2, \varepsilon, \lambda)), \varepsilon]\, d\sigma = 0 \qquad (8.48)$$

Equation (8.48) is a pair of scalar equations in the variables c_2, ε, λ. If we were to apply the procedure used in earlier work, we would try to solve for c_2 and λ in terms of ε, that is, to find $c_2(\varepsilon)$ and $\lambda = \lambda(\varepsilon)$. Having obtained such solutions we would use (8.44) to express

$$c_j(c_2, \varepsilon, \lambda) \qquad\qquad (j = 3, \dots, n)$$

in terms of ε. Instead, in this case, we solve for ε and λ in terms of c_2. To do this, we first divide equation (8.42) by c_2. More precisely, we define the 2-vector

$$F(c_2, \varepsilon, \lambda) = \begin{bmatrix} F_1(c_2, \varepsilon, \lambda) \\ F_2(c_2, \varepsilon, \lambda) \end{bmatrix}$$

as follows. If $c_2 \neq 0$, let

$$F(c_2, \varepsilon, \lambda) = [e^{(C+\varepsilon H(\varepsilon))2\pi\lambda} - I]\begin{bmatrix} 0 \\ 1 \end{bmatrix}$$

$$+ \frac{1}{c_2}\int_0^{2\pi\lambda} e^{(C+\varepsilon H(\varepsilon))(2\pi\lambda-\sigma)} f[x(\sigma, c_2, \varepsilon), \varepsilon]\, ds$$

where f denotes the vector $\begin{bmatrix} f_1 \\ f_2 \end{bmatrix}$ and $x(\sigma, c_2, \varepsilon)$ denotes

$$x(\sigma, c_2, c_3(c_2, \varepsilon, \lambda), \dots, c_n(c_2, \varepsilon, \lambda), \varepsilon)$$

If $c_2 = 0$, let

$$F(0, \varepsilon, \lambda) = [e^{(c+\varepsilon H(\varepsilon))2\pi\lambda} - I]\begin{bmatrix} 0 \\ 1 \end{bmatrix}$$

Now we have: if $c_2 \neq 0$

$$|F(c_2, \varepsilon, \lambda) - F(0, \varepsilon, \lambda)|$$

$$= \left|\frac{1}{c_2}\int_0^{2\pi\lambda} e^{[c+\varepsilon H(\varepsilon)](2\pi\lambda-\sigma)} f[x(\sigma, c_2, \varepsilon), \varepsilon]\, d\sigma\right| \qquad (8.49)$$

Since $x(\sigma, 0, \varepsilon) = 0$, then there exists $M > 0$ such that

$$|x(\sigma, c_2, \varepsilon)| = |x(\sigma, c_2, \varepsilon) - x(\sigma, 0, \varepsilon)| < M|c_2|$$

By the hypothesis on f, it follows that

$$|f[x(\sigma, c_2, \varepsilon), \varepsilon]| < \{\eta[M|c_2|]\}M|c_2| \qquad (8.50)$$

From (8.49) and (8.50), we conclude that

$$\lim_{c_2 \to 0} F(c_2, \varepsilon, \lambda) = F(0, \varepsilon, \lambda)$$

Thus there exists an open neighborhood of $(0, 0, 1)$ in R^3 such that $F(c_2, \varepsilon, \lambda)$ is continuous in the neighborhood. Next we compute $\frac{\partial F}{\partial \varepsilon}$ and $\frac{\partial F}{\partial \lambda}$ at $c_2 = 0$, $\varepsilon = 0$, $\lambda = 1$.

$$\frac{\partial F}{\partial \varepsilon} = \frac{\partial F(0, \varepsilon, 1)}{\partial \varepsilon} = \frac{\partial}{\partial \varepsilon} \left\{ [e^{[C + \varepsilon H(\varepsilon)]2\pi}] \begin{bmatrix} 0 \\ 1 \end{bmatrix} \right\}$$

$$= \frac{\partial}{\partial \varepsilon} [e^{2\pi C} e^{\varepsilon H(\varepsilon)2\pi}] \begin{bmatrix} 0 \\ 1 \end{bmatrix}$$

$$= \frac{\partial}{\partial \varepsilon} \left\{ \begin{bmatrix} \cos 2\pi & \sin 2\pi \\ -\sin 2\pi & \cos 2\pi \end{bmatrix} [e^{2\pi \varepsilon h(\varepsilon)}] \begin{bmatrix} 0 \\ 1 \end{bmatrix} \right\}$$

$$= \frac{\partial}{\partial \varepsilon} \begin{bmatrix} e^{2\pi \varepsilon h(\varepsilon)} & \\ & e^{2\pi \varepsilon h(\varepsilon)} \end{bmatrix} \begin{bmatrix} 0 \\ 1 \end{bmatrix}$$

$$= \begin{bmatrix} 0 \\ e^{2\pi \varepsilon h(\varepsilon)}[2\pi h(\varepsilon) + 2\pi \varepsilon h'(\varepsilon)] \end{bmatrix}$$

Thus at $c_2 = 0$, $\varepsilon = 0$, $\lambda = 1$

$$\frac{\partial F}{\partial \varepsilon} = \begin{bmatrix} 0 \\ 2\pi \end{bmatrix} \tag{8.51}$$

A similar calculation shows that at $c_2 = 0$, $\varepsilon = 0$, $\lambda = 1$,

$$\frac{\partial F}{\partial \lambda} = \begin{bmatrix} 2\pi \\ 0 \end{bmatrix} \tag{8.52}$$

Then using (8.51) and (8.52) we may use the implicit function theorem to show that there exist $\delta_1 > 0$, $\delta_2 > 0$, $\delta_3 > 0$ such that if $|c_2| < \delta_1$ the equation

$$F(c_2, \varepsilon, \lambda) = 0$$

may be solved uniquely for λ and ε as functions of c_2, that is, $\lambda(c_2)$ and $\varepsilon(c_2)$, such that

$$|\lambda(c_2) - 1| < \delta_2$$

and

$$|\varepsilon(c_2)| < \delta_3$$

Since equation (8.48) may be written as

$$c_2 F(c_2, \varepsilon, \lambda) = 0$$

we have thus solved (8.48) for ε and λ in terms of c_2. From (8.45), we obtain the corresponding values of $c_3, \ldots, c_n$ and we conclude that if $|c_2|, \ldots, |c_n|$, $|\varepsilon|$, and

$|\lambda - 1|$ are small enough, then equation (8.38) has a unique solution of period $2\pi\lambda$ with initial value $c_1 = 0$.

Finally we show that each solution of (8.38) with sufficiently small $|c|$, $|\varepsilon|$, $|\lambda - 1|$ which is periodic of period $2\pi\lambda$ can be parameterized so that its initial value c_1 is zero. Thus it follows from the conclusion of the preceding paragraph that if $|c_1|, \ldots, |c_n|$, ε, and $|\lambda - 1|$ are sufficiently small there is a unique solution of period $2\pi\lambda$ of (8.38).

Lemma 8.3 *If $x(t, c, \varepsilon)$ is a periodic solution of (8.38) of period $2\pi\lambda$ and if $|c|$, $|\varepsilon|$ and $|\lambda - 1|$ are sufficiently small, then there exists t_0 such that if $x_1(t, c, \varepsilon)$ is the first component of $x(t, c, \varepsilon)$ then*

$$x_1(t_0, c, \varepsilon) = 0$$

That is, the solution $x(t, c, \varepsilon)$ can be reparameterized (to

$$x(t + t_0, c, \varepsilon))$$

so that the first component of the initial value at $t = 0$ is itself zero.

Proof Assume that the conclusion of the lemma is not true. That is, assume that given $\delta_1, \delta_2, \delta_3$, arbitrarily small positive numbers, then there exists a solution $x(t, c, \varepsilon)$ of period $2\pi\lambda$ such that $|c| < \delta_1$, $|\varepsilon| < \delta_2$, $|\lambda - 1| < \delta_3$ and $t \in [0, 2\pi\lambda]$

$$x_1(t, c, \varepsilon) \neq 0$$

For definiteness, assume that for all t,

$$x_1(t, c, \varepsilon) > 0$$

We consider the projection of the orbit of $x(t, c, \varepsilon)$ on the (x_1, x_2)-plane, that is, the point set described by

$$P_{n-r}[x(t, c, \varepsilon)] = (x_1(t, c, \varepsilon), x_2(t, c, \varepsilon)) \qquad t \in [0, 2\pi\lambda] \qquad (8.53)$$

This point set is a closed curve.

By hypothesis, there exist $\tilde{c}_1$ and t_0 such that

$$\max_t \ x_1(t, c, \varepsilon) = x_1(t_0, c, \varepsilon) = \tilde{c}_1 > 0$$

Let

$$x_2(t_0, c, \varepsilon) = \tilde{c}_2$$

Reparameterize $x(t, c, \varepsilon)$ so that $t_0 = 0$. Then the first two components of the initial condition c are $c_1 = \tilde{c}_1$ and $c_2 = \tilde{c}_2$ and

$$P_{n-r}x(\pi, c, \varepsilon) = e^{\pi(C + \varepsilon H(\varepsilon))} \begin{bmatrix} c_1 \\ c_2 \end{bmatrix} + o(|c|)$$

$$= \begin{bmatrix} \cos \pi & \sin \pi \\ -\sin \pi & \cos \pi \end{bmatrix} \begin{bmatrix} e^{\pi \varepsilon h(\varepsilon)} c_1 \\ e^{\pi \varepsilon h(\varepsilon)} c_2 \end{bmatrix} + o(|c|)$$

$$= \begin{bmatrix} -e^{\pi \varepsilon h(\varepsilon)} c_1 \\ -e^{\pi \varepsilon h(\varepsilon)} c_2 \end{bmatrix} + o(|c|)$$

Hence if $|c|$ and $|\varepsilon|$ are small enough, then $P_{n-r}x(\pi, c, \varepsilon)$ is very close to

$$\begin{bmatrix} -c_1 \\ -c_2 \end{bmatrix}$$

Since $c_1 = \tilde{c}_1 > 0$, then $P_1[x(\pi, c, \varepsilon)]$ is close to $-c_1 < 0$. This is a contradiction to the hypothesis. This completes the proof of Lemma 8.3. $\qquad\square$

Thus we have completed the proof of:

Hopf Bifurcation Theorem

Suppose the equation

$$\frac{dx}{dt} = Ax + \varepsilon L(\varepsilon)x + f(x, \varepsilon) \tag{8.38}$$

where

$$A = \begin{bmatrix} 0 & 1 & \\ 1 & 0 & \\ & & D \end{bmatrix}$$

where C is an $(n-2) \times (n-2)$ constant matrix each of whose eigenvalues has nonzero real part, and $L(\varepsilon)$ is an $n \times n$ matrix each of whose entries is a differentiable function of ε for $\varepsilon \in I$, an interval on R with midpoint 0, such that

$$L(\varepsilon) = \begin{bmatrix} h(\varepsilon) & 0 & \\ 0 & h(\varepsilon) & \\ & & \mathcal{D}(\varepsilon) \end{bmatrix}$$

where $h(0) = 1$ and $\mathcal{D}(\varepsilon)$ is a differentiable $(n-2) \times (n-2)$ matrix. Also suppose that $f(x, \varepsilon)$ has continuous second derivatives in x and ε for $(x, \varepsilon) \in R^n \times I$ and that

$$|f(x, \varepsilon)| = o(|x|)$$

uniformly in ε, that is, for $(x, \varepsilon) \in R^n \times I$

$$|f(x, \varepsilon)| < \left\{ \eta(1 \times 1) \right\} |x|$$

where η is a continuous monotonic increasing function with domain $\{r \in R / r \geq 0\}$ and $\eta(0) = 0$.

Conclusion: There is an open interval J with midpoint 0 and a neighborhood N in R^n of $(1, 0, 0, \ldots, 0)$ such that for each $c_2 \in J$, there exists a unique periodic solution

$$x(t, \lambda, \varepsilon)$$

of period $2\pi\lambda$ such that $c_1 = 0$ and

$$(\lambda, \varepsilon, c_3, \ldots, c_n) \in N.$$

Moreover, any periodic solution satisfying the conditions that c_1 is sufficiently small and $(\lambda, \varepsilon, c_3, \ldots, c_n) \in N$ can be reparameterized so that $c_1 = 0$. Thus the conclusion can be formulated as: If $|c_1|$ is sufficiently small then for each $c_2 \in J$, there exists unique periodic solution

$$x(t, \lambda, c_1, c_2, \ldots, c_n)$$

of period $2\pi\lambda$ such that

$$(\lambda, \varepsilon, c_3, \ldots, c_n) \in N$$

Remark

In the proof, we obtain ε as a function of c_2. If $\varepsilon(c_2) \geq 0$ for all $c_2 \in J$, then there exists no periodic solution for $\varepsilon < 0$, and we have the case which describes the intuitive picture of periodic solutions appearing as ε increases above zero.

Chapter 9

Using the Averaging Method:
An Introduction

Introduction

The principle of averaging and the averaging method are somewhat ill-defined terms used to refer to certain approximating methods which originated in work of Clairant, Lagrange, and Laplace. However, although these techniques were widely developed and used, a rigorous treatment was not undertaken until the twentieth century. For a history of the subject, see Arnold [1988, Vol. 3, Chapter 5], Sanders and Verhulst [1985], and other references in these books.

Here we use the term, *"averaging method"* to mean the mathematical theory developed by Kryloff and Bogoliuloff [1937] and Bogoliublov and Mitropolskii [1961]. The basic idea is to consider a differential equation which is called the standard form, then determine an "averaged" version of this standard form, and finally show that the solutions of the averaged equation yield information about the solutions of the standard form.

The standard form is the n-dimensional equation

$$\frac{dx}{dt} = \varepsilon f(t, x, \varepsilon) \tag{9.1}$$

where ε is a small parameter and f is such that there exists

$$F(x) = \lim_{T \to \infty} \int_0^T f(t, x, 0)dt$$

The averaged equation of (9.1) is

$$\frac{dx}{dt} = \varepsilon F(x) \tag{9.2}$$

A rather natural question arises here: Because of the factor ε which multiplies f, equation (9.1) seems to represent a peculiarly limited class of equations. Why would anyone want to study the solutions of (9.1)? The answer is that large classes of important problems can be transformed into equations in standard form: problems in classical mechanics (see Bogoliubov and Mitropolskii [1961] and Arnold [1988, Vol. 3, Chapter 5]), existence of periodic solutions and almost periodic solutions (see Bogoliubov and Mitropolskii [1961], Hale [1963, 1969], Malkin [1959], and

Roseau [1966]) and approximations to solutions of equations over finite and infinite intervals of the t-axis. See Bogoliubov and Mitropolskii [1961] and Malkin [1959].

In this chapter, we illustrate two uses of the averaging method, first in the study of periodic solutions and then almost periodic solutions. The great power of the method is revealed in this second application which provides, as we shall see, the solution of a "small divisors" problem.

The treatment in this chapter is intended merely as a brief introduction to a large subject. Many important topics will go unmentioned, and we will omit all proofs except for the study of periodic solutions. (A more complete discussion would be lengthy; the theory of almost periodic functions is a subject in itself and some of the proofs we omit require considerable "hard" analysis.)

Periodic Solutions

As pointed out before, large classes of important problems can be transformed into equations in standard form. As an example of such a transformation we show how a problem about periodic solutions can be recast in this form. We consider

$$\frac{dx}{dt} = Ax + \varepsilon f(t, x, \varepsilon) \tag{9.3}$$

where A is a constant $n \times n$ matrix, and

$$f: R \times R^n \times I \to R^n$$

where I is an interval on R with midpoint at 0, and for all t, x, ε

$$f(t + T, x, \varepsilon) = f(t.x.\varepsilon)$$

For brevity we consider the simplest resonance case, that is, we assume that

$$e^{TA} - I = 0.$$

Let y be defined by

$$x = e^{tA} y$$

Substituting in (9.3), we have

$$A e^{tA} y + e^{tA} \frac{dy}{dt} = A e^{tA} y + \varepsilon f(t, e^{tA} y, \varepsilon)$$

Multiplication by e^{-tA} yields the standard form

$$\frac{dy}{dt} = \varepsilon e^{-tA} f(t, e^{tA} y, \varepsilon) \tag{9.4}$$

We seek a solution $y(t, \varepsilon)$ of period T for equation (9.4). (Since e^{tA} has period T, $x(t, \varepsilon) = e^{tA} y(t, \varepsilon)$ is a solution of (9.3) of period T.)

The corresponding averaged equation for equation (9.4) is

$$\frac{dy}{dt} = \varepsilon \int_0^T e^{-sA} f(s, e^{sA} y, 0) \, ds \tag{9.5}$$

Now suppose that y_0 is an equilibrium point of (9.5), that is, suppose

$$\int_0^T e^{-sA} f(s, e^{sA} y_0, 0) \, ds = 0$$

and suppose further that y_0 is a nondegenerate equilibrium point, that is, the Jacobian at y_0 of

$$\int_0^T e^{-sA} f(s, e^{sA} y, 0) \, ds$$

is nonzero. Then it is proved in studies of the averaging method that if $|\varepsilon|$ is sufficiently small, equation (9.4) (and hence equation (9.3)) has a solution of period T.

However, it is easy to see that the hypothesis about y_0 is the Poincaré condition for equation (9.3) which was studied in Chapter 7, Part A. See especially the first example of resonance. Also the further analysis in Chapter 7 about the number of periodic solutions and their stability properties is applicable here. Thus use of the averaging method to study periodic solutions is logically contained in the Poincaré method. The Poincaré method has the advantage of being more direct: The equation which is the periodicity condition is analyzed. The averaging method involves recasting the given equation into a standard form and then proving that nondegenerate equilibrium points of the averaged equation give rise to periodic solutions of the standard form.

Almost Periodic Solutions

Almost periodic functions arise easily and naturally in the study of oscillations. In the studies in Chapters 7 and 8, the perturbation term or forcing term is periodic, but now suppose there are two forcing terms which have different periods, say T_1 and T_2. If T_1 and T_2 are rationally related, for example, if

$$T_2 = \frac{m}{n} T_1$$

where m, n are integers, then the sum of the two forcing terms is periodic of period $nT_2 = mT_1$, and we could apply the results in Chapter 7 to seek solutions of period mT_1. But if T_1, T_2 are not rationally related, then the sum of the two forcing terms is said to be quasiperiodic. In this case, we cannot expect to obtain a periodic solution of the perturbed equation. However, one would expect intuitively to obtain a solution with some kind of oscillatory properties. The property of quasiperiodicity is a special case of almost periodicity (see Chapter 6), and it turns out that the generalization to almost periodicity yields an extensive and natural generalization of the study of periodic solutions.

However, developing the theory of almost periodic solutions is much more difficult than finding periodic solutions. The first reason for this is that the necessary and sufficient condition for periodicity used in Chapters 7 and 8 (which consists of solving a system of n real equations in n real variables) has no counterpart in the study of almost periodicity. Almost periodicity is too complicated a condition to be characterized by a system of n equations in n variables. Indeed, a crucial step in establishing the existence of almost periodic solutions of nonlinear equation is the use of successive approximations to solve a functional equation. That is, the problem of studying solutions of equations in R^n in the search for periodic solutions is replaced by the problem of studying solutions of functional equations (equations in an infinite-dimensional space) in the search for almost periodic solutions.

The second reason for the difficulty in the theory of almost periodic solutions appears in attempts to use power series to represent the almost periodic solution. A classical way to study periodic solutions of

$$\frac{dx}{dt} = f(t, x, \varepsilon)$$

for the case in which f has period T in t and f is analytic is to represent the sought-after periodic solution as a power series in ε in which the coefficients are periodic in t. If a formal procedure for determining the coefficients in the power series (such as equating coefficients of like powers of ε) can be applied and if the resulting power series converges, then the power series is a periodic solution of the equation. Or if a periodic solution is known to exist, then it follows that the formally obtained power series converges to a periodic solution. (See Coddington and Levinson [1955, pp. 350–351].)

It is natural to try to use this method in the search for almost periodic solutions. But a serious difficulty intervenes: Even in simple cases, the formally obtained power series diverges. This is an example of the problem of "small divisors." (The power series does not converge because the terms have small denominators.) See Malkin [1959, pp. 269–273] for an example of such small divisors.

Our purpose here is simply to state some of the major results concerning almost periodic solutions. (Basic properties of almost periodic functions are summarized in Chapter 6.)

Theorem 9.1   *Given the equation*

$$\frac{dx}{dt} = Ax + f(t) \tag{9.6}$$

in which

(1) The eigenvalues of the $n \times n$ constant matrix A all have nonzero real parts.

(2) $f(t)$ is an almost periodic function.

Then equation (9.6) has a unique almost periodic solution $x(t)$ and

$$\text{lub}_t |x(t)| < \alpha \, \text{lub}_t |f(t)|$$

where α is a constant which depends only on A.

For a proof of Theorem 9.1, see Roseau [1966], Malkin [1959], and Hale [1963].

Now we state the central result for nonlinear systems with a small parameter which is due to Kryloff and Bogoliuboff.

Theorem 9.2 *Consider the system of differential equations*

$$\frac{dx}{dt} = Ax + \varepsilon f(x, y, t, \varepsilon)$$
$$\frac{dy}{dt} = \varepsilon g(x, y, t, \varepsilon) \tag{9.7}$$

where $x \in R^n$, $y \in R^n$, $\mu \in I$, where I is an interval in R with midpoint 0, and A is a constant real $m \times m$ matrix, and we impose the following hypotheses:

(i) f, g are almost periodic for each fixed (x, y, ε).

(ii) f, g have continuous partial derivatives on an open set $U^m \times U^n \times R \times I \subset R^m \times R^n \times R \times R$.

(iii) There exists $b \in U^n$ such that

$$\lim_{T \to \infty} \frac{1}{T} \int_0^T g(0, b, t, 0) dt = 0$$

(iv) Let

$$S = \lim_{T \to \infty} \frac{1}{T} \int_0^T \frac{\partial g}{\partial y}(0, b, t, 0) dt$$

We require that all the eigenvalues of S and all the eigenvalues of S have nonzero real parts.

Conclusion: If $|\varepsilon|$ is sufficiently small, then there exists an almost periodic solution of (9.7), $(x(t, \varepsilon), y(t, \varepsilon))$ such that

$$\lim_{\mu \to 0} (x(t, \varepsilon), y(t, \varepsilon)) = (0, b)$$

and the convergence is uniform in t.

There are a number of proofs of versions of the theorem. See Bogoliubov and Mitropolskii [1961, Chapter 6], Hale [1963], and, for a particularly clear and succinct account, Roseau [1966, Chapter 16].

The case in which eigenvalues of matrix S may be zero real parts has been treated by Malkin [1959] and is discussed by Roseau [1966].

Other Uses of the Averaging Method

We have indicated that the averaging method is used to study existence of periodic and almost periodic solutions. But the averaging method can be used to study other wide classes of problems. An especially important study is how to compare the solution of the standard form and the solution of the averaged equation if $|\varepsilon|$ is small enough and if they have the same initial values (or initial values which are very close together). There are two kinds of comparison that can be made. The first is an estimate of the difference between the solutions over a finite interval of the t-axis.

Definition Let $D \subset R^n$ and $r > 0$ be given. Then the *r-neighborhood of D*, denoted by $\mathcal{N}_\rho(D)$, is the set

$$\left\{ p \in R^n \, \Big/ \, \underset{g \in D}{g\ell b} \, |p - q| < r \right\}$$

Theorem 9.3 *Suppose that $f(t, x)$ satisfies the following conditions:*

(i) *There exists an open set $D \subset R^n$ such that $f(t, x)$ is bounded on D, that is, there exists $M > 0$ such that if $t \geq 0$ and $x \in D$, then*

$$|f(t, x)| < M$$

(ii) *There exists $m > 0$ such that if $t \geq 0$ and $x^{(1)}$, $x^{(2)}$ are elements of D, then*

$$|f(t, x^{(1)}) - f(t, x^{(2)})| < m|x^{(1)} - x^{(2)}|$$

(iii) *If $x \in D$, then there exists*

$$\lim_{T \to 0} \int_0^T f(t, x)dx = F(x)$$

Conclusion: Given $\rho > 0$, $\eta > 0$, and $L > 0$ (where ρ and η are small and L is large), then there exists $\varepsilon_0 > 0$ such that if

$$0 < \varepsilon < \varepsilon_0$$

and if $\bar{x}(t)$ is a solution of

$$\frac{dx}{dt} = \varepsilon F(x)$$

which is defined for all $t > 0$ and such that

$$\{\bar{x}(t)/t > 0\} \subset \mathcal{N}_\rho(D)$$

and if $x(t)$ is the solution of

$$\frac{dx}{dt} = \varepsilon f(t, x)$$

such that

$$x(0) = \bar{x}(0)$$

then it follows that for

$$t \in \left(0, \frac{L}{\varepsilon} \right)$$

we have

$$|x(t) - \bar{x}(t)| < \eta$$

Proof   See Bogoliubov and Mitropolskii [1961, pp. 429–435] and Roseau [1966, pp. 102–106].

The problem of comparing the solutions of the standard form and the averaged equation over an infinite interval on the *t*-axis is considerably more difficult and requires much more effort. Indeed, the remainder of Bogoliubov and Mitropolskii [1961, Chapter 6] is devoted to the problem. See also Roseau [1966].

For later work on this problem, see Mitropolskii and Homa [1983].   ⬜

Appendix

Ascoli's Theorem

First we remind the reader of the following definition:

Definition   Suppose $\{f_n\}$ is a sequence of functions each defined on an interval $[a, b]$. Sequence $\{f_n\}$ is *equicontinuous at a point* $x_0 \in [a, b]$ if: given $\varepsilon > 0$ there exists $\delta > 0$ such that if $|x - x_0| < \delta$ and $x \in [a, b]$, then for each n

$$|f_n(x) - f_n(x_0)| < \varepsilon.$$

The sequence $\{f_n\}$ is *equicontinuous on* $[a, b]$ if $\{f_n\}$ is equicontinuous at each point of $[a, b]$.

Ascoli's Theorem

Let $\{f_n\}$ be an equicontinuous sequence of funcitons on $[a, b]$, and suppose there exists $M > 0$ such that for all n and for all $x \in [a, b]$,

$$|f_n(x)| < M$$

Then $\{f_n\}$ contains a subsequence which is uniformly convergent on $[a, b]$.

This is the simplest form of Ascoli's Theorem and it is sufficient for our purposes. For a more general version of the theorem, see, for example, Royden [1968, p. 179].

Principle of Contraction Mappings

Definition   Let S be an arbitrary metric space with metric d. A mapping

$$F : S \to S$$

is a *contraction mapping* if there exists a number $k \in (0, 1)$ such that for all $x, y \in S$,

$$d(F(x), F(y)) < kd(x, y)$$

Definition A metric space S is *complete* if each Cauchy sequence in S has a limit. That is, if $\{x_n\} \subset S$ is such that for $\varepsilon > 0$ there is a number $N(\varepsilon)$ so that if $m, n > N(\varepsilon)$, we have

$$d(x_m, x_n) < \varepsilon$$

then there exists an $x_0 \in S$ such that

$$\lim_{n \to \infty} d(x_n, x_0) = 0$$

Principle of Contraction Mappings: If F is a contraction mapping from a complete metric space into itself, then F has a unique fixed point.

Proof Let x be an arbitrary point of S, and let

$$x_n = F^n(x)$$

where $F^n(x)$ is defined inductively by: $F^1(x) = F(x)$ and $F^n(x) = F[F^{n-1}(x)]$, $n = 2, 3, \ldots$. Then $\{x_n\}$ is a Cauchy sequence because

$$
\begin{aligned}
d(x_m, x_n) &= (F^m(x), F^n(x)) \\
&< k^m d(x, F^{n-m}(x)) \\
&\quad \text{(where, for definiteness, we assume } n > m) \\
&< k^m \{d(x, x_1) + d(x_1, x_2) + \cdots + d(x^{n-m-1}, x^{n-m})\} \\
&< k^m \{d(x, x_1) + k d(x, x_1) + \cdots + k^{n-m-1} d(x, x_1)\} \\
&< k^m \frac{1}{1-k} d(x, x_1)
\end{aligned}
$$

Since $k \in (0, 1)$, then if m is sufficiently large, the quantity

$$\frac{k^m}{1-k} d(x, x_1)$$

can be made arbitrarily small.

Since S is complete, the Cauchy sequence $\{x_n\}$ has a limit x_0. Then x_0 is a fixed point because

$$
\begin{aligned}
F(x_0) &= F\left[\lim_{n \to \infty} F^n x \right] \\
&= \lim F^{n+1}(x) \\
&= \lim_{n \to \infty} x_{n+1} \\
&= x_0
\end{aligned}
$$

To prove the uniqueness of the fixed point x_0, suppose that there exist two fixed points x and y. Then

$$d(x, y) = d(Fx, Fy) < k d(x, y)$$

Since $k < 1$, then $d(x, y) = 0$ and $x = y$. $\qquad\qquad \square$

The Weierstrass Preparation Theorem

The classical Weierstrass preparation theorem, which is all that is used in this book, may be stated as follows:

Weierstrass Preparation Theorem. *Let $f(x_1, \ldots, x_n)$ be a real or complex power series in $x_1, \ldots, x_n$ which converges in a neighborhood of the origin and is such that*

$$f \text{ is zero at the origin and } f(x_1, 0, \ldots, 0) \neq 0$$

Let p be the least degree in x_1 of $f(x_1, 0, \ldots, 0)$. Then

$$f(x_1, \ldots, x_n) = F(x_1, x_2, \ldots, x_n)E(x_1, \ldots, x_n)$$

where $E(x_1, \ldots, x_n)$ is a power series such that

$$E(0, \ldots, 0) \neq 0$$

and $F(x_1, \ldots, x_n)$ is polynomial of degree p in x_1, that is,

$$F(x_1, \ldots, x_n) = x_1^p + a_1(x_2, \ldots, x_n)x_1^{p-1} + \cdots + a_p(x_2, \ldots, x_n)$$

and each coefficient $a_r(x_2, \ldots, x_n)$, $r = 1, \ldots, p$ is such that

$$a_r(0, \ldots, 0) = 0$$

and $a_r(x_2, \ldots, x_n)$ is a power series in $x_2, \ldots, x_n$.

For a proof, see Fuks [1962]. A more general version of the theorem for smooth functions, that is, C^∞ functions, is given by Golubitsky and Guillemin [1973, Chapter 4].

Topological Degree

Topological degree is a geometric notion which has proved extremely useful in qualitative studies of various kinds of nonlinear functional equations: ordinary differential equations, partial differential equations, integral equations, and integro-differential equations. Although degree theory can be used to obtain only qualitative results of a general nature (existence theorems and some limited results about stability), use of degree theory is an important method in analysis because the results obtained are significant and have not so far been obtained by other methods. Here we will merely sketch the definition of topological degree of mappings in Euclidean spaces (the Brouwer degree), and normed linear spaces (the Leray-Schauder degree) and list those properties of degree which will be useful for the studies in this book.

Our sketch of the definition of degree is based on the definition given by Nagumo [1951]. For a more general and elegant version of the definition, see Nirenberg [1974].

We use the Nagumo definition because it requires the least technical language and is suggestive of methods for computing the degree of particular classes of mappings. Descriptions of the use of topological degree in other studies of functional equations may be found in Cronin [1964], Krasnosel'skii [1964], Mawhin and Rouche [1973], and Nirenberg [1974]. Definitions of topological degree for more general classes of mappings in normed linear spaces have been given by Nussbaum [1971, 1972, 1974] and Browder and Petryshyn [1969].

Let $\bar{U}$ be the closure of a bounded open set U in R^n and let f be a continuous mapping with domain $\bar{U}$ and range contained in R^n, that is,

$$f : \bar{U} \to R^n$$

Mapping f can be described by

$$f : (x_1, \ldots, x_n) \to (f_1(x_1, \ldots, x_n), \ldots, f_n(x_1, \ldots, x_n))$$

where $f_1, \ldots, f_n$ are real-valued functions. We will assume that the partial derivatives

$$\frac{\partial f_i}{\partial x_j} \qquad (i, j = 1, \ldots, n)$$

exist and are continuous at each point of U. Let $p \in R^n$ be such that

$$p \notin f(\bar{U} - U)$$

that is, p is not an image point of the boundary of U. Suppose that $q \in U$ is such that $f(q) = p$. We will say that q is a point of multiplicity $+1$ if

$$\det \left[\frac{\partial f_i}{\partial x_j} (q) \right] > 0$$

that is, the Jacobian of f at q is positive, and q is a point of multiplicity -1 if

$$\det \left[\frac{\partial f_i}{\partial x_j} \right] (q) < 0$$

Let us assume that if $q \in f^{-1}(p)$, the Jacobian of f at q is nonzero. It follows easily then that $f^{-1}(p)$ is finite. The topological degree or Brouwer degree of f at p and relative to the set $\bar{U}$, which we denote by $\deg[f, \bar{U}, p]$, is defined to be:

$$\deg[f, \bar{U}, p] = [\text{Number of points } q \text{ in } f^{-1}(p) \text{ of multiplicity } +1]$$
$$- [\text{Number of points } q \text{ in } f^{-1}(p) \text{ of multiplicity } -1]$$

That is, to determine $\deg[f, \bar{U}, p]$ one counts the number of points in $f^{-1}(p)$ that have multiplicity $+1$ and subtracts from it the number of points in $f^{-1}(p)$ that have multiplicity -1.

This definition has two serious deficiencies. First, it is not generally true that the Jacobian of f is nonzero at each point of $f^{-1}(p)$. Indeed, the set $f^{-1}(p)$ is not, in general, finite. Also, it is desirable (for aesthetic and practical reasons) to define deg[$f, \bar{U}, p$] for mappings f which are merely continuous rather than restricting our definition to differentiable mappings.

It turns out that if $\bar{f}$ is a continuous mapping from $\bar{U}$ into R^n and if $p \notin \bar{f}(\bar{U} - U)$, then $\bar{f}$ can be approximated arbitrarily closely by a mapping f to which our definition is applicable, that is, a mapping f which is differentiable and is such that if $p \notin f(\bar{U}, \bar{U})$, the set $f^{-1}(p)$ is finite, and

$$\det \left[\frac{\partial f_i}{\partial x_j} (q) \right] \neq 0$$

for each $q \in f^{-1}(p)$. Moreover, if $f^{(1)}$ and $f^{(2)}$ are two such mappings which approximate $\bar{f}$ sufficiently well, then

$$\deg \left[f^{(1)}, \bar{U}, p \right] = \deg \left[f^{(2)}, \bar{U}, p \right]$$

Hence if f is a differentiable mapping such that $f^{-1}(p)$ is finite and

$$\det \left| \frac{\partial f_i}{\partial x_j} (q) \right| \neq 0$$

for each $q \in f^{-1}(p)$ and if f approximates $\bar{f}$ sufficiently well, we may define the topological degree or Brouwer degree of $\bar{f}$ at p and relative to $\bar{U}$, denoted by deg[$\bar{f}, \bar{U}, p$], as

$$\deg[\bar{f}, \bar{U}, p] = \deg[f, \bar{U}, p]$$

Note that we have omitted entirely the question of: (i) proving that a continuous mapping $\bar{f}$ can be approximated by a differentiable map with the desired properties; (ii) proving that if $f^{(1)}$ and $f^{(2)}$ are two such (sufficiently fine) approximations, then

$$\deg[f^{(1)}, \bar{U}, p] = \deg[f^{(2)}, \bar{U}, p]$$

Such proofs are well-known (Nagumo [1951], Alexandroff and Hopf [1935], Cronin [1964]) but they do not seem to provide any help in applying topological degree to problems in functional equations. Hence, we omit them. The proofs use Sard's theorem which we state later when we need it for showing how the degree measures the number of solutions.

We add just one remark concerning the geometric meaning of the condition

$$\det \left[\frac{\partial f_i}{\partial x_j} (q) \right] > 0$$

Since f is differentiable, then for points $\bar{q}$ in a neighborhood of q, the mapping

$$\bar{q} - q \rightarrow f(\bar{q}) - f(q)$$

can be approximated by the linear mapping described by the matrix

$$\mathcal{M} = \left[\frac{\partial f_i}{\partial x_j}(q) \right]$$

The condition

$$\det \left[\frac{\partial f_i}{\partial x_j}(q) \right] > 0$$

implies that the mapping described by $\mathcal{M}$ is orientation-preserving. (In the two-dimensional case, if $\mathcal{M}$ is applied to a triangle whose vertices are oriented in counterclockwise order, the images of the vertices will also be oriented in counterclockwise order. A complete description of "orientation-preserving," which we will not give, requires a rigorous definition of "counterclockwise" and an analogous definition in the n-dimensional case where $n > 2$. See Alexandroff and Hopf [1935] and Cronin [1964].)

It is occasionally convenient to use the following special case of the degree. Suppose that mapping f takes $\bar{U}$ into R^n and suppose that $p \in R^n$ is such that $p \notin f(\bar{U} - U)$ and $f^{-1}(p)$ is a single point $q \in U$. From the definition of degree, it follows that if V is any open set in R^n such that $q \in V$ and $\bar{V} \subset \bar{U}$, then

$$\deg[f, \bar{V}, p] = \deg[f, \bar{U}, p]$$

Definition   The *topological index of f at q is*

$$\deg[f, \bar{V}, p]$$

where V is any open set in R^n such that $q \in V$ and $\bar{V} \subset \bar{U}$.

Example 1.   *If I is the identity mapping from R^n into R^n and U is a bounded open set, then*

$$\deg[I, \bar{U}, p] = +1 \quad \text{if} \quad p \in U$$
$$\deg[I, \bar{U}, p] = 0 \quad \text{if} \quad p \in \bar{U}^c$$

This statement follows at once from the definition of Brouwer degree.

Example 2.   *If A is a linear homogeneous mapping from R^n into R^n and A is described by a nonsingular matrix $\mathcal{M}$, then*

$$\deg[A, \bar{U}, p] = \text{sgn} \det \mathcal{M} \quad \text{if } p \in A(U)$$
$$\deg[A, \bar{U}, p] = 0 \quad \quad \text{if } p \in [\overline{A(U)}]^c$$

This statement also follows at once from the definition of Brouwer degree.

Example 3. *If g is a mapping from $\bar{V}$, the closure of a bounded open set V in R^q, into R^q, that is,*

$$g(x_1 \ldots, x_q) \rightarrow (g_1(x_1 \ldots, x_q), \ldots, g_q(x_1, \ldots, x_q))$$

and if f is a mapping from $\bar{U}$, the closure of a bounded open set in R^n, into R^n, where $n > q$, and $\bar{U} \cap R^q = \bar{V}$ and f is defined by

$$f : (x_1, \ldots, x_q, \ldots, x_n) \rightarrow (g_1(x_1, \ldots, x_q), \ldots, g_q(x_1, \ldots, x_q), x_{q+1}, \ldots, x_n)$$

Then if

$$p = \left(x_1^0, \ldots, x_q^0, x_{q+1}^0, \ldots, x_n^0\right)$$

and $p \notin f(\bar{U} - U)$ and if $p_0 = (x_1^0, \ldots, x_q^0)$ and $p_0 \notin g(\bar{V} - V)$, then

$$\deg[f, U, p] = \deg[g, \bar{V}, p_0]$$

The proof of this result follows from a careful examination of the definition of degree. The mapping f is called a *suspension* of mapping g.

Example 4. *If the mapping f is a constant mapping, that is, for all $q \in \bar{U}$, $f(q) = \tilde{p}$, then if $p \neq \tilde{p}$,*

$$\deg[f, \bar{U}, p] = 0$$

Now we describe some of the properties of topological degree which make it useful in analysis. The most fundamental of these properties, which is indeed at the basis of applications, is:

Property 1. If $\deg[f, \bar{U}, p] \neq 0$, then the set $f^{-1}(p)$ is nonempty. In other words, if $\deg[f, \bar{U}, p] \neq 0$, the equation

$$f(x) = p$$

has a solution $x \in U$.

The proof of Property 1 for differentiable approximating mapping f is trivial. To extend the proof to all continuous f involves only arguments of a routine nature.

Property 1 shows that the problem of solving an equation can be translated into the problem of computing the degree of a mapping. However, if we look at the definition of $\deg[f, \bar{U}, p]$ and at Property 1, there seems to be a kind of circularity involved. In order to compute $\deg[f, \bar{U}, p]$, we look at the set $f^{-1}(p)$. Then if $\deg[f, \bar{U}, p]$ is shown to be nonzero, we conclude from Property 1 that the equation

$$f(x) = p$$

has a solution, that is, that the set $f^{-1}(p)$ is nonempty. It seems that in order to establish the mere fact that $f^{-1}(p)$ is nonempty, we must study in some detail the properties of the points in $f^{-1}(p)$! Such a procedure is clearly unreasonable. Actually, we proceed by various indirect methods to show that $\deg[f, \bar{U}, p] \neq 0$ and then invoke Property 1 to conclude that the equation

$$f(x) = p$$

has a solution.

The basis for the most important method for computing $\deg[f, \bar{U}, p]$ is the fact that $\deg[f, \bar{U}, p]$ is constant under a continuous deformation of f or, in more standard terminology, that $\deg[f, \bar{U}, p]$ is invariant under homotopy. The precise statement is:

Property 2. Suppose that F denotes a continuous mapping from $\bar{U} \times [0, 1]$ into R^n. If $t \in [0, 1]$, the mapping $F/\bar{U} \times \{t\}$ will be denoted by F_t. Then the mapping F_t can be identified with the mapping g from $\bar{U}$ into R^n defined by:

$$g : q \to F_t(q)$$

Let $p(\tau)$ be a continuous curve, that is, a continuous mapping from the interval $0 \leq \tau \leq 1$ into R^n such that for all $t \in [0, 1]$ and all $\tau \in [0, 1]$,

$$p(\tau) \notin F_t(\bar{U} - U)$$

Then for each $t \in [0, 1]$ and each $\tau \in [0, 1]$, the topological degree

$$\deg[f_t, \bar{U}, p(\tau)]$$

is defined, and for all $t \in [0, 1]$ and $\tau \in [0, 1]$, it has the same value.

The proof of Property 2 is not particularly difficult. But since it seems to shed no light on how to go about applying degree theory in analysis, we omit it.

Definition   The mappings F_0 and F_1 in Property 2, regarded as mappings from $\bar{U}$ into R^n, are said to be *homotopic* in

$$R^n - \{p(\tau) \mid \tau \in [0, 1]\}.$$

Mapping F is a *homotopy* in $R^n - \{p(\tau) \mid \tau \in [0, 1]\}$.

Property 2 is applied in the following way. Suppose we wish to compute $\deg[f, \bar{U}, p]$. We look for a mapping F from $\bar{U} \times [0, 1]$ into R^n such that

(i) $F_0 = f$;

(ii) $p \notin F_t(\bar{U} - U)$ for all $t \in [0, 1]$;

(iii) $\deg[F_1, \bar{U}, p]$ can be easily computed.

It follows from Property 2 that

$$\deg[f, \bar{U}, p] = \deg[F_0, \bar{U}, p] = \deg[F_1, \bar{u}, p]$$

Since we have so far computed the degrees of only a couple of very simple mappings, we have little hope that such a search for a mapping F would be successful. Finding such a mapping F is, in general, very difficult. Nevertheless, the mapping F can be found in a large number of cases which yield significant results in analysis. Now we illustrate the use of Property 2 with some examples.

Example 5. *Suppose that $f(\bar{U})$ is contained in an m-dimensional subspace R^m of R^n where $m < n$. Then if $p \in R^n$ is such that $p \notin f(\bar{U} - U)$,*

$$\deg[f, \bar{U}, p] = 0$$

Proof Since $p \notin f(\bar{U} - U)$, there is a connected neighborhood N of p such that

$$N \cap f(\bar{U} - U) = \emptyset$$

But if $\bar{p} \in N$, there is a continuous curve in N which joins p and $\bar{p}$. Hence, by Property 2, if $\bar{p} \in N$, then

$$\deg[f, \bar{U}, p] = \deg[f, \bar{U}, \bar{p}]$$

But there is a point $\tilde{p} \in N$ such that $\tilde{p} \notin R^m$ and hence such that $\tilde{p} \notin f(\bar{U})$. Since $\tilde{p} \notin f(U)$, then by Property 1, it follows that

$$\deg[f, \bar{U}, \tilde{p}] = 0$$

Thus

$$\deg[f, \bar{U}, p] = \deg[f, \bar{U}, \tilde{p}] = 0$$

$\Box$

Example 6. *If the mapping*
$$f : \bar{U} \to R^n$$

is described by

$$f : (x_1, \ldots, x_n) \to (f_1(x_1, \ldots, x_n), \ldots, f_n(x_1, \ldots, x_n))$$

and one of the functions, say $f_j(x_1, \ldots, x_n)$, is nonnegative or nonpositive, and if $0 \notin f(\bar{U} - U)$, then
$$\deg[f, \bar{U}, 0] = 0$$

Proof If $f_j(x_1, \ldots, x_n)$ is nonnegative, then if $(\bar{x}_1, \ldots, \bar{x}_n)$ is such that $\bar{x}_j < 0$,

$$(\bar{x}_1, \ldots, \bar{x}_n) \notin f(\bar{U})$$

The remainder of the proof is very similar to that for Example 5.

$\Box$

Example 7. *Let f be the mapping from R^2 into R^2 described in terms of a complex variable by:*
$$f : z \to z^n$$

where n is a positive integer. Let U be a bounded open set in R^2 such that $0 \in U$. Then

$$\deg[f, \bar{U}, 0] = n$$

Proof   Let $z_0 \neq 0$ be sufficiently close to 0 so that $f^{-1}(z_0) \subset U$. From Property 2, it follows that

$$\deg[f, \bar{U}, 0] = \deg[f, \bar{U}, z_0]$$

(To apply Property 2, let the set $\{p(\tau) \mid \tau \in [0, 1]\}$ be the line segment joining 0 and z_0.) To compute $\deg[f, U, z_0]$, note that $f^{-1}(z_0)$ is the set of nth roots of z_0 which is a set of n distinct complex numbers $w_1, \dots, w_n$ each of which is nonzero. To find the sign of the Jacobian of f at these points, we notice that if

$$f(z) = u(x, y) + iv(x, y)$$

then the determinant whose sign must be found is

$$\det \begin{bmatrix} u_x & v_x \\ u_y & v_y \end{bmatrix} = u_x v_y - v_x u_y$$

But by the Cauchy-Riemann equations (which certainly hold for the simple analytic function $f(z) = z^n$)

$$u_x v_y - v_x u_y = u_x^2 + v_x^2$$

But

$$u_x^2 + v_x^2 = |f'(x)|^2 = |nz^{n-1}|^2$$

and for $j = 1, \dots, n$

$$|nw_i^{n-1}|^2 > 0$$

Hence at each point $w_j (j = 1, \dots, n)$ the Jacobian is positive, which completes the proof. (The kind of argument used in this example is applicable to a wide class of mappings as we will show later.)   ∎

We prove now a generalization of a well-known theorem in complex variable.

Rouché's Theorem *Let $\bar{U}$ be the closure of a bounded open set U in R^n, and let f and g be continuous mappings from $\bar{U}$ into R^n. Suppose that $p \in f(\bar{U} - U)$ and that for all $x \in \bar{U} - U$,*

$$|g(x) - f(x)| < |f(x) - p|$$

Then

$$\deg[g, \bar{U}, p] = \deg[f, \bar{U}, p]$$

Proof Since $p \notin f(\bar{U} - U)$, then $\deg[f, \bar{U}, p]$ is defined. The mapping

$$f(x) + t[g(x) - f(x)]$$

is a homotopy in $R^n - \{p\}$ because if $x \in \bar{U} - U$,

$$\begin{aligned} |f(x) + t[g(x) - f(x)] - p| \\ \geq |f(x) - p| - t|g(x) - f(x)| \\ \geq |f(x) - p| - |g(x) - f(x)| > 0 \end{aligned}$$

Hence f and g are homotopic in $R^n - \{p\}$ and by Property 2, the conclusion of the theorem holds. □

Example 8. *Suppose mapping f is defined on $\bar{U}$ where U is an open neighborhood of the origin 0 and f has the form*

$$f(x) = \mathcal{M}x + h(x)$$

where $\mathcal{M}$ is a nonsingular matrix and

$$h(x) = o(x)$$

that is,

$$|h(x)| < \eta(|x|)|x|$$

where $\eta(r)$ is a continuous monotonic increasing function from the set of nonnegative real numbers into the nonnegative real numbers such that

$$\eta(0) = 0$$

Then there exists a number $r_0 > 0$ such that if V is an open neighborhood of the origin such that $V \subset B_{r_0}$, the ball of radius r_0 and center 0, then

$$\deg[f, \bar{V}, 0] = sgn \det \mathcal{M}$$

Proof Since $\mathcal{M}$ is nonsingular, there is a positive number m such that if $|x| \neq 0$, then

$$|\mathcal{M}x| > m|x|$$

Let r_0 be such that $\eta(r_0) < m$. Then if $0 < |x| < r_0$

$$|f(x) - \mathcal{M}x| = |\mathcal{M}x + h(x) - \mathcal{M}x| = |h(x)| < \eta(r_0)|x| < m|x| < |\mathcal{M}x|$$

By Example 2

$$\deg[\mathcal{M}, \bar{V}, 0] = sgn \det \mathcal{M}$$

Hence the conclusion follows from Rouché's theorem. □

Example 9. *Suppose mapping f is defined on R^n and f has the form*

$$f(x) = \mathcal{M}x + k(x)$$

where $\mathcal{M}$ is a nonsingular matrix and

$$|k(x)| < N(|x|)|x|$$

where $N(r)$ is a continuous function from the set of positive numbers into the positive numbers and

$$\varlimsup_{r\to\infty} N(r) < m$$

where m has the same meaning as in Example 8. Then there exists a number $r_1 > 0$ such that if W is an open neighborhood of 0 such that $B_{r_1} \subset W$, then

$$\deg[f, \bar{W}, 0] = \text{sgn} \det \mathcal{M}$$

Proof   Let r_1 be such that if $r \geq r_1$, then $N(r) < m$. It follows that if $x \in \bar{W} - W$,

$$|f(x) - \mathcal{M}x| = |\mathcal{M}x + k(x) - \mathcal{M}x| = |k(x)| < N(r_1)|x| < m|x| < |\mathcal{M}x|$$

Again the conclusion follows from Rouché's Theorem.   ▯

Another property of topological degree which is useful for computing the degree is the product theorem, that is, the statement which says roughly that the product of the degrees is the degree of the product. The precise statement is:

Product Theorem *Let $\bar{U}$ be the closure of a bounded open set in R^n and f a continuous mapping from $\bar{U}$ into R^n. Let g be a continuous mapping from $f[\bar{U}]$ into R^n and suppose $p \in R^n$ is such that*

$$p \notin gf(\bar{U} - U)$$

Assume further that $g^{-1}(p)$ is a finite set of points $q_1, \ldots, q_m$ such that for $j = 1, \ldots, m$,

$$q_j \in \text{Int } f[U]$$

Let $V_1, \ldots, V_m$ be a collection of pairwise disjoint open sets such that for $j = 1, \ldots, m$,

$$q_j \in V_j$$

and

$$\bar{V}_j \subset f[U]$$

Then

$$\deg[gf, \bar{U}, p] = \sum_{j=1}^{n} \deg[f, \bar{U}, q_j] \deg[g, \bar{V}_j, p]$$

The proof of this theorem is not difficult and can be based on a careful examination of the definition of degree.

Example 10. *Let f be a mapping from R^{2m} into R^{2m}, where $m > 1$, defined as follows. Regard R^{2m} as complex Euclidean m-space and let f be defined by:*

$$f : (z_1, \ldots, z_m) \to (z_1^{k_1}, \ldots, z_m^{k_m})$$

Then f is the product of the mappings

$$f_j : (z_1, \ldots, z_m) \to (z_1, \ldots, z_j^{k_j}, \ldots, z_m) \quad (j = 1, \ldots, m)$$

Hence by Example 3, Example 7, and the product theorem, if B^{2m} is a ball in R^{2m} with center 0, then

$$\deg[f, B^{2m}, 0] = \prod_{j=1}^{m} k_j$$

(We note that $\deg[f, B^{2m}, 0]$ can also be computed directly from the definition of topological degree by using the technique used in Example 7. Showing that the Jacobians are positive is rather tedious.)

Example 11. *Let f be a mapping from R^{2m} into R^{2m}, where $m > 1$, defined as follows. Regard R^{2m} as complex Euclidean m-space and let f be defined by:*

$$f : (z_1, \ldots, z_m) \to (f_1(z_1, \ldots, z_m), \ldots, f_m(z_1, \ldots, z_m))$$

where $f_j(z_1, \ldots, z_m)$ is a polynomial homogeneous of degree k_j in $z_1, \ldots, z_m$ for $j = 1, \ldots, m$. if $R(f_1, \ldots, f_m)$ is the resultant of the polynomials $f_1, \ldots, f_m$, then a classical theorem (see Macaulay [1916] or van der Waerden [1940]) states that $R(f_1, \ldots, f_m) = 0$ if and only if the equations

$$f_1(z_1, \ldots, z_m) = 0$$
$$f_m(z_1, \ldots, z_m) = 0$$

have a nonzero solution. In the language of degree theory, we may state this theorem as: If B^{2m} is the ball with radius one and center 0, in R^{2m}, then $\deg[f, B^{2m}, 0]$ is defined if and only if $R(f_1, \ldots, f_m) \neq 0$. Suppose $R(f_1, \ldots, f_m) \neq 0$. Then

$$\deg[f, B^{2m}, 0] = \prod_{j=1}^{m} k_j$$

Proof See Cronin [1964].

Example 12. *Let f be a mapping from R^{2m} into R^{2m}, where $m > 1$, defined by:*

$$f : (z_1, \ldots, z_m) \to (f_1(z_1, \ldots, z_m)$$
$$+ h_1(z_1, \ldots, z_m), \ldots, f_m(z_1, \ldots, z_m) + h_m(z_1, \ldots, z_m))$$

where $f_j(z_1, \ldots, z_m)$ is a polynomial homogeneous of degree k_j in $z_1, \ldots, z_m$, $h_j(z_1, \ldots, z_m)$ is a continuous function of order $k_j + 1$, that is, if $r^2 = \sum_{j=1}^{m} |z_j|^2$, then

$$\lim_{r \to 0} \frac{|h_j(z_1, \ldots, z_m)|}{r^{k_j}} = 0$$

If the resultant $R(f_1 m, \ldots, f_m)$ is nonzero then there is a positive number b such that if B^{2m} is a ball with center 0 and radius less than b then

$$\deg[f, B^{2m}, 0] = \prod_{j=1}^{m} k_j$$

If $R(f_1, \ldots, f_m) = 0$ and if $\deg[f, B^{2m}, 0]$ is defined and if the radius of B^{2m} is less than some positive number b, then

$$\deg[f, B^{2m}, 0] \geq \prod_{j=1}^{m} k_j$$

Proof   For the case in which $R(f_1, \ldots, f_m) \neq 0$, the proof is obtained by a straightforward application of Rouché's theorem. For the proof of the second case, see Cronin [1964].   ∎

Example 13. *Let f be defined by:*

$$f : (z_1, \ldots, z_m) \to (f_1(z_1, \ldots, z_m)$$
$$+ s_1(z_1, \ldots, z_m), \ldots, f_m(z_1, \ldots, z_m) + s_m(z_1, \ldots, z_m))$$

where $f_1, \ldots, f_m$ are as in Example 12, and $R(f_1, \ldots, f_m) \neq 0$ and $s_j(z_1, \ldots, z_m)$ is a continuous function of order less than k_j, that is, if $r^2 = \sum_{j=1}^{m} |z_j|^2$, then

$$\lim_{r \to \infty} \frac{|s_j(z_1, \ldots, z_m)|}{r^{k_j}} = 0$$

Then there is a positive number b_1 such that if B^{2m} is a ball with center 0 and radius greater than b_1, then

$$\deg[f, B^{2m}, 0] = \prod_{j=1}^{m} k_j$$

Proof   By an argument similar to that for Example 12.   ∎

Example 14. *Let h be a homeomorphism from $\bar{U}$, the closure of a bounded connected open set U in R^n, into R^n. Then if $p \in h(U)$,*

$$\deg[h, \bar{U}, p] = \pm 1$$

Proof Apply the product theorem with $f = h$ and $g = h^{-1}$, and use the fact that the topological degree of the identity mapping at points in the image set is $+1$ (Example 1). Note that in order to be certain that $\deg[h^{-1}, h(\bar{U}), h^{-1}(p)]$ is defined, one must know that $h(\bar{U})$ is the closure of a bounded open set. This follows from the invariance of domain theorem (see, for example, Hurewicz and Wallman [1948]). At the cost of extra labor, the invariance of domain theorem itself can be proved by using the product theorem. See Leray [1935].

The definition of topological degree that we have described is the "covering number" definition which was given originally by Brouwer [1912]. From the point of view of the analyst who is interested in solving equations, it is a very natural definition. There is, however, an equivalent definition which is equally important. The second definition, which is sometimes called the "intersection number" definition, is obtained as follows. Suppose, as before, that $\bar{U}$ is the closure of a bounded open set $U \subset R^n$, that f is a continuous mapping from $\bar{U}$ into R^n and that

$$p \notin f(\bar{U} - U)$$

Let L be a half-ray in R^n emanating from the point p, that is,

$$L = \{p + \lambda \vec{v} \mid \lambda \geq 0\}$$

where $\vec{v}$ is a fixed nonzero n-vector. Let K be the set of points of intersection of L and $f(\bar{U} - U)$, that is,

$$K = L \cap \{f(\bar{U} - U)\}$$

Suppose that $p_i \in K$ and $q_i = f^{-1}(p_i)$. By standard but tedious considerations of orientation, an intersection multiplicity of $+1$ or -1 can be assigned to the point p_i. This is illustrated for the case $n = 2$ in Figure A.1. (The "counterclockwise orientation" in R^2 is assumed to be the positive orientation.) The point p_1 has intersection multiplicity $+1$ because r_1, r_2, q_1 and $f(r_1), f(r_2), p_1$ both have the same orientation. The point p_2 has intersection multiplicity -1 because s_1, s_2, q_2 and $f(s_1), f(s_2), p_2$ have opposite orientations. Then $\deg[f, \bar{U}, p]$ is defined to be

$$\deg[f, \bar{U}, p] = [\text{Number of points in } K \text{ with intersection multiplicity } +1]$$
$$- [\text{Number of points in } K \text{ with intersection multiplicity } -1]$$

As with our presentation of the covering number definition, this definition has many "holes" in it. In the first place, we have only indicated the "orientation considerations" that must be made. Secondly, we have not indicated how the half-ray L should be chosen. A fairly lengthy investigation shows that the definition is independent of the half-ray L. "Well-behaved" approximations g of f are used, and it is shown that for sufficiently fine approximations g, the value of $\deg[g, \bar{U}, p]$ is always the same.

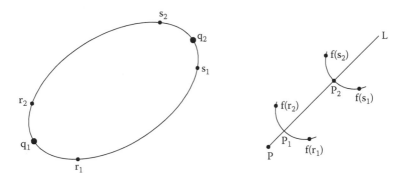

Figure 1

At the cost of considerable labor, it can be shown that the covering number definition and the winding number definition just described are equivalent. A detailed discussion is given in Alexandroff and Hopf [1935] and Cronin [1964]. Notice that the fact that the two definitions are equivalent implies that the topological degree depends only on the values of f on $\bar{U} - U$, that is, on the function $f/(\bar{U} - U)$. Indeed the function f need only be defined on $\bar{U} - U$.

There are two important reasons for introducing the intersection number definition. First, the intersection number definition contains as a special case (the case in which $n = 2$) the concept of the winding number, familiar from complex variable. Second, use of the intersection number definition of topological degree makes possible the computation of the topological degree of some further classes of mappings as we now show. $\square$

Example 15. *Let*

$$f : (x, y) \rightarrow (\bar{x}, \bar{y})$$

be a mapping from R^2 into R^2 defined by:

$$\bar{x} = P(x, y)$$
$$\bar{y} = Q(x, y)$$

where P, Q are polynomials homogeneous in x and y of degrees p and q, respectively. Let

$$B^2 = \{(x, y) \mid x^2 + y^2 \leq 1\}$$

We assume that $\deg[f, B^2, 0]$ is defined, that is, we assume that there is no point $p \in B^2 - \{Int\ B^2\}$ such that

$$f(p) = 0$$

This means we assume that P and Q have no common real linear factors. Now we write P and Q in factored form, that is,

$$P(x, y) = K_1 \prod_{k=1}^{m} (y - a_i x)^{p_i}$$

$$Q(x, y) = K_2 \prod_{j=1}^{n} (y - b_j x)^{q_j}$$

where K_1, K_2 are constants. (We include the possibility that some $a_i = \infty$ or some $b_j = \infty$; equivalently, that the factor $y - a_i x$ is equal to $-x$ or the factor $y - b_j x$ is equal to $-x$.) It follows from the definition of topological degree that if

$$g: (x, y) \to (\tilde{x}, \tilde{y})$$

is the mapping from R^2 into R^2 defined by

$$\tilde{x} = \prod_{i=1}^{n} (y - a_i x)^{p_i}$$

$$\tilde{y} = \prod_{j=1}^{n} (y - b_j x)^{q_j}$$

(1)

then

$$\deg[g, B^2, 0] = (\deg[f, B^2, 0])(\text{sign } K_1 K_2)$$

In order to compute $\deg[g, B^2, 0]$ we use the intersection number definition and investigate the image under g of $B^2 - (\text{Int } B^2)$ by studying the changes of sign of $\tilde{x}$ and $\tilde{y}$ as (x, y) varies over $B^2 - (\text{Int } B^2)$. By using Property 2 (the invariance under homotopy of the degree) it can be easily shown (Cronin [1964, pp. 39–40]) that the value of $\deg[g, B^2, 0]$ is not affected if the following factors in the products on the right-hand side of (1) above are omitted.

(1) *Pairs of factors $(y - a_{i_1} x)$, $(y - a_{i_2} x)$ in which a_{i_1} and a_{i_2} are complex conjugates.*

(2) *Factors $(y - a_{i_1} x)^{p_i}$ where a_{i_1} is real and p_i is even.*

(3) *Pairs of factors $(y - a_{i_1} x)$, $(y - a_{i_2} x)$ such that $a_{i_1} < a_{i_2}$ and there is no b_j or a_{i_3} such that*

$$a_{i_1} < b_j < a_{i_2}$$

or

$$a_{i_1} < a_{i_3} < a_{i_2}$$

(4) *Pairs of factors $(y - \tilde{a}_1 x)$, $(y - \tilde{a}_m x)$ where $\tilde{a}_1$ and $\tilde{a}_m$ are the smallest and largest of all the numbers $a_1, \ldots, a_m, b_1, \ldots, b_n$.*

(5) *The factors listed above with the a_i's replaced by b_j's and the b_j replaced by a_i.*

If all the factors in

$$\prod_{i=1}^{m}(y - a_i x)^{p_i}$$

and in

$$\prod_{j=1}^{n}(y - b_j x)^{q_j}$$

are included in the above four classifications, then $\deg[g, B_1, 0]$ *is zero because g is homotopic in* $R^n - \{0\}$ *to a mapping*

$$M : (x, y) \rightarrow (x_1, y_1)$$

where x_1 *or* y_1 *is a constant. To complete the study we need only consider a mapping* M_1 *such that*

$$a_1 < b_1 < a_2 < \cdots < a_m < b_m \tag{2}$$

or

$$b_1 < a_1 < b_2 < \cdots < b_m < a_m \tag{3}$$

and with each factor $(y - a_i x)$ *and* $(y - b_j x)$ *having exponent one. It is easily shown (Cronin [1964, p. 40]) by using the intersection number definition of topological degree that if (2) holds, then* $\deg[g, B_1, 0]$ *is m. If (3) holds, then* $\deg[g, B_q, 0]$ *is* $-m$.

Property 1 states that if the degree of the mapping is nonzero, then the corresponding equation has at least one solution. Since the degree is an integer, it is natural to raise the question of whether the degree or its absolute value measures or counts the number of solutions of the equation. The answer to this question is not obvious. By Example 7, the degree of the mapping

$$f : z \rightarrow z^n$$

relative to the closed unit disc and at 0 is n. But the equation

$$f(z) = 0$$

has just one solution $z = 0$. So the degree (or its absolute value) is not, in general, a lower bound for the number of solutions. On the other hand, suppose the mapping h with domain

$$B^2 = \{(x, y) \mid x^2 + y^2 \leq 1\} = \{(r, \theta) \mid 0 \leq r \leq 1\}$$

is defined by

$$h(x, y) = (\lambda(r)x, \lambda(r)y)$$

where

$$\lambda(r) = 0 \quad \text{if} \quad 0 \le r \le \frac{1}{2}$$

$$\lambda(r) = 2\left(r - \frac{1}{2}\right) \quad \text{if} \quad r \ge \frac{1}{2}$$

Mapping h is the identity mapping on the circle $r = 1$. Hence

$$\deg[h, B^2, 0] = +1$$

But the equation

$$h(r, \theta) = (0, 0)$$

has an infinite set of solutions, that is, the set

$$\left\{ (r, \theta) \mid 0 \le r \le \frac{1}{2} \right\}$$

So the degree (on its absolute value) is certainly not an upper bound for number of solutions.

However, we can show that if the mapping f is differentiable, then the degree does yield a kind of count of the number of solutions. To prove this we need a special case of a well-known result in analysis.

Definition   Let f be a differentiable mapping of an open set $G \subset R^n$ into R^n. A point $x \in G$ at which the Jacobian of f is zero is a *critical point* of f.

Sard's Theorem   (Special case) *Let f be a differentiable mapping of a bounded open set $G \subset R^n$ into R^n. Let D be an open set such that $\bar{D} \subset G$. Then the image under f of the set of critical points in D has measure zero.*

For a proof of Sard's theorem, see Sard [1942] or Golubitsky and Guillemin [1973].

Sard's theorem yields almost immediately the following theorem which shows that if f is differentiable, then the absolute value of the degree of f is essentially a lower bound for the number of solutions of the corresponding equation.

Theorem 1.   *If f is a differentiable mapping of a bounded open set $G \subset R^n$ into R^n and if $G \supset \bar{U}$, where $\bar{U}$ is the closure of a bounded open set U, and if $\deg[f, \bar{U}, p]$ exists and $\deg[f, \bar{U}, p] = m$, then there is an open neighborhood N of p and a set $S \subset N$ such that S has n-measure zero and such that:*

(i) *For all $\bar{p} \in N$,*

$$\deg[f, \bar{U}, \bar{p}] = \deg[f, \bar{U}, p];$$

(ii) *if $\bar{p} \in N - S$, then $[f^{-1}(\bar{p})] \cap U$ is a finite set of points;*

(iii) *if $\bar{p} \in N - S$ and $q \in [f^{-1}(\bar{p})] \cap U$, then the Jacobian of f at q is nonzero.*

(iv) *if $\bar{p} \in N - S$ and n is the number of points in $[f^{-1}(\bar{p})] \cap U$, then $n \geq |m|$.*

(Sard's theorem is used as a basis for one way of setting up the definition of topological degree. See Nagumo [1951].)

Mappings defined by analytic functions have the useful property that the degree of the mapping is positive if and only if the corresponding equation

$$f(z) = p$$

has a solution. Moreover the degree is equal to the number of solutions of the equation

$$f(z) = \bar{p}$$

for all $\bar{p}$ in a neighborhood of p except for a finite set. The underlying reason for this is quite simple. Suppose that

$$f(z) = u(x, y) + iv(x, y)$$

is a function analytic in an open set $\mathcal{U}$ in the complex plane. We consider the corresponding mapping

$$f: (x, y) \rightarrow (u(x, y), v(x, y))$$

Suppose that

$$f(x, y) = 0$$

has a solution $(\bar{x}, \bar{y})$, that is,

$$u(\bar{x}, \bar{y}) = 0$$
$$v(\bar{x}, \bar{y}) = 0$$

From the Cauchy-Riemann equations, it follows that the Jacobian of f at $(\bar{x}, \bar{y})$ is

$$\det \begin{bmatrix} u_x & u_y \\ v_x & v_y \end{bmatrix} = \det \begin{bmatrix} u_x & -v_x \\ v_x & u_x \end{bmatrix} = u_x^2 + v_x^2 \tag{4}$$

Since $\partial f / \partial z$ is an analytic function and

$$\frac{\partial f}{\partial z} = u_x(x, y) + iv_x(x, y)$$

then from the identity theorem for analytic functions, it follows that in a bounded set in R^2, there is only a finite set of points $(\bar{x}_1, \bar{y}_1), \ldots, (\bar{x}_m, \bar{y}_m)$ such that

$$u_x(\bar{x}_j, \bar{y}_j) + iv_x(\bar{x}_j, \bar{y}_j) = 0 \qquad (j = 1, \ldots, m)$$

Hence if $\bar{U}$ is the closure of a bounded open set in R^2 such that $\bar{U} \subset \mathcal{U}$, then if

$$p \notin f\left[\left\{\sum_{j=1}^{m}(\bar{x}_j, \bar{y}_j)\right\} \cup [\bar{U} - U]\right]$$

it follows from (4) that $\deg[f, \bar{U}, p]$ is nonnegative and is equal to the number of solutions in the set U of the equation

$$f(z) = p$$

If C^n denotes complex Euclidean n-space, extension of the reasoning above and use of Theorem 1 yields the following result.

Theorem 2. *Suppose that*

$$f_j(z_1, \ldots, z_n) \qquad (j = 1, \ldots, n)$$

is analytic (i.e., can be represented by a power series in $z_1, \ldots, z_n$) in an open set $\mathcal{U}$ in C^n. Let $\bar{U}$ be the closure of a bounded open set U in R^{2n} (where R^{2n} is identified with C^n) such that $\bar{U} \subset \mathcal{U}$. Let f be the mapping of R^{2n} into R^{2n} described in terms of points in C^n by

$$f : (z_1, \ldots, z_n) \rightarrow (f_1(z_1, \ldots, z_n), \ldots, f_n(z_1 \ldots, z_n))$$

Then if $p \notin f(\bar{U} - U)$ and $p \in f(U)$,

$$\deg[f, \bar{U}, p] > 0$$

Also there is a set $S \subset R^{2n}$ such that S has $2n$-measure zero and such that if

$$p \in f(U) - S$$

the number of solutions of the equation

$$f(z) = p$$

is equal to $\deg[f, \bar{U}, p]$.

Further considerations of the same kind yield the following theorem:

Theorem 3. *Suppose that*

$$f_j(z_1, \ldots, z_n) \qquad (j = 1, \ldots, n)$$

is analytic (i.e., can be represented by a power series in $z_1, \ldots, z_n$) in an open set $\mathcal{U}$ in C^n, and suppose further that

$$f_j(\bar{z}_1, \ldots, \bar{z}_n) = \overline{f_j(z_1, \ldots, z_n)} \tag{5}$$

where $\bar{z}_i$ denotes the conjugate of z_i (condition (5) will be satisfied, for example, if the coefficients in the power series expansion of f_j ($j = 1, \ldots, n$) are real.) Suppose $\bar{U}$ is the closure of a bounded open set U in R^n such that $\bar{U} \subset \mathcal{U}$ where R^n is identified with the subspace of the points

$$(z_1, \ldots, z_n) = (x_1 + iy_1, \ldots, x_n + iy_n)$$

in C^n that consists of points of the form $(x_1, \ldots, x_n)$. Let h be the mapping from R^n into R^n defined by

$$h: (x_1, \ldots, x_n) \to (f_1(x_1, \ldots, x_n), \ldots, f_n(x_1, \ldots, x_n))$$

Suppose further that $\mathcal{V}$ is an open set in C^n such that $\bar{\mathcal{V}} \subset \mathcal{U}$ and $\mathcal{V} \cap R^n = U$. Then there is a set $M \subset R^n$ of n-measure zero such that if

$$p \in R^n - M - h(\bar{U} - U)$$

(so that $\deg[f, \bar{\mathcal{V}}, p]$ is defined) and

$$\deg[h, \bar{U}, p] \neq 0$$

then the number n_h of solutions of the equation

$$h(x) = p$$

is finite and

$$n_h \leq \deg[f, \bar{\mathcal{V}}, p]$$

and

$$n_h = \deg[f, \bar{\mathcal{V}}, p] \quad (\text{mod } 2)$$

Proof   See Cronin [1960, 1971].   ☐

Example 16.   *Let f be a mapping from C^n (or R^{2n}) into C^n (or R^{2n}) defined by*

$$f: (z_1, \ldots, z_n) = (f_1(z_1, \ldots, z_n), \ldots, f_n(z_1, \ldots, z_n))$$

where for $j = 1, \ldots, n$,

$$f_j(z_1, \ldots, z_n) = P_{k_j}(z_1, \ldots, z_n) + H_{k_j}(z_1, \ldots, z_n)$$

where P_{k_j} is a polynomial homogeneous of degree k_j in $z_1, \ldots, z_n$ and $H_{k_j}(z_1, \ldots, z_n)$ is of order $k_j + 1$ in the sense that if

$$r^2 = \sum_{i=1}^{n} |z_i|^2$$

$$\lim_{r \to 0} \frac{|H_{k_j}(z_1, \ldots, z_n)|}{r^{k_j}} = 0$$

If B_r is a ball of radius r and center 0 in R^{2n} and if r is sufficiently small, then

$$\deg[f, B_r, 0] \geq \prod_{j=1}^{n} k_j \qquad (5')$$

Proof   This is proved by using Sard's theorem. See Cronin [1953].   ⬚

We have already seen numerous examples of mappings with degree zero. It is natural to raise the question of whether the fact that the degree is zero yields information about the existence of solutions. First, if the degree is zero, the corresponding equation may nevertheless have solutions. By Example 6, the mapping

$$f : (x, y) \rightarrow (x^2, y^2)$$

on the unit disc B^2 is such that

$$\deg[f, B^2, 0] = 0$$

But the equation

$$f(x, y) = (0, 0)$$

has the solution: $x = 0$, $y = 0$. Thus the fact that the degree is zero seems an inconclusive observation. However, this fact can be utilized in some cases to obtain useful results. In order to see how it is used, we state one more basic property of the topological degree.

Property 3. Suppose that f is a continuous mapping from $\bar{U} \cup \bar{V}$, where U and V are bounded open sets in R^n and $U \cap V = \emptyset$, into R^n. Let $p \in R^n$ be such that

$$p \notin [f(\bar{U} - U)] \cup [f(\bar{V} - V)]$$

Then

$$\deg[f, \bar{U}, p] + \deg[f, \bar{V}, p] = \deg[f, \overline{U \cup V}, p]$$

Property 3 follows easily from the covering number definition of degree and we will not give a detailed proof.

Using Property 3, we have immediately the following theorem.

Theorem 4.   *Suppose U is a bounded open set in R^n and W is an open set such that $W \subset U$. If f is a mapping from $\bar{U}$ into R^n and*

$$p \in R^n - f(\bar{W} - W) - f(\bar{U} - U)$$

and if

$$\deg[f, \bar{W}, p] \neq \deg[f, \bar{U}, p] \tag{6}$$

then the equation

$$f(x) = p$$

has a solution $x \in U - \bar{W}$.

Proof   By Property 3,

$$\deg[f, \bar{W}, p] + \deg[f, \overline{U - \bar{W}}, p] = \deg[f, \bar{U}, p]$$

Hence by (6),

$$\deg[f, \overline{U - \bar{W}}, p] \neq 0$$

and the theorem follows from Property 1.   ▯

Theorem 4 does not seem a very impressive statement, but it is frequently useful in analysis. It is often possible to show that

$$\deg[f, \bar{U}, p] = 0$$

or $\deg[f, \bar{U}, p]$ is even and to show that f is a homeomorphism in an open neighborhood $\mathcal{N}$ of some point in $f^{-1}(p)$ or that the differential of f at some point in $f^{-1}(p)$ is nonsingular. Then by Example 14, $\deg[f, \mathcal{N}, p] = \pm 1$ and the hypotheses of Theorem 4 are satisfied.

For work in differential equations, it is often convenient to think in terms of a vector field. Analytically, the concept of vector field is, in a simple way, equivalent to the concept of a mapping.

Definition   Let E be a subset of R^n. A *vector field* on E, denoted by $V(x)$ where $x \in E$, is a mapping from E into the collection of real n-vectors. (Geometrically, one thinks of the vector $V(x)$ attached to the point x so that the initial point of $V(x)$ is the point x.) From the analytic viewpoint, the vector field $V(x)$ is simply a mapping M_v from E into R^n, that is, the mapping from x into the head of the vector $V(x)$ if the initial point of $V(x)$ is placed at the origin. We say that the vector field $V(x)$ is *continuous* or *differentiable* if M_v is continuous or differentiable.

Notice that an n-dimensional autonomous system

$$x' = f(x)$$

can be regarded geometrically simply as a vector field $f(x)$. We make use of this viewpoint only briefly (Exercise 4 in Chapter 6) but it is of basic importance in some studies of differential equations.

Definition   A *singularity* of vector field $V(x)$ is a point $\bar{x} \in E$ such that $V(\bar{x}) = 0$. If vector field $V(x)$ has no singularities on E, then $V(x)$ is *nonsingular*.

Definition   Suppose that $E = \bar{U}$ where $\bar{U}$ is the closure of a bounded open set U in R^n, and suppose that if $y \in \bar{U} - U$, then $V(y) \neq 0$. The *index* of $V(x)$ is $\deg[M_v, \bar{U}, 0]$.

Definition   Suppose $E = \bar{U}$ and $p \in U$ is such that $V(p) = 0$ and there is a neighborhood N of p such that $V(x)$ is nonzero at each point $x \in \bar{N} - \{p\}$. The *index of the vector field $V(x)$ at the singularity p* is the topological index of M_v at p.

Now let $\bar{U}$ be the closure of a bounded open set U in R^n such that at each point q of the boundary of U there is a tangent hyperplane H_q. This will hold if, for example, there is a neighborhood N of q in $\bar{U} - U$ such that N is the image of an open set in R^{n-1} under a one-to-one mapping of the form

$$(x_1, \ldots, x_{n-1}) \to (f_1(x_1, \ldots, x_{n-1}), \ldots, f_{n-1}(x_1, \ldots, x_{n-1}))$$

where each f_j $(j = 1, \ldots, n-1)$ is a differentiable function. (In the language of differential geometry, the condition holds if $\bar{U} - U$ is a differentiable manifold.) Then if L_q is the line through q which is normal to H_q, there exists a fixed n-vector V_q of unit length such that

$$L_q = \{p(\lambda) \mid p(\lambda) = q + \lambda V_q, \ \lambda \text{ real}\}$$

and such that there exists a number $\eta > 0$ with the property that if $\lambda \in (0, \eta)$, then $p(\lambda) \in U$ and if $\lambda \in (-\eta, 0)$ then $p(\lambda) \in \bar{U}^c$, the complement of $\bar{U}$ relative to R^n.

Theorem 5.   *Let $V(x)$ be a nonsingular vector field on $\bar{U} - U$ such that for each $q \in \bar{U} - U$, the vector $V(q)$ is such that*

$$(V(q), V_q) \geq 0 \tag{7}$$

Then the index of $V(x)$ is equal to the index of the vector field V_x.

Proof   The index of $V(x)$ is, by definition, $\deg[M_v, \bar{U}, 0]$. (Remember that the degree depends only on $M_v/(\bar{U} - U)$.) Consider the homotopy

$$(1 - t)V(x) + tV_x$$

defined on $[\bar{U} - U] \times [0, 1]$. Suppose there is a point $p_0 \in \bar{U} - U$ and a number $t_0 \in (0, 1)$ such that

$$(1 - t_0)V(p_0) + t_0 v_{p_0} = 0$$

Then

$$V(p_0) = -\frac{t_0}{1 - t_0} V_{p_0} \tag{8}$$

Take the innerproduct of V_{p_0} with (8) and obtain

$$(V(p_0), V_{p_0}) = -\frac{t_0}{1 - t_0} \| V_{p_0} \|^2 \tag{9}$$

Since $\| V_{p_0} \|^2 = 1$ and $-t_0/1 - t_0 < 0$, equation (9) contradicts (7). $\quad\square$

Corollary 1 *If $\bar{U}$ is a ball of unit radius with center at the origin, then the index of $V(x)$ is $+1$.*

Proof The index of the vector field V_x is $\deg[I, \bar{U}, 0]$ where I is the identity mapping. $\quad\square$

Corollary 2 *If for each $q \in \bar{u} - U$, the vector $V(q)$ is such that*

$$(V(q), V_q) < 0$$

then the index of $V(q)$ is equal to $(-1)^n$ times the index of V_q. $\quad\square$

Proof First, $(-V(q), V_q) > 0$ and hence by the theorem, the index of $[-V(q)]$ is equal to the index of V_q. But the index of $[-V(q)]$ is $(-1)^n$ times the index of $V(q)$. $\quad\square$

Corollary 3 *If the hypotheses of Corollary 1 or Corollary 2 are satisfied, then the vector field $V(p)$ has a singularity in U.*

Proof Follows from the definition of index of a vector field and Property 1 of topological degree. $\quad\square$

Brouwer Fixed Point Theorem
Let $\bar{U}$ be the closure of a bounded open set U in R^n such that $\bar{U}$ is homeomorphic to the unit ball B^n in R^n. Suppose f is a continuous mapping from $\bar{U}$ into $\bar{U}$. Then f has a fixed point, that is, there is a point $p \in \bar{U}$ such that $f(p) = p$.

Proof Let
$$h : \bar{U} \rightarrow B_n$$
be the homeomorphism from $\bar{U}$ onto B^n given the hypothesis. We show that $g = hfh^{-1}$ has a fixed point. Suppose g does not have a fixed point on ∂B^n. Since g takes B^n into itself, then the vector field

$$G(p) = \overrightarrow{p, g(p)}$$

has index $(-1)^n$ by Corollary 2 and hence has a singularity, that is, there exists $p \in B^n$ such that $p = g(p)$. $\quad\square$

By using the topological degree theory previously developed, we can give the above quick proof of the Brouwer fixed point theorem. However, the Brouwer fixed point theorem is independent of the degree theory in the sense that an "elementary" proof, which requires no "topological machinery," can also be given. See Alexandroff-Hopf [1935, p. 376].

Jordan Curve Theorem. *Let h be a 1 − 1 continuous mapping of*

$$S = \{(x, y) \mid x^2 + y^2 = 1\}$$

into R^2. Then

$$R^2 - h(S) = C_1 \cup C_2$$

where C_1, C_2 are disjoint connected sets; the set C_1 is bounded and the boundary of C_1 is $h(S)$; the set C_2 is unbounded and the boundary of C_2 is $h(S)$. If $p \in C_1$, then

$$\deg[h, B_2, p] = \pm 1 \tag{10}$$

the sign depending on the orientation of $h(S)$. If $p \in C_2$, then

$$\deg[h, B_2, p] = 0 \tag{11}$$

Proof See Dieudonné [1960, p. 251 ff.]. ⬚

The conditions (10) and (11) in the Jordan curve theorem as stated here are often omitted from the statement of the theorem. However, we need these conditions in the proof of the Poincaré-Bendixson theorem.

Next we define a topological degree for certain classes of mappings in infinite-dimensional spaces, that is, Banach spaces. (Actually for the definition of this topological degree, it is sufficient to consider a linear normed space. But the spaces that occur in applications are usually Banach spaces.) It turns out that it is not possible to define a topological degree for all continuous mappings as in the finite-dimensional case. It can be shown with examples (see Cronin [1964, pp. 124–130]) that if it is assumed that a topological degree with Properties 1 and 2 has been defined for continuous mappings from a Banach space into itself, then the identity mapping (which must have topological degree ±1) is homotopic to a constant mapping (which must have topological degree 0). This, of course, contradicts the invariance under homotopy (Property 2).

However a topological degree has been defined for a special class of mappings which often occur in analysis.

Definition Let X be a Banach space, and let f be a continuous mapping from X into itself such that if B is a bounded set in X (i.e., there exists a positive number M such that if $x \in B$, then $\|x\| < M$) then the set $F(B)$ is compact in X (i.e., each infinite subset of $F(B)$ has a limit point in X). Then F is said to be a *compact mapping* of X into itself.

It is not difficult to show that a mapping of the form $I + F$ where I is the identity and F is compact can be approximated by a sequence of mappings in finite-dimensional spaces and all mappings in this sequence which are fine enough approximations of $I + F$ have the same topological degree (i.e., Brouwer degree). This topological degree is defined to be the topological degree of $I + F$. It is usually called the Leray-Schauder degree after the two mathematicians who introduced the definition and is denoted by $\deg_{LS}[I + F, \bar{W}, p]$ where W is a bounded open set in X and $p \notin (I + F)(\bar{W} - W)$. The Leray-Schauder degree has properties exactly analogous to the Properties 1, 2, and 3 of the Brouwer degree. Most important, it turns out that many functional equations can be formulated as the study of an equation in a Banach space of X of the form

$$(I + F)x = y \tag{12}$$

and hence the problem can be approached by investigating the Leray-Schauder degree of $I + F$. For example, it is not difficult to show that the integral equation

$$x(t) + \int_0^1 K[s, t, x(s)] \, ds = y(t)$$

where $y(t)$ is a continuous real-valued function on $[0, 1]$ and $K(s, t, \xi)$ is a continuous real-valued function on

$$[0, 1] \times [0, 1] \times R$$

where R is the set of real numbers, is of the form (12), that is, the mapping

$$x(t) \rightarrow \int_0^1 K[s, t, x(s)] \, ds$$

is a compact mapping from $C[0, 1]$ into itself.

Many important and useful results in nonlinear analysis have been obtained by use of the Leray-Schauder degree. For a summary of some of the results up to 1963, see Cronin [1964]. Numerous applications have been obtained since then (see the *Mathematical Reviews*). Also the definition of Leray-Schauder degree has been extended to include larger classes of mappings. Some references to these definitions are given at the beginning of our discussion of topological degree.

Property 1. Let $\bar{W}$ be the closure of a bounded open set W in a Banach space X and F a compact mapping from $\bar{W}$ into X. Suppose $p \in X$ is such that

$$p \notin (I + F)(\bar{W} - W)$$

and suppose

$$\deg_{LS}\left[I + F, \bar{W}, p\right] \neq 0$$

Then there exists $q \in W$ such that $(I + F)q = p$.

Property 2. Let $\bar{W}$ be the closure of a bounded open set W in a Banach space X and suppose that $\mathcal{F}$ is a continuous mapping from $\bar{W} \times [0, 1]$ into X such that for each $\mu \in [0, 1]$ the mapping

$$\mathcal{F}/(\bar{W} \times \{\mu\})$$

regarded as a mapping from $\bar{W}$ into X, is a compact mapping and such that $\mathcal{F}$ is uniformly continuous in μ, that is, given $\varepsilon > 0$ then there exists a $\delta > 0$ such that if

$$|\mu_1 - \mu_2| < \delta$$

then for all $x \in \bar{W}$,

$$\|\mathcal{F}(x, \mu_1) - \mathcal{F}(x, \mu_2)\| < \varepsilon$$

Let $p \in X$ be such that

$$p \notin I(\bar{W} - W) + \mathcal{F}\{(\bar{W} - W) \times [0, 1]\}$$

Then for each $\mu \in [0, 1]$ the Leray-Schauder degree of $I + \mathcal{F}/\bar{W} \times \{\mu\}$ is defined and for all $\mu \in [0, 1]$, the Leray-Schauder degree of $I + \mathcal{F}/\bar{W} \times \{\mu\}$ has the same value.

Finally we state the infinite-dimensional analog of the Brouwer fixed point theorem.

Schauder Fixed Point Theorem. *Let K be a bounded convex closed set in a Banach space X and F a compact mapping from K into X such that $F(K) \subset K$. Then F has a fixed point, that is, there exists an $x \in K$ such that $F(x) = x$.*

References

Aggarwal, J. K. and Vidyasagar, M. [1977]: *Nonlinear systems stability analysis*, Dowden, Hutchinson and Ross, Inc., Stroudsberg, Pennsylvania.

Ahlfors, Lars V. [1966]: *Complex analysis, an introduction to the theory of analytic functions of one complex variable*, 2nd edition, McGraw-Hill, New York.

Alexandroff, P. and Hopf, H. [1935]: *Topologie*, I, Springer, Berlin (reprinted by Edwards Brothers, Ann Arbor, Michigan, 1945).

Andronov, A. A. [1929]: Les cycles limites de Poincaré et la théorie des oscillations autoentretenues, *Comptes Rendus*, Volume 189, p. 559.

Andronov, A. A. and Chaikin, C. E. [1949]: *Theory of oscillations*, English Language Edition edited under the direction of Solomon Lefschetz, Princeton University Press, Princeton, New Jersey.

Andronov, A. A. and Leontovich, E. A. [1937]: Some cases of dependence of limit cycles on parameters, *Uchenie Zapiski Gor'kovskovo Gos. Univ.* 6, 3–24.

Anosov, D. V. and Arnold, V. I. (editors) [1988]: *Encyclopedia of Mathematical Sciences,* Volume 1, Dynamical Systems I (translated by E. R. Dawson and D. O'Shea) Springer-Verlag, New York.

Arnold, V. I. [1963]: "Proof of a theorem of A. N. Kolmogorov on the invariance of quasi-periodic motions under small perturbations of the Hamiltonian," *Russian Mathematical Surveys*, 18, 9–36.

Arnold, V. I. (editor) [1988]: *Encyclopedia of Mathematical Sciences,* Volume 3, Dynamical Systems III (translated by A. Iacob), Springer-Verlag, New York.

———— [1973]: *Ordinary differential equations,* translated and edited by Richard S. Silverman, M.I.T. Press, Cambridge, Massachusetts.

Bendixson, I. [1901]: "Sur les courbes définies par des équations différentielle," *Acta Math.*, Volume 24, pp. 1–88.

Besicovitch, A. S. [1954]: *Almost periodic functions*, Dover Publications, Inc., New York.

Beuter, A., Glass, L., Mackey, M. C., and Titcombe, M. S. [2003]: *Nonlinear dynamics in physiology and medicine*, Springer, New York.

Bogoliubov, N. N. and Mitropolskii, Y. A. [1961]: *Asymptotic Methods in the theory of non-linear oscilaltions* (translated from Russian), Hindustan Publishing, Delhi-6.

Boyce, W. and DiPrima, R. [1986]: *Elementary differential equations and boundary value problems*, 4th edition, John Wiley & Sons, New York.

Brouwer, L. E. J. [1912]: "Uber Abbildung von Mannigfaltigkeiten," *Mathematische Annalen* 71, 97–115.

Browder, Felix E. and Petryshyn, W. [1969]: "Approximation methods and the generalized topological degree for nonlinear mappings in Banach spaces," *Journal of Functional Analysis* 3, 217–245.

Cesari, L. [1971]: *Asymptotic behavior and stability problems in ordinary differential equations*, 3rd edition, Springer-Verlag, New York.

Churchill, Ruel V. and Brown, James W. [1987]: *Fourier series and boundary value problems*, 4th edition, McGraw-Hill, New York.

Coddington, Earl A. and Levinson, N. [1952]: "Perturbations of linear systems with constant coefficients passing periodic solutions," Contributions to the *Theory of Nonlinear Oscillations*, Volume II, Annals of Mathematics Studies, Number 29, Princeton University Press, Princeton.

Coddington, Earl A. and Levinson, N. [1955]: *Theory of ordinary differential equations*, McGraw-Hill, New York.

Courant, R. and Hilbert, D. [1953]: *Methods of mathematical physics*, Volume 1, Interscience, New York.

Cronin, J. [1953]: "Analytic functional mappings," *Annals of Mathematics* 58, 175–181.

———— [1956]: "Some mappings with topological degree zero," *Proceedings of the American Mathematical Society* 7, 1139–1145.

———— [1960]: "An upper bound for the number of periodic solutions of a perturbed system," *Journal of Mathematical Analysis and Applications*, I, 334–341.

———— [1964]: *Fixed Points and topological degree in nonlinear analysis*, Mathematical Surveys, Number 11, American Mathematical Society, Rhode Island.

———— [1971]: "A stability criterion," *Applicable Analysis* I, 25–30.

———— [1971]: "Topological degree and the number of solutions of equations," *Duke Mathematical Journal* 38, 531–538.

———— [1987]: *Mathematical aspects of Hodgkin-Huxley neural theory*, Cambridge Studies in Mathematical Biology, Cambridge University Press, Cambridge.

Dickson, L. E. [1939]: *New first course in the theory of equations*, John Wiley & Sons, New York.

Dieudonne, J. [1960]: *Foundations of modern analysis*, Academic Press, New York.

Dulac, H. [1923]: Sur les cycles limites, *Bull. Soc. France*, Vol. 51, pp. 45–108.

Dumortier, F. [1977]: Singularities of vector fields in the plane. *Journal of Differential Equations* 23, 53–106.

Edwards, C. H. and Penney, David E. [1989]: *Elementary differential equations with boundary value problems*, 2nd edition, Prentice Hall, Englewood Cliffs, New Jersey.

Farkas, M. [1994]: *Periodic motions*, Springer-Verlag, New York.

Field, R. J., Körös, E., and Noyes, R. M. [1972]: "Oscillations in chemical systems, II. Thorough analysis of temporal oscillations in the bromate-cerium-malonic acid system," *Journal of the American Chemical Society*, 94, 8649–8664.

Field, R. J. and Noyes, R. M. [1974]: "Oscillations in chemical systems, IV. Limit cycle behavior in a model of a real chemical reaction," *Journal of Chemical Physics*, 60, 1877–1884.

FitzHugh, R. [1969]: "Mathematical models of excitation and propagation in nerve," Chapter I, *Biological engineering*, edited by Herman P. Schwan, McGraw-Hill, New York.

Friedrichs, K. O. [1965]: *Lectures on advanced ordinary differential equations*, Gordon and Breach, New York.

Fuks, B. A. [1962]: *Introduction to the Theory of Analytic Functions of Several Complex Variables* (translated from Russian), American Mathematical Society, Providence, Rhode Island.

Golubitsky, M. and Guillemin, V. [1973]: *Stable mappings and their singularities*, Springer-Verlag, New York.

Gomory, R. [1956]: "Critical points at infinity and forced oscillations," *Contributions to the Theory of Nonlinear Oscillations*, Volume III, *Annals of Mathematics Studies*, No. 36, Princeton University Press, Princeton, New Jersey.

Goodwin, B. C. [1963]: *Temporal organization in cells*, Academic Press, New York.

———— [1965]: "Oscillatory behavior in enzymetric control processes," *Advances in Enzyme Regulation* (G. Weber, Ed.) 3.

Goursat, E. [1904]: *A course in mathematical analysis*, Vol. I (translated by E. R. Hendrick), Ginn and Co., New York.

Greenberg, Michael D. [1978]: *Foundations of applied mathematics*, Prentice Hall, Englewood Cliffs, New Jersey.

Guckenheimer, J. and Holmes, P. [1983]: *Nonlinear oscillations, dynamical systems, and bifurcations of vector fields*, Springer-Verlag, New York.

Hahn, W. [1967]: *Stability of motion*, Springer-Verlag, New York.

Hale, Jack K. [1963]: *Oscillations in nonlinear systems*, McGraw-Hill, New York.

———— [1969]: *Ordinary differential equations*, Wiley-Interscience, New York.

Halmos, Paul R. [1958]: *Finite-dimensional vector spaces*, 2nd edition, Van Nostrand, Princeton, New Jersey.

Hartman, P. [1964]: *Ordinary differential equations*, John Wiley & Sons, New York.

Hastings, J. W. [1973]: "Are circadian rhythms conditioned?," Abstract, *Proceedings of the XI Conference of the International Society for Chronobiology*, Hanover, Germany.

Hastings, S. P. and Murray, J. D. [1975]: "The existence of oscillatory solutions in the Field-Noyes model for the Belousov-Zhabotinski reaction," *SIAM Journal of Applied Mathematics* 28, 678–688.

Hirsch, Morris W. and Smale, S. [1974]: *Differential equations, dynamical systems, and linear algebra*, Academic Press, New York.

Hodgkin, A. L. and Huxley, A. F. [1952]: "A quantitative description of membrane current and its applications to conduction and excitation in nerve," *Journal of Physiology* 117, 500–544.

Hopf, E. [1942]: "Abzweigung einer periodischen Lösung von einer stationären Lösung eines differentialsystems," *Berichten der Mathematisch-Physischen*

Klasse der Sachsischen Akademie der Wissenschaften zu Leipzig, XCIV, Band Sitzung vom. 19 Januar.

Hurewicz, W. and Wallman, H. [1948]: *Dimension theory*, revised, Princeton University Press, Princeton, New Jersey.

Il'yashenko, Yu. S. [1991]: *Finite Theorems for Limit Cycles*, Translations of Mathematical Monographs, Volume 94, American Mathematical Society, Providence, Rhode Island.

Ize, J. [1976]: "Bifurcation theory for Fredholm operators," *Memoirs of the American Mathematical Society*, Vol. 174, American Mathematical Society, Providence, Rhode Island.

Jack, J. J. B., Noble, D., and Tsien, R. W. [1975]: *Electric current flow in excitable cells*, Oxford University Press, London.

Keener, J. and Sneyd, J. [1998]: *Mathemtical Physiology*, Springer-Verlag, New York.

Kolmogorov, A. N. and Formin, S. V. [1957]: *Elements of the theory of functions and functional analysis* (translated by Leo F. Boron), Graylock Press, Rochester, New York.

Kopell, N. and Ermentrout, G. B. [1986]: Symmetry and phaselocking in chains of weakly coupled oscillators, *Com. Pure Applied Math.*, Vol. XXXIX, 623–660.

Kopell, N. and Howard, L. N. [1973]: "Plane wave solutions to reaction-diffusion equations," *Studies in Applied Mathematics*, Vol. LII, 291–328.

Krasnosel'ski, M. A. [1964]: *Topological methods in the theory of nonlinear integral equations* (translated from Russian by A. H. Armstrong; translation edited by J. Burlak), Macmillan Company, New York.

Kryloff, N. and Bogoliuboff, N. [1937]: *Introduction to non-linear mechanics* (Russian) Kieff. English version by S. Lefschetz, Princeton University Press, Princeton, New Jersey [1943].

Lakshmikantham, V., Leela, S., and Martynyuk [1990]: *Practical stabiity of nonlinear systems*, World Scientific Publishing Co. Pte Ltd., Singapore.

LaSalle, Joseph and Lefschetz, S. [1961]: *Stability by Liapunov's direct method with applications*, Academic Press, New York.

Lefschetz, S. [1962]: *Differential equations: geometric theory*, 2nd edition, Interscience Publishers, New York. Reprinted by Dover Publications, Inc., Mineola, New York.

Leontovich, A. M. [1962]: "On the stability of the Lagrangian periodic solutions of the restricted three-body problem," *Doklady Akademii Nauk* SSSR 143, 525–528. (English translation: *Soviet Mathematics*, Doklady 3, 425–429).

Leray, J. [1935]: "Note, Topologie des espaces abstracts de M. Banach," *Comptes Rendus Hebdomadaires des Séances de l'Académie des Sciences* (Paris) 200, 1082–1084.

Lyapunov (Liapunov), A. M. [1947]: *Problème général de la stabilité du mouvement* (Reproduction of the French translation in 1907 of a Russian memoire dated 1892), *Annals of Mathematics Studies*, No. 17, Princeton University Press, Princeton, New Jersey.

Macaulay, F. S. [1916]: *The algebraic theory of modular systems*, Cambridge University Press.

Malkin, I. G. [1959]: *Some problems in the theory of nonlinear oscillations*, 2 volumes (AEC-tr-3766 book 1-2), U.S. Atomic Energy Commission Technical Information Service.

Marden, M. [1949]: *The geometry of the zeros of a polynomial in a complex variable, Mathematical Surveys*, No. 3, American Mathematical Society, Providence, Rhode Island.

Marden, M. [1966]: *Geometry of polynomials*, 2nd edition, Mathematical Survey, American Mathematical Society, Providence, Rhode Island.

Marsden, J. and McCracken, M. [1976]: *The Hopf bifurcation and its applications,* Applied Mathematical Sciences, Volume 19, Springer-Verlag, New York.

Mawhin, J. and Rouche, N. [1973]: *Équations différentielles ordinaires*, Tome 1, Tome 2, Masson et cie, Paris.

May, Robert W. [1973]: *Stability and complexity in model ecosystems*, Princeton University Press, Princeton.

Maynard Smith, J. [1974]: *Models in ecology*, Cambridge University Press, Cambridge.

Mitropolskii, Yu. A. and Homa, G. P. [1983]: *The mathematical foundation of the asymptotic methods of nonlinear mechanics*, Naukova Dumka, Kiev (Russian).

Moser, J. [1973]: *Stable and random motions in dynamical systems,* Annals of Mathematics Studies Number 77, Princeton University Press, Princeton, New Jersey.

Murray, J. D. [1974]: "On a model for the temporal oscillations in the Belousov-Zhabotinsky reaction," *Journal of Chemical Physics* 61, pp. 3610–3613.

Murray, J. D. [2003]: *Mathematical Biology* I, II, 3rd edition, Springer-Verlag, New York.

Nagumo, M. [1951]: "A theory of degree of mapping based on infinitesimal analysis," *American Journal of Mathematics* 73, 485–496.

Nemytskii, V. V. and Stepanov, V. V. [1960]: *Qualitative theory of differential equations*, Princeton University Press, Princeton, New Jersey.

Nirenberg, L. [1974]: *Topics in nonlinear functional analysis*, Courant Institute of Mathematical Sciences, New York.

Nussbaum, Roger [1971]: "The fixed point index for local condensing maps," *Annali di Matematica* 89, 217–258.

———— [1972]: "Degree theory for local condensing maps," *Journal of Mathematical Analysis and Applications* 37, 741–766.

———— [1974]: "On the uniqueness of the topological degree for k-set contractions," *Mathematische Zeitschrift* 137, 1–6.

Oatley, K. and Goodwin, B. C. [1971]: "The explanation and investigation of biological rhythms," Chapter I in *Biological rhythms and human performance*, edited by W. P. Colquhoun, Academic Press, New York.

Poincaré, H. [1886]: Mémoire sur les courbes définies par une équation différentielle, *Journal de mathématiques Pures et Appliquées* (3) 7 (1881), 375–442; 8 (1882), 251–296; *Journal de Mathématiques Pures et Appliquées* (4) 1 (1885), 167–244; 2, 151–217.

Poincaré, H. [1892–1899]: *Les méthodes nouvelles de la mécanique céleste*, 3 volumes, Gauthiers-Villars, Paris. Reprinted in 1958 by Dover Publications, New York.

Roseau, M. [1966]: *Vibrations non-linéaires et théorie de la stabilité*, Springer-Verlag, Berlin.

Royden, H. L. [1968]: *Real analysis*, 2nd edition, Macmillan Company, London.

Ruelle, D. and Takens, F. [1962]: "On the nature of turbulence," *Communications in Mathematical Physics* 20, 1971.

Sanders, J. A. and Verhulst, F. [1985]: *Averaging methods in nonlinear dynamical systems*, Springer-Verlag, New York.

Sard, A. [1942]: "The measure of the critical values of differentiable maps," *Bulletin of the American Mathematical Society* 48, 883–890.

Schwartz, A. J. [1963]: "A generalization of a Poincaré-Bendixson theorem to closed two-dimensional manifolds," *American Journal of Mathematics* 85, 453–458.

Scott, Alwyn C. [1975]: "The electrophysics of a nerve fiber," *Reviews of Modern Physics* 47, 487–553.

Sell, G. [1966]: "Periodic solutions and asymptotic stability," *Journal of Differential Equations* 2, 143–157.

Siegel, C. R. [1969]: *Topics in complex function theory*, Volume I, *Elliptic functions and uniformizaton theory,* Wiley-Interscience, New York.

Traub, Roger D. and Miles, R. [1991]: *Neuronal networks of the hippocampus*, Cambridge University Press, Cambridge.

Tyson, John J. [1976]: *The Belousov-Zhatotinski reaction,* Lecture Notes in Biomathematics, Springer-Verlag, Berlin, New York.

Van der Waerdon, B. [1940]: *Moderne algebra*, Volume II, 2nd edition, Springer, Berlin. Translation into English by Fred Blum, Ungar Publishing Co., New York, 1949.

Verhulst, F. [1990]: *Nonlinear differential equations and dynamical systems*, Springer-Verlag, New York.

Winfree, Arthur T. [1967]: "Biological rhythms and the behavior of populations of coupled oscillators," *Journal of Theoretical Biology* 16, 15–42.

Wintner, A. [1947]: *The analytical foundations of celestial mechanics*, Princeton University Press, Princeton, New Jersey.

Index